797,885 Books

are available to read at

www.ForgottenBooks.com

Forgotten Books' App
Available for mobile, tablet & eReader

ISBN 978-1-330-41564-1
PIBN 10057621

This book is a reproduction of an important historical work. Forgotten Books uses state-of-the-art technology to digitally reconstruct the work, preserving the original format whilst repairing imperfections present in the aged copy. In rare cases, an imperfection in the original, such as a blemish or missing page, may be replicated in our edition. We do, however, repair the vast majority of imperfections successfully; any imperfections that remain are intentionally left to preserve the state of such historical works.

Forgotten Books is a registered trademark of FB &c Ltd.
Copyright © 2015 FB &c Ltd.
FB &c Ltd, Dalton House, 60 Windsor Avenue, London, SW19 2RR.
Company number 08720141. Registered in England and Wales.

For support please visit www.forgottenbooks.com

1 MONTH OF FREE READING

at

www.ForgottenBooks.com

By purchasing this book you are eligible for one month membership to ForgottenBooks.com, giving you unlimited access to our entire collection of over 700,000 titles via our web site and mobile apps.

To claim your free month visit:

www.forgottenbooks.com/free57621

* Offer is valid for 45 days from date of purchase. Terms and conditions apply.

English
Français
Deutsche
Italiano
Español
Português

www.forgottenbooks.com

Mythology Photography **Fiction** Fishing Christianity **Art** Cooking Essays Buddhism Freemasonry Medicine **Biology** Music **Ancient Egypt** Evolution Carpentry Physics Dance Geology **Mathematics** Fitness Shakespeare **Folklore** Yoga Marketing **Confidence** Immortality Biographies Poetry **Psychology** Witchcraft Electronics Chemistry History **Law** Accounting **Philosophy** Anthropology Alchemy Drama Quantum Mechanics Atheism Sexual Health **Ancient History** **Entrepreneurship** Languages Sport Paleontology Needlework Islam **Metaphysics** Investment Archaeology Parenting Statistics Criminology **Motivational**

AN ELEMENTARY TREATISE

ON

QUATERNIONS

BY

P. G. TAIT, M. A.

FORMERLY FELLOW OF ST. PETER'S COLLEGE, CAMBRIDGE
PROFESSOR OF NATURAL PHILOSOPHY IN THE UNIVERSITY OF EDINBURGH

. . . . τεκρακτύν,
παγὰν ἀενάου φύσεως ῥιζώματ' ἔχουσαν.

SECOND EDITION, ENLARGED

Cambridge
AT THE UNIVERSITY PRESS

[*All Rights reserved.*]

LIBRARY
JUN 11 1968
UNIVERSITY OF TORONTO
QA
257
T3
1875

PREFACE.

To THE first edition of this work, published in 1867, the following was prefixed :—

'THE present work was commenced in 1859, while I was a Professor of Mathematics, and far more ready at Quaternion analysis than I can now pretend to be. Had it been then completed I should have had means of testing its teaching capabilities, and of improving it, before publication, where found deficient in that respect.

'The duties of another Chair, and Sir W. Hamilton's wish that my volume should not appear till after the publication of his *Elements*, interrupted my already extensive preparations. I had worked out nearly all the examples of Analytical Geometry in Todhunter's Collection, and I had made various physical applications of the Calculus, especially to Crystallography, to Geometrical Optics, and to the Induction of Currents, in addition to those on Kinematics, Electrodynamics, Fresnel's Wave Surface, &c., which are reprinted in the present work from the *Quarterly Mathematical Journal* and the *Proceedings of the Royal Society of Edinburgh*.

'Sir W. Hamilton, when I saw him but a few days before his death, urged me to prepare my work as soon as possible, his being almost ready for publication. He then expressed, more strongly perhaps than he had ever done before, his profound conviction of the importance of Quaternions to the progress of physical science; and his desire that a really elementary treatise on the subject should soon be published.

'I regret that I have so imperfectly fulfilled this last request of my revered friend. When it was made I was already engaged, along with Sir W. Thomson, in the laborious work of preparing a large Treatise on Natural Philosophy. The present volume has thus been written under very disadvantageous circumstances, espe cially as I have not found time to work up the mass of materials which I had originally collected for it, but which I had not put into a fit state for publication. I hope, however, that I have to some extent succeeded in producing a thoroughly elementary work, intelligible to any ordinary student; and that the numerous examples I have given, though not specially chosen so as to display the full merits of Quaternions, will yet sufficiently shew their admirable simplicity and naturalness to induce the reader to attack the *Lectures* and the *Elements;* where he will find, in profusion, stores of valuable results, and of elegant yet powerful analytical investigations, such as are contained in the writings of but a very few of the greatest mathematicians. For a succinct account of the steps by which Hamilton was led to the invention of Quaternions, and for other interesting information regarding that remarkable genius, I may refer to a slight sketch of his life and works in the *North British Review* for September 1866.

'It will be found that I have not servilely followed even so great a master, although dealing with a subject which is entirely his own. I cannot, of course, tell in every case what I have gathered from his published papers, or from his voluminous correspondence, and what I may have made out for myself. Some theorems and processes which I have given, though wholly my own, in the sense of having been made out for myself before the publication of the *Elements*, I have since found there. Others also may be, for I have not yet read that tremendous volume completely, since much of it bears on developments unconnected with Physics. But I have endeavoured throughout to point out to the reader all the more important parts of the work which I know to be wholly due to Hamilton. A great part, indeed, may be said to be obvious to any one who has mastered the preliminaries; still I think that, in the two last Chapters especially, a good deal of original matter will be found.

'The volume is essentially a *working* one, and, particularly in the later Chapters, is rather a collection of examples than a detailed treatise on a mathematical method. I have constantly aimed at avoiding too great extension; and in pursuance of this object have omitted many valuable elementary portions of the subject. One of these, the treatment of Quaternion logarithms and exponentials, I greatly regret not having given. But if I had printed all that seemed to me of use or interest to the student, I might easily have rivalled the bulk of one of Hamilton's volumes. The beginner is recommended merely to *read* the first five Chapters, then to *work* at Chapters VI, VII, VIII (to which numerous easy Examples are appended). After this he may work at the first five, with their (more difficult) Examples; and the remainder of the book should then present no difficulty.

'Keeping always in view, as the great end of every mathematical method, the physical applications, I have endeavoured to treat the subject as much as possible from a geometrical instead of an analytical point of view. Of course, if we premise the properties of i, j, k merely, it is possible to construct from them the whole system*; just as we deal with the imaginary of Algebra, or, to take a closer analogy, just as Hamilton himself dealt with Couples, Triads, and Sets. This may be interesting to the pure analyst, but it is repulsive to the physical student, who should be led to look upon i, j, k from the very first as geometric realities, not as algebraic imaginaries.

'The most striking peculiarity of the Calculus is that *multiplication is not generally commutative*, i.e. that qr is in general different from rq, r and q being quaternions. Still it is to be remarked that something similar is true, in the ordinary coördinate methods, of operators and functions: and therefore the student is not wholly unprepared to meet it. No one is puzzled by the fact that log.cos.x

* This has been done by Hamilton himself, as one among many methods he has employed; and it is also the foundation of a memoir by M. Allégret, entitled *Essai sur le Calcul des Quaternions* (Paris, 1862).

is not equal to cos.log.x, or that $\sqrt{\frac{dy}{dx}}$ is not equal to $\frac{d}{dx}\sqrt{y}$. Sometimes, indeed, this rule is most absurdly violated, for it is usual to take $\cos^2 x$ as equal to $(\cos x)^2$, while $\cos^{-1}x$ is not equal to $(\cos x)^{-1}$. No such incongruities appear in Quaternions; but what is true of operators and functions in other methods, that they are not generally commutative, is in Quaternions true in the multipli cation of (vector) cöordinates.

'It will be observed by those who are acquainted with the Cal culus that I have, in many cases, not given the shortest or simplest proof of an important proposition. This has been done with the view of including, in moderate compass, as great a variety of methods as possible. With the same object I have endeavoured to supply, by means of the Examples appended to each Chapter, hints (which will not be lost to the intelligent student) of farther developments of the Calculus. Many of these are due to Hamilton, who, in spite of his great originality, was one of the most excellent examiners any University can boast of.

'It must always be remembered that Cartesian methods are mere particular cases of Quaternions, where most of the distinctive features have disappeared; and that when, in the treatment of any particular question, scalars have to be adopted, the Quaternion solution becomes identical with the Cartesian one. Nothing therefore is ever lost, though much is generally gained, by employing Quaternions in preference to ordinary methods. In fact, even when Quaternions degrade to scalars, they give the solution of the most general statement of the problem they are applied to, quite independent of any limitations as to choice of particular cöordinate axes.

'There is one very desirable object which such a work as this may possibly fulfil. The University of Cambridge, while seeking to supply a real want (the deficiency of subjects of examination for mathematical honours, and the consequent frequent introduction of the wildest extravagance in the shape of data for "Problems"), is in danger of making too much of such elegant trifles as Trilinear

Cöordinates, while gigantic systems like Invariants (which, by the way, are as easily introduced into Quaternions as into Cartesian methods) are quite beyond the amount of mathematics which even the best students can master in three years' reading. One grand step to the supply of this want is, of course, the introduction into the scheme of examination of such branches of mathematical physics as the Theories of Heat and Electricity. But it appears to me that the study of a mathematical method like Quaternions, which, while of immense power and comprehensiveness, is of extraordinary simplicity, and yet requires constant thought in its applications, would also be of great benefit. With it there can be no "shut your eyes, and write down your equations," for mere mechanical dexterity of analysis is certain to lead at once to error on account of the novelty of the processes employed.

'The Table of Contents has been drawn up so as to give the student a short and simple summary of the chief fundamental formulae of the Calculus itself, and is therefore confined to an analysis of the first five [and the two last] chapters.

'In conclusion, I have only to say that I shall be much obliged to any one, student or teacher, who will point out portions of the work where a difficulty has been found; along with any inaccuracies which may be detected. As I have had no assistance in the revision of the proof-sheets, and have composed the work at irregular intervals, and while otherwise laboriously occupied, I fear it may contain many slips and even errors. Should it reach another edition there is no doubt that it will be improved in many important particulars.'

To this I have now to add that I have been equally surprised and delighted by so speedy a demand for a second edition—and the more especially as I have had many pleasing proofs that the work has had considerable circulation in America. There seems now at last to be a reasonable hope that Hamilton's grand invention will soon find its way into the working world of science, to which it is certain to render enormous services, and not be laid aside to be unearthed some centuries hence by some grubbing antiquary.

It can hardly be expected that one whose time is mainly engrossed by physical science, should devote much attention to the purely analytical and geometrical applications of a subject like this; and I am conscious that in many parts of the earlier chapters I have not fully exhibited the simplicity of Quaternions. I hope, however, that the corrections and extensions now made, especially in the later chapters, will render the work more useful for my chief object, the Physical Applications of Quaternions, than it could have been in its first crude form.

I have to thank various correspondents, some anonymous, for suggestions as well as for the detection of misprints and slips of the pen. The only absolute error which has been pointed out to me is a comparatively slight one which had escaped my own notice: a very grave blunder, which I have now corrected, seems not to have been detected by any of my correspondents, so that I cannot be quite confident that others may not exist.

I regret that I have not been able to spare time enough to re-write the work; and that, in consequence of this, and of the large additions which have been made (especially to the later chapters), the whole will now present even a more miscellaneously jumbled appearance than at first.

It is well to remember, however, that it is quite possible to make a book *too easy* reading, in the sense that the student may read it through several times without feeling those difficulties which (except perhaps in the case of some rare genius) must attend the acquisition of really useful knowledge. It is better to have a rough climb (even cutting one's own steps here and there) than to ascend the dreary monotony of a marble staircase or a well-made ladder. Royal roads to knowledge reach only the particular locality aimed at—and there are no views by the way. It is not on them that pioneers are trained for the exploration of unknown regions.

But I am happy to say that the possible repulsiveness of my

early chapters cannot long be advanced as a reason for not attacking this fascinating subject. A still more elementary work than the present will soon appear, mainly from the pen of my colleague Professor KELLAND. In it I give an investigation of the properties of the linear and vector function, based directly upon the Kinematics of Homogeneous Strain, and therefore so different in method from that employed in this work that it may prove of interest to even the advanced student.

Since the appearance of the first edition I have managed (at least partially) to effect the application of Quaternions to line, surface, and volume integrals, such as occur in Hydrokinetics, Electricity, and Potentials generally. I was first attracted to the study of Quaternions by their promise of usefulness in such applications, and, though I have not yet advanced far in this new track, I have got far enough to see that it is certain in time to be of incalculable value to physical science. I have given towards the end of the work all that is necessary to put the student on this track, which will, I hope, soon be followed to some purpose.

One remark more is necessary. I have employed, as the positive direction of rotation, that of the earth about its axis, or about the sun, as seen in our northern latitudes, i.e. that *opposite* to the direction of motion of the hands of a watch. In Sir W. Hamilton's great works the opposite is employed. The student will find no difficulty in passing from the one to the other; but, without previous warning, he is liable to be much perplexed.

With regard to notation, I have retained as nearly as possible that of Hamilton, and where new notation was necessary I have tried to make it as simple and as little incongruous with Hamilton's as possible. This is a part of the work in which great care is absolutely necessary; for, as the subject gains development, fresh notation is inevitably required; and our object must be to make each step such as to defer as long as possible the revolution which must ultimately come.

Many abbreviations are possible, and sometimes very useful in private work; but, as a rule, they are unsuited for print. Every analyst, like every short-hand writer, has his own special contractions; but, when he comes to publish his results, he ought invariably to put such devices aside. If all did not use a common mode of public expression, but each were to print as he is in the habit of writing for his own use, the confusion would be utterly intolerable.

Finally, I must express my great obligations to my friend M. M. U. Wilkinson of Trinity College, Cambridge, for the care with which he has read my proofs, and for many valuable suggestions.

P. G. TAIT.

College, Edinburgh,
October 1873.

CONTENTS.

ERRATUM.

Page 102, line 20, *for* $\phi\psi\rho-\psi\phi\rho$ *read* $\phi\psi'\rho-\psi\phi'\rho$.

QUATERNIONS.

CHAPTER I.

VECTORS, AND THEIR COMPOSITION.

1.] For more than a century and a half the geometrical representation of the negative and imaginary algebraic quantities, -1 and $\sqrt{-1}$, or, as some prefer to write them, $-$ and $-^{\frac{1}{2}}$, has been a favourite subject of speculation with mathematicians. The essence of almost all of the proposed processes consists in employing such expressions to indicate the *direction*, not the *length*, of lines.

2.] Thus it was long ago seen that if positive quantities were measured off in one direction along a fixed line, a useful and lawful convention enabled us to express negative quantities of the same kind by simply laying them off on the same line in the opposite direction. This convention is an essential part of the Cartesian method, and is constantly employed in Analytical Geometry and Applied Mathematics.

3.] Wallis, towards the end of the seventeenth century, proposed to represent the impossible roots of a quadratic equation by going *out of* the line on which, if real, they would have been laid off. His construction is equivalent to the consideration of $\sqrt{-1}$ as a directed unit-line perpendicular to that on which real quantities are measured.

4.] In the usual notation of Analytical Geometry of two dimensions, when rectangular axes are employed, this amounts to reckoning each unit of length along Oy as $+\sqrt{-1}$, and on Oy' as $-\sqrt{-1}$; while on Ox each unit is $+1$, and on Ox' it is

-1. If we look at these four lines in circular order, i.e. in the order of positive rotation (opposite to that of the hands of a watch), they give

$$1, \quad \sqrt{-1}, \quad -1, \quad -\sqrt{-1}.$$

In this series each expression is derived from that which precedes it by multiplication by the factor $\sqrt{-1}$. Hence we may consider $\sqrt{-1}$ as an operator, analogous to a handle perpendicular to the plane of xy, whose effect on any line is to make it rotate (positively) about the origin through an angle of 90°.

5.] In such a system, a point is defined by a single imaginary expression. Thus $a+b\sqrt{-1}$ may be considered as a single quantity, denoting the point whose cöordinates are a and b. Or, it may be used as an expression for the line joining that point with the origin. In the latter sense, the expression $a+b\sqrt{-1}$ implicitly contains the *direction*, as well as the *length*, of this line; since, as we see at once, the direction is inclined at an angle $\tan^{-1}\frac{b}{a}$ to the axis of x, and the length is $\sqrt{a^2+b^2}$.

6.] Operating on this symbol by the factor $\sqrt{-1}$, it becomes $-b+a\sqrt{-1}$; and now, of course, denotes the point whose x and y cöordinates are $-b$ and a; or the line joining this point with the origin. The length is still $\sqrt{a^2+b^2}$, but the angle the line makes with the axis of x is $\tan^{-1}\left(-\frac{a}{b}\right)$; which is evidently 90° greater than before the operation.

7.] De Moivre's Theorem tends to lead us still farther in the same direction. In fact, it is easy to see that if we use, instead of $\sqrt{-1}$, the more general factor $\cos\alpha+\sqrt{-1}\sin\alpha$, its effect on any line is to turn it through the (positive) angle α in the plane of x, y. [Of course the former factor, $\sqrt{-1}$, is merely the particular case of this, when $\alpha=\frac{\pi}{2}$.]

Thus $(\cos\alpha+\sqrt{-1}\sin\alpha)(a+b\sqrt{-1})$

$$= a\cos\alpha-b\sin\alpha+\sqrt{-1}\,(a\sin\alpha+b\cos\alpha),$$

by direct multiplication. The reader will at once see that the new form indicates that a rotation through an angle α has taken place, if he compares it with the common formulæ for turning the cöordinate axes through a given angle. Or, in a less simple manner, thus—

$$\text{Length} = \sqrt{(a\cos\alpha-b\sin\alpha)^2+(a\sin\alpha+b\cos\alpha)^2}$$
$$= \sqrt{a^2+b^2} \qquad \text{as before.}$$

Inclination to axis of x

$$= \tan^{-1}\frac{a\sin\alpha + b\cos\alpha}{a\cos\alpha - b\sin\alpha} = \tan^{-1}\frac{\tan\alpha + \frac{b}{a}}{1 - \frac{b}{a}\tan\alpha}$$

$$= \alpha + \tan^{-1}\frac{b}{a}.$$

8.] We see now, as it were, *why* it happens that

$$(\cos\alpha + \sqrt{-1}\sin\alpha)^m = \cos m\alpha + \sqrt{-1}\sin m\alpha.$$

In fact, the first operator produces m successive rotations in the same direction, each through the angle α; the second, a single rotation through the angle $m\alpha$.

9.] It may be interesting, at this stage, to anticipate so far as to state that a Quaternion can, in general, be put under the form

$$N(\cos\theta + \varpi\sin\theta),$$

where N is a numerical quantity, θ a real angle, and

$$\varpi^2 = -1.$$

This expression for a quaternion bears a very close analogy to the forms employed in De Moivre's Theorem; but there is the essential difference (to which Hamilton's chief invention referred) that ϖ is not the algebraic $\sqrt{-1}$, but may be *any directed unit-line* whatever in space.

10.] In the present century Argand, Warren, and others, extended the results of Wallis and De Moivre. They attempted to express as a line the product of two lines each represented by a symbol such as $a + b\sqrt{-1}$. To a certain extent they succeeded, but simplicity was not gained by their methods, as the terrible array of radicals in Warren's Treatise sufficiently proves.

11.] A very curious speculation, due to Servois and published in 1813 in Gergonne's *Annales*, is the only one, so far as has been discovered, in which the slightest trace of an anticipation of Quaternions is contained. Endeavouring to extend to *space* the form $a + b\sqrt{-1}$ for the plane, he is guided by analogy to write for a directed unit-line in space the form

$$p\cos\alpha + q\cos\beta + r\cos\gamma,$$

where α, β, γ are its inclinations to the three axes. He perceives easily that p, q, r must be *non-reals*: but, he asks, "seraient-elles *imaginaires* réductibles à la forme générale $A + B\sqrt{-1}$?" This he could not answer. In fact they are the i, j, k of the Quaternion Calculus. (See Chap. II.)

12.] Beyond this, few attempts were made, or at least recorded, in earlier times, to extend the principle to space of three dimensions;

and, though many such have been made within the last forty years, none, with the single exception of Hamilton's, have resulted in simple, practical methods; all, however ingenious, seeming to lead at once to processes and results of fearful complexity.

For a lucid, complete, and most impartial statement of the claims of his predecessors in this field we refer to the Preface to Hamilton's *Lectures on Quaternions*.

13.] It was reserved for Hamilton to discover the use of $\sqrt{-1}$ as a *geometric reality*, tied down to no particular direction in space, and this use was the foundation of the singularly elegant, yet enormously powerful, Calculus of Quaternions.

While all other schemes for using $\sqrt{-1}$ to indicate direction make one direction in space expressible by real numbers, the remainder being imaginaries of some kind, leading in general to equations which are heterogeneous; Hamilton makes all directions in space equally imaginary, or rather equally real, thereby ensuring to his Calculus the power of dealing with space indifferently in all directions.

In fact, as we shall see, the Quaternion method is independent of axes or any supposed directions in space, and takes its reference lines solely from the problem it is applied to.

14.] But, for the purpose of elementary exposition, it is best to begin by assimilating it as closely as we can to the ordinary Cartesian methods of Geometry of Three Dimensions, which are in fact a mere particular case of Quaternions in which most of the distinctive features are lost. We shall find in a little that it is capable of soaring above these entirely, after having employed them in its establishment; and, indeed, as the inventor's works amply prove, it can be established, *ab initio*, in various ways, without even an allusion to Cartesian Geometry. As this work is written for students acquainted with at least the elements of the Cartesian method, we keep to the first-mentioned course of exposition; especially as we thereby avoid some reasoning which, though rigorous and beautiful, might be apt, from its subtlety, to prove repulsive to the beginner.

We commence, therefore, with some very elementary geometrical ideas.

15.] Suppose we have two points A and B in *space*, and suppose A given, on how many numbers does B's relative position depend?

If we refer to Cartesian cöordinates (rectangular or not) we find that the data required are the excesses of B's three cöordinates over those of A. Hence *three* numbers are required.

Or we may take polar cöordinates. To define the moon's position with respect to the earth we must have its Geocentric Latitude and Longitude, *or* its Right Ascension and Declination, and, in addition, its distance or radius-vector. *Three* again.

16.] Here it is to be carefully noticed that nothing has been said of the *actual* cöordinates of either A or B, or of the earth and moon, in space; it is only the *relative* cöordinates that are contemplated.

Hence any expression, as $\overline{AB}$, denoting a line considered with reference to direction as well as length, contains implicitly *three* numbers, and all lines parallel and equal to AB depend in the same way upon the same three. Hence, *all lines which are equal and parallel may be represented by a common symbol, and that symbol contains three distinct numbers.* In this sense a line is called a VECTOR, since by it we pass from the one extremity, A, to the other, B; and it may thus be considered as an instrument which *carries* A to B: so that a vector may be employed to indicate a definite *translation* in space.

17.] We may here remark, once for all, that in establishing a new Calculus, we are at liberty to give any definitions whatever of our symbols, provided that no two of these interfere with, or contradict, each other, and in doing so in Quaternions *simplicity* and (so to speak) *naturalness* were the inventor's aim.

18.] Let $\overline{AB}$ be represented by α, we know that α depends on *three* separate numbers. Now if CD be equal in length to AB and if these lines be parallel, we have evidently $\overline{CD} = \overline{AB} = \alpha$, where it will be seen that the sign of *equality* between vectors contains implicitly *equality in length* and *parallelism in direction.* So far we have *extended* the meaning of an algebraical symbol. And it is to be noticed that an equation between vectors, as

$$\alpha = \beta,$$

contains *three* distinct equations between mere numbers.

19.] We must now define + (and the meaning of − will follow) in the new Calculus. Let A, B, C be any three points, and (with the above meaning of =) let

$$\overline{AB} = \alpha, \quad \overline{BC} = \beta, \quad \overline{AC} = \gamma.$$

If we define + (in accordance with the idea (§ 16) that a vector represents a *translation*) by the equation

$$\alpha + \beta = \gamma,$$

or
$$\overline{AB}+\overline{BC}=\overline{AC},$$
we contradict nothing that precedes, but we at once introduce the idea that *vectors are to be compounded, in direction and magnitude, like simultaneous velocities*. A reason for this may be seen in another way if we remember that by *adding* the differences of the Cartesian cöordinates of A and B, to those of the cöordinates of B and C, we get those of the cöordinates of A and C. Hence these cöordinates enter *linearly* into the expression for a vector.

20.] But we also see that if C and A coincide (and C may be *any* point)
$$\overline{AC}=0,$$
for no vector is then required to carry A to C. Hence the above relation may be written, in this case,
$$\overline{AB}+\overline{BA}=0,$$
or, introducing, and by the same act defining, the symbol $-$,
$$\overline{BA}=-\overline{AB}.$$

Hence, *the symbol* $-$, *applied to a vector, simply shows that its direction is to be reversed.*

And this is consistent with all that precedes; for instance,
$$\overline{AB}+\overline{BC}=\overline{AC},$$
and
$$\overline{AB}=\overline{AC}-\overline{BC},$$
or
$$=\overline{AC}+\overline{CB},$$
are evidently but different expressions of the same truth.

21.] In any triangle, ABC, we have, of course,
$$\overline{AB}+\overline{BC}+\overline{CA}=0;$$
and, in any closed polygon, whether plane or gauche,
$$\overline{AB}+\overline{BC}+\ldots\ldots+\overline{YZ}+\overline{ZA}=0.$$

In the case of the polygon we have also
$$\overline{AB}+\overline{BC}+\ldots\ldots+\overline{YZ}=\overline{AZ}.$$

These are the well-known propositions regarding composition of velocities, which, by the second law of motion, give us the geometrical laws of composition of forces.

22.] If we compound any number of *parallel* vectors, the result is obviously a numerical multiple of any one of them.

Thus, if A, B, C are in one straight line,
$$\overline{BC}=x\,\overline{AB};$$
where x is a number, positive when B lies between A and C, otherwise negative: but such that its numerical value, independent of sign, is the ratio of the length of BC to that of AB. This is

at once evident if AB and BC be commensurable; and is easily extended to incommensurables by the usual *reductio ad absurdum.*

23.] An important, but almost obvious, proposition is that *any vector may be resolved, and in one way only, into three components parallel respectively to any three given vectors, no two of which are parallel, and which are not parallel to one plane.*

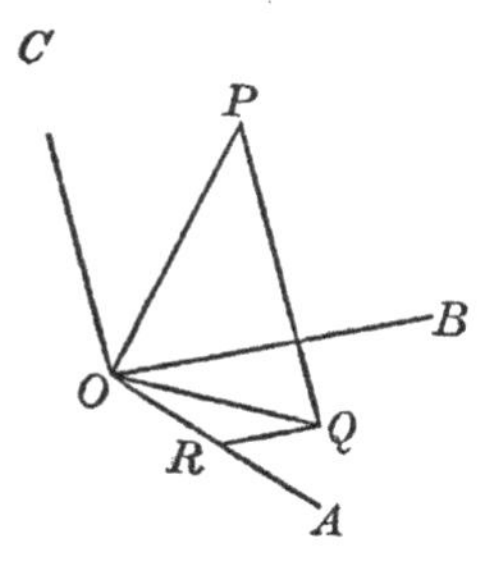

Let OA, OB, OC be the three fixed vectors, OP any other vector. From P draw PQ parallel to CO, meeting the plane BOA in Q. [There must be a definite point Q, else PQ, and therefore CO, would be parallel to BOA, a case specially excepted.] From Q draw QR parallel to BO, meeting OA in R. Then we have $\overline{OP} = \overline{OR} + \overline{RQ} + \overline{QP}$ (§ 21),

and these components are respectively parallel to the three given vectors. By § 22 we may express $\overline{OR}$ as a numerical multiple of $\overline{OA}$, $\overline{RQ}$ of $\overline{OB}$, and $\overline{QP}$ of $\overline{OC}$. Hence we have, generally, for any vector in terms of three fixed non-coplanar vectors, α, β, γ,

$$\overline{OP} = \rho = x\alpha + y\beta + z\gamma,$$

which exhibits, in one form, the *three* numbers on which a vector depends (§ 16). Here x, y, z are perfectly definite, and can have but single values.

24.] Similarly any vector, as $\overline{OQ}$, in the same plane with OA and OB, can be resolved into components $\overline{OR}$, $\overline{RQ}$, parallel respectively to $\overline{OA}$ and $\overline{OB}$; so long, at least, as these two vectors are not parallel to each other.

25.] There is particular advantage, in certain cases, in employing a series of three *mutually perpendicular unit-vectors* as lines of reference. This system Hamilton denotes by i, j, k.

Any other vector is then expressible as

$$\rho = xi + yj + zk.$$

Since i, j, k are unit-vectors, x, y, z are here the lengths of conterminous edges of a rectangular parallelepiped of which ρ is the vector-diagonal; so that the length of ρ is, in this case,

$$\sqrt{x^2 + y^2 + z^2}. \quad = T\rho$$

Let

$$\varpi = \xi i + \eta j + \zeta k$$

be any other vector, then (by the proposition of § 23) the vector equation

$$\rho = \varpi$$

obviously involves the following three equations among numbers,

$$x = \xi, \qquad y = \eta, \qquad z = \zeta.$$

Suppose i to be drawn eastwards, j northwards, and k upwards, this is equivalent merely to saying that *if two points coincide, they are equally to the east (or west) of any third point, equally to the north (or south) of it, and equally elevated above (or depressed below) its level.*

26.] It is to be carefully noticed that it is only when α, β, γ are not coplanar that a vector equation such as

$$\rho = \varpi,$$

or

$$x\alpha + y\beta + z\gamma = \xi\alpha + \eta\beta + \zeta\gamma,$$

necessitates the three numerical equations

$$x = \xi, \quad y = \eta, \quad z = \zeta.$$

For, if α, β, γ be coplanar (§ 24), a condition of the following form must hold

$$\gamma = a\alpha + b\beta.$$

Hence

$$\rho = (x + za)\alpha + (y + zb)\beta,$$
$$\varpi = (\xi + \zeta a)\alpha + (\eta + \zeta b)\beta,$$

and the equation

$$\rho = \varpi$$

now requires only the *two* numerical conditions

$$x + za = \xi + \zeta a, \quad y + zb = \eta + \zeta b.$$

27.] *The Commutative and Associative Laws hold in the combination of vectors by the signs + and −.* It is obvious that, if we prove this for the sign +, it will be equally proved for −, because before a vector (§ 20) merely indicates that it is to be reversed before being considered positive.

Let A, B, C, D be, in order, the corners of a parallelogram; we have, obviously,

$$\overline{AB} = \overline{DC}, \quad \overline{AD} = \overline{BC}.$$

And

$$\overline{AB} + \overline{BC} = \overline{AC} = \overline{AD} + \overline{DC} = \overline{BC} + \overline{AB}.$$

Hence the commutative law is true for the addition of any two vectors, and is therefore generally true.

Again, whatever four points are represented by A, B, C, D, we have

$$\overline{AD} = \overline{AB} + \overline{BD} = \overline{AC} + \overline{CD},$$

or substituting their values for $\overline{AD}, \overline{BD}, \overline{AC}$ respectively, in these three expressions,

$$\overline{AB} + \overline{BC} + \overline{CD} = \overline{AB} + (\overline{BC} + \overline{CD}) = (\overline{AB} + \overline{BC}) + \overline{CD}.$$

And thus the truth of the associative law is evident.

28.] The equation

$$\rho = x\beta,$$

where ρ is the vector connecting a variable point with the origin, β a definite vector, and x an indefinite number, represents the straight line drawn from the origin parallel to β (§ 22).

The straight line drawn from A, where $\overline{OA} = \alpha$, and parallel to β, has the equation

$$\rho = \alpha + x\beta. \qquad (1)$$

In words, we may pass directly from O to P by the vector $\overline{OP}$ or ρ; or we may pass first to A, by means of $\overline{OA}$ or α, and then to P along a vector parallel to β (§ 16).

Equation (1) is one of the many useful forms into which Quaternions enable us to throw the general equation of a straight line in space. As we have seen (§ 25) it is equivalent to *three* numerical equations; but, as these involve the indefinite quantity x, they are virtually equivalent to but *two*, as in ordinary Geometry of Three Dimensions.

29.] A good illustration of this remark is furnished by the fact that the equation

$$\rho = y\alpha + x\beta,$$

which contains two indefinite quantities, is virtually equivalent to only one numerical equation. And it is easy to see that it represents the plane in which the lines α and β lie; or the surface which is formed by drawing, through every point of OA, a line parallel to OB. In fact, the equation, as written, is simply § 24 in symbols.

And it is evident that the equation

$$\rho = \gamma + y\alpha + x\beta$$

is the equation of a plane passing through the extremity of γ, and parallel to α and β.

It will now be obvious to the reader that the equation

$$\rho = p_1 \alpha_1 + p_2 \alpha_2 + \ldots\ldots = \Sigma p \alpha,$$

where α_1, α_2, &c. are given vectors, and p_1, p_2, &c. numerical quantities, *represents a straight line* if p_1, p_2, &c. be linear functions of *one* indeterminate number; and a *plane*, if they be linear expressions containing *two* indeterminate numbers. Later (§ 31 (l)), this theorem will be much extended.

Again, the equation

$$\rho = x\alpha + y\beta + z\gamma$$

refers to *any* point whatever in space, provided α, β, γ are not coplanar. (*Ante*, § 23).

30.] The equation of the line joining any two points A and B, where $\overline{OA} = \alpha$ and $\overline{OB} = \beta$, is obviously

$$\rho = \alpha + x(\beta - \alpha),$$

or

$$\rho = \beta + y(\alpha - \beta).$$

These equations are of course identical, as may be seen by putting $1 - y$ for x.

The first may be written

$$\rho+(x-1)\alpha-x\beta = 0;$$

or $$p\rho+q\alpha+r\beta = 0,$$

subject to the condition $p+q+r = 0$ identically. That is—A homogeneous linear function of three vectors, equated to zero, expresses that the extremities of these vectors are in one straight line, *if the sum of the coefficients be identically zero.*

Similarly, the equation of the plane containing the extremities A, B, C of the three non-coplanar vectors α, β, γ is

$$\rho = \alpha+x(\beta-\alpha)+y(\gamma-\beta),$$

where x and y are each indeterminate.

This may be written

$$p\rho+q\alpha+r\beta+s\gamma = 0,$$

with the identical relation

$$p+q+r+s = 0.$$

which is the condition that four points may lie in one plane.

31.] We have already the means of proving, in a very simple manner, numerous classes of propositions in plane and solid geometry. A very few examples, however, must suffice at this stage; since we have hardly, as yet, crossed the threshold of the subject, and are dealing with mere linear equations connecting two or more vectors, and even with them *we are restricted as yet to operations of mere addition.* We will give these examples with a painful minuteness of detail, which the reader will soon find to be necessary only for a short time, if at all.

(*a.*) *The diagonals of a parallelogram bisect each other.*

Let $ABCD$ be the parallelogram, O the point of intersection of its diagonals. Then

$$\overline{AO}+\overline{OB} = \overline{AB} = \overline{DC} = \overline{DO}+\overline{OC},$$

which gives $$\overline{AO}-\overline{OC} = \overline{DO}-\overline{OB}.$$

The two vectors here equated are parallel to the diagonals respectively. Such an equation is, of course, absurd unless

(1) The diagonals are parallel, in which case the figure is not a parallelogram;

(2) $\overline{AO} = \overline{OC}$, and $\overline{DO} = \overline{OB}$, the proposition.

(*b.*) *To show that a triangle can be constructed, whose sides are parallel, and equal, to the bisectors of the sides of any triangle.*

Let ABC be any triangle, Aa, Bb, Cc the bisectors of the sides.

Then
$$\overline{Aa} = \overline{AB} + \overline{Ba} = \overline{AB} + \tfrac{1}{2}\overline{BC},$$
$$\overline{Bb} \;\;-\;-\;-\;\; = \overline{BC} + \tfrac{1}{2}\overline{CA},$$
$$\overline{Cc} \;\;-\;-\;-\;\; = \overline{CA} + \tfrac{1}{2}\overline{AB}.$$

Hence
$$\overline{Aa} + \overline{Bb} + \overline{Cc} = \tfrac{3}{2}(\overline{AB} + \overline{BC} + \overline{CA}) = 0;$$
which (§ 21) proves the proposition.

Also
$$\begin{aligned}\overline{Aa} &= \overline{AB} + \tfrac{1}{2}\overline{BC} \\ &= \overline{AB} - \tfrac{1}{2}(\overline{CA} + \overline{AB}) \\ &= \tfrac{1}{2}(\overline{AB} - \overline{CA}) = \tfrac{1}{2}(\overline{AB} + \overline{AC}),\end{aligned}$$
results which are sometimes useful. They may be easily verified by producing Aa to twice its length and joining the extremity with B.

(b'.) *The bisectors of the sides of a triangle meet in a point, which trisects each of them.*

Taking A as origin, and putting α, β, γ for vectors parallel, and equal, to the sides taken in order BC, CA, AB; the equation of Bb is (§ 28 (1))
$$\rho = \gamma + x\left(\gamma + \frac{\beta}{2}\right) = (1+x)\gamma + \frac{x}{2}\beta.$$
That of Cc is, in the same way,
$$\rho = -(1+y)\beta - \frac{y}{2}\gamma.$$
At the point O, where Bb and Cc intersect,
$$\rho = (1+x)\gamma + \frac{x}{2}\beta = -(1+y)\beta - \frac{y}{2}\gamma.$$
Since γ and β are not parallel, this equation gives
$$1 + x = -\frac{y}{2}, \text{ and } \frac{x}{2} = -(1+y).$$
From these
$$x = y = -\tfrac{2}{3}.$$
Hence
$$\overline{AO} = \tfrac{1}{3}(\gamma - \beta) = \tfrac{2}{3}\overline{Aa}. \quad \text{(See Ex. } (b)\text{)}$$
This equation shows, being a vector one, that Aa passes through O, and that $AO : Oa :: 2 : 1$.

(c) If
$$\overline{OA} = \alpha,$$
$$\overline{OB} = \beta,$$
$$\overline{OC} = a\alpha + b\beta,$$

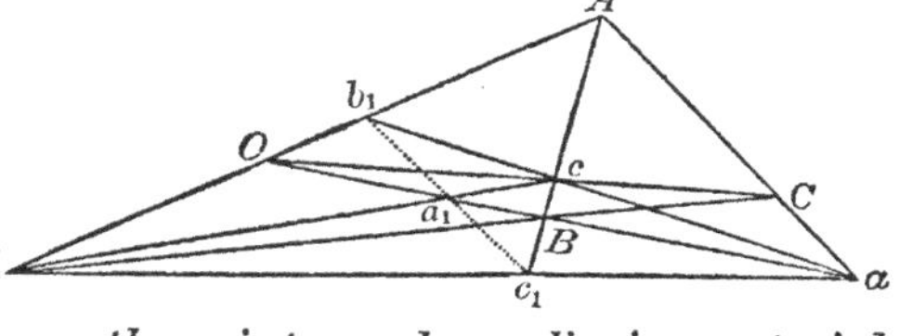

be three given co-planar vectors, and the lines indicated in the figure be drawn, the points a_1, b_1, c_1 lie in a straight line.

We see at once, by the process indicated in § 30, that

$$\overline{Oc} = \frac{a\alpha + b\beta}{a+b}, \qquad \overline{Ob} = \frac{a\alpha}{1-b}, \qquad \overline{Oa} = \frac{b\beta}{1-a}.$$

Hence we easily find

$$\overline{Oa_1} = -\frac{b\beta}{1-a-2b}, \quad \overline{Ob_1} = -\frac{a\alpha}{1-2a-b}, \quad \overline{Oc_1} = \frac{-a\alpha+b\beta}{b-a}.$$

These give

$$-(1-a-2b)\overline{Oa_1} + (1-2a-b)\overline{Ob_1} - (b-a)\overline{Oc_1} = 0.$$

But $\quad -(1-a-2b)+(1-2a-b)-(b-a) = 0$ identically.

This, by § 30, proves the proposition.

(*d.*) Let $\overline{OA} = \alpha$, $\overline{OB} = \beta$, be any two vectors. *If MP be parallel to OB; and OQ, BQ, be drawn parallel to AP, OP respectively; the locus of Q is a straight line parallel to OA.*

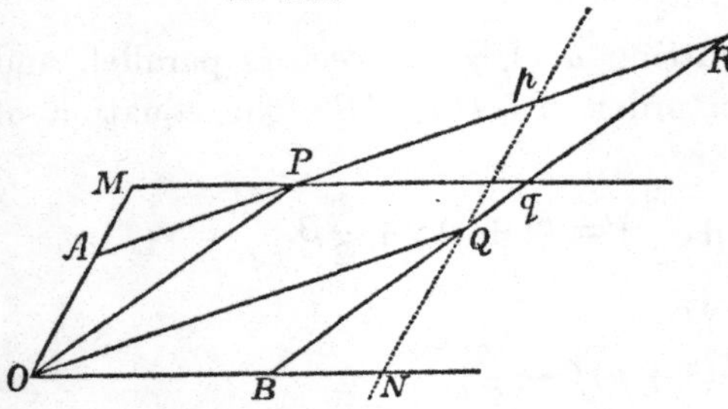

Let $\overline{OM} = e\alpha$.

Then

$$\overline{AP} = \overline{e-1}\,\alpha + x\beta.$$

Hence the equation of OQ is

$$\rho = y(\overline{e-1}\,\alpha + x\beta);$$

and that of BQ is

$$\rho = \beta + z(e\alpha + x\beta).$$

At Q we have, therefore,

$$\left.\begin{aligned} xy &= 1 + zx, \\ y(e-1) &= ze. \end{aligned}\right\}$$

These give $xy = e$, and the equation of the locus of Q is

$$\rho = e\beta + y'\alpha,$$

i.e. a straight line parallel to OA, drawn through N in OB produced, so that $\quad ON : OB :: OM : OA.$

Cor. If BQ meet MP in q, $\overline{Pq} = \beta$; and if AP meet NQ in p, $\overline{Qp} = \alpha$.

Also, for the point R we have $\overline{pR} = \overline{AP}$, $\overline{QR} = \overline{Bq}$.

Hence, *if from any two points, A and B, lines be drawn intercepting a given length Pq on a given line Mq; and if, from R their point of intersection, Rp be laid off = PA, and RQ = qB; Q and p lie on a fixed straight line, and the length of Qp is constant.*

(*e.*) *To find the centre of inertia of any system.*

If $\overline{OA} = \alpha$, $\overline{OB} = \alpha_1$, be the vector sides of any triangle, the vector from the vertex dividing the base AB in C so that

$$BC : CA :: m : m_1 \text{ is}$$
$$\frac{m\alpha + m_1\alpha_1}{m + m_1}.$$

For $\overline{AB}$ is $\alpha_1 - \alpha$, and therefore $\overline{AC}$ is

$$\frac{m_1}{m + m_1}(\alpha_1 - \alpha).$$

Hence
$$\begin{aligned}\overline{OC} &= \overline{OA} + \overline{AC} \\ &= \alpha + \frac{m_1}{m + m_1}(\alpha_1 - \alpha) \\ &= \frac{m\alpha + m_1\alpha_1}{m + m_1}.\end{aligned}$$

This expression shows how to find the centre of inertia of two masses; m at the extremity of α, m_1 at that of α_1. Introduce m_2 at the extremity of α_2, then the vector of the centre of inertia of the three is, by a second application of the formula,

$$\frac{(m + m_1)\left(\frac{m\alpha + m_1\alpha_1}{m + m_1}\right) + m_2\alpha_2}{(m + m_1) + m_2} = \frac{m\alpha + m_1\alpha_1 + m_2\alpha_2}{m + m_1 + m_2}.$$

For any number of masses, expressed generally by m at the extremity of the vector α, we have the vector of the centre of inertia

$$\beta = \frac{\Sigma(m\alpha)}{\Sigma(m)}.$$

This may be written $\qquad \Sigma m(\alpha - \beta) = 0.$

Now $\alpha_1 - \beta$ is the vector of m_1 with respect to the centre of inertia. Hence the theorem, *If the vector of each element of a mass, drawn from the centre of inertia, be increased in length in proportion to the mass of the element, the sum of all these vectors is zero.*

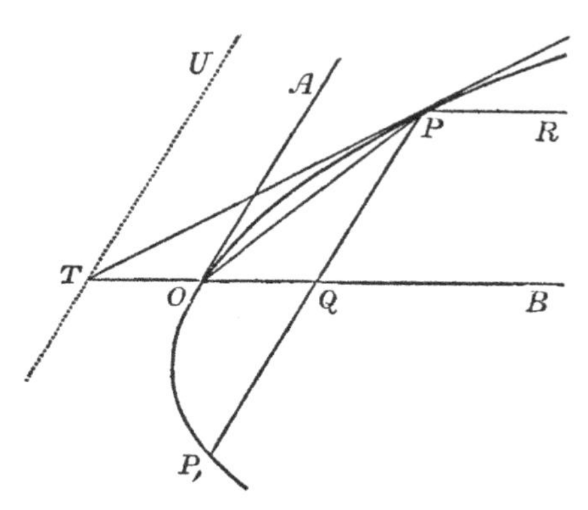

(*f*.) We see at once that the equation

$$\rho = \alpha t + \frac{\beta t^2}{2},$$

where t is an indeterminate number, and α, β given vectors, represents a parabola. The origin, O, is a point on the curve, β is parallel to the axis, i. e. is the diameter OB drawn from the origin, and α is OA the tangent at the origin. In the figure

$$\overline{QP} = \alpha t, \qquad \overline{OQ} = \frac{\beta t^2}{2}.$$

The secant joining the points where t has the values t and t' is represented by the equation

$$\rho = \alpha t + \frac{\beta t^2}{2} + x\left(\alpha t' + \frac{\beta t'^2}{2} - \alpha t - \frac{\beta t^2}{2}\right) \quad (\S\ 30)$$

$$- \alpha t + \frac{\beta t^2}{2} + x(t'-t)\left\{\alpha + \beta\frac{t'+t}{2}\right\}.$$

Put $t' = t$, and write x for $x(t'-t)$ [which may have any value] and the equation of the tangent at the point (t) is

$$\rho = \alpha t + \frac{\beta t^2}{2} + x(\alpha + \beta t).$$

Put $x = -t$, $\qquad \rho = -\frac{\beta t^2}{2}$,

or the intercept of the tangent on the diameter is $-$the abscissa of the point of contact.

Otherwise: the tangent is parallel to the vector $\alpha + \beta t$ or $\alpha t + \beta t^2$ or $\alpha t + \frac{\beta t^2}{2} + \frac{\beta t^2}{2}$ or $\overline{OQ} + \overline{OP}$. But $\overline{TP} = \overline{TO} + \overline{OP}$, hence $\overline{TO} = \overline{OQ}$.

(*g*.) Since the equation of any tangent to the parabola is

$$\rho = \alpha t + \frac{\beta t^2}{2} + x(\alpha + \beta t),$$

let us find the tangents which can be drawn from a given point. Let the vector of the point be

$$\rho = p\alpha + q\beta \quad (\S\ 24).$$

Since the tangent is to pass through this point, we have, as conditions to determine t and x, $\quad t + x = p$,

$$\frac{t^2}{2} + xt = q;$$

by equating respectively the coefficients of α and β.

Hence $\qquad t = p \pm \sqrt{p^2 - 2q}$.

Thus, in general, *two* tangents can be drawn from a given point. These coincide if

$$p^2 = 2q;$$

that is, if the vector of the point from which they are to be drawn is

$$\rho = p\alpha + q\beta = p\alpha + \frac{p^2}{2}\beta,$$

i.e. if the point lies *on* the parabola. They are imaginary if $2q > p^2$, i.e. if the point be

$$\rho = p\alpha + \left(\frac{p^2}{2} + r\right)\beta,$$

r being *positive*. Such a point is evidently *within* the curve, as at R, where $\overline{OQ} = \frac{p^2}{2}\beta$, $\overline{QP} = p\alpha$, $\overline{PR} = r\beta$.

(*h.*) Calling the values of t for the two tangents found in (*g*) t_1 and t_2 respectively, it is obvious that the vector joining the points of contact is

$$\alpha t_1 + \frac{\beta t_1^2}{2} - \alpha t_2 - \frac{\beta t_2^2}{2},$$

which is parallel to

$$\alpha + \beta \frac{t_1 + t_2}{2};$$

or, by the values of t_1 and t_2 in (*g*),

$$\alpha + p\beta.$$

Its direction, therefore, does not depend on q. In words, *If pairs of tangents be drawn to a parabola from points of a diameter produced, the chords of contact are parallel to the tangent at the vertex of the diameter.* This is also proved by a former result, for we must have $\overline{OT}$ for *each* tangent equal to $\overline{QO}$.

(*i.*) The equation of the chord of contact, for the point whose vector is

$$\rho = p\alpha + q\beta,$$

is thus

$$\rho = \alpha t_1 + \frac{\beta t_1^2}{2} + y(\alpha + p\beta).$$

Suppose this to pass always through the point whose vector is

$$\rho = a\alpha + b\beta.$$

Then we must have

$$\left.\begin{aligned} t_1 + y &= a, \\ \frac{t_1^2}{2} + py &= b; \end{aligned}\right\}$$

or

$$t_1 = p \pm \sqrt{p^2 - 2pa + 2b}.$$

Comparing this with the expression in (*g*), we have

$$q = pa - b;$$

that is, the point from which the tangents are drawn has the vector

$$\begin{aligned} \rho &= p\alpha + (pa - b)\beta \\ &= -b\beta + p(\alpha + a\beta), \text{ a straight line } (\S\ 28\ (1)). \end{aligned}$$

The mere form of this expression contains the proof of the usual properties of the pole and polar in the parabola; but, for the sake of the beginner, we adopt a simpler, though equally general, process.

Suppose $a = 0$. This merely restricts the pole to the particular diameter to which we have referred the parabola. Then the pole is Q, where

$$\rho = b\beta,$$

and the polar is the line TU, for which

$$\rho = -b\beta + p\alpha.$$

Hence *the polar of any point is parallel to the tangent at the extremity of the diameter on which the point lies, and its intersection with that diameter is as far beyond the vertex as the pole is within, and vice versâ.*

(*j.*) As another example let us prove the following theorem. *If a triangle be inscribed in a parabola, the three points in which the sides are met by tangents at the angles lie in a straight line.*

Since O is any point of the curve, we may take it as one corner of the triangle. Let t and t_1 determine the others. Then, if $\varpi_1, \varpi_2, \varpi_3$ represent the vectors of the points of intersection of the tangents with the sides, we easily find

$$\varpi_1 = \frac{t_1^2}{2t_1 - t}\left(a + \frac{t}{2}\beta\right),$$

$$\varpi_2 = \frac{t^2}{2t - t_1}\left(a + \frac{t_1}{2}\beta\right),$$

$$\varpi_3 = \frac{tt_1}{t_1 + t}\,a.$$

These values give

$$\frac{2t_1 - t}{t_1}\varpi_1 - \frac{2t - t_1}{t}\varpi_2 - \frac{t_1^2 - t^2}{tt_1}\varpi_3 = 0.$$

Also $$\frac{2t_1 - t}{t_1} - \frac{2t - t_1}{t} - \frac{t_1^2 - t^2}{tt_1} = 0 \text{ identically.}$$

Hence, by § 30, the proposition is proved.

(*k.*) Other interesting examples of this method of treating curves will, of course, suggest themselves to the student. Thus

$$\rho = a \cos t + \beta \sin t$$

or $$\rho = ax + \beta\sqrt{1 - x^2}$$

represents an ellipse, of which the given vectors a and β are semi-conjugate diameters.

Again, $\rho = at + \frac{\beta}{t}$ or $\rho = a \tan x + \beta \cot x$

evidently represents a hyperbola referred to its asymptotes.

But, so far as we have yet gone with the explanation of the calculus, as we are not prepared to determine the lengths or inclinations of vectors, we can investigate only a very small class of the properties of curves, represented by such equations as those above written.

(*l.*) We may now, in extension of the statement in § 29, make the obvious remark that

$$\rho = \Sigma p \alpha$$

is the equation of a *curve* in space, if the numbers p_1, p_2, &c. are functions of *one* indeterminate. In such a case the equation is sometimes written

$$\rho = \phi(t).$$

But, if p_1, p_2, &c. be functions of *two* indeterminates, the locus of the extremity of ρ is a *surface*; whose equation is sometimes written

$$\rho = \phi(t, u).$$

(*m.*) Thus the equation

$$\rho = \alpha \cos t + \beta \sin t + \gamma t$$

belongs to a helix.

Again,
$$\rho = p\alpha + q\beta + r\gamma$$
with a condition of the form

$$ap^2 + bq^2 + cr^2 = 1$$

belongs to a central surface of the second order, of which α, β, γ are the directions of conjugate diameters. If a, b, c be all positive, the surface is an ellipsoid.

32.] In Example (*f*) above we performed an operation equivalent to the differentiation of a vector with reference to a single *numerical* variable of which it was given as an explicit function. As this process is of very great use, especially in quaternion investigations connected with the motion of a particle or point; and as it will afford us an opportunity of making a preliminary step towards overcoming the novel difficulties which arise in quaternion differentiation; we will devote a few sections to a more careful exposition of it.

33.] It is a striking circumstance, when we consider the way in which Newton's original methods in the Differential Calculus have been decried, to find that Hamilton was *obliged* to employ them, and not the more modern forms, in order to overcome the characteristic difficulties of quaternion differentiation. Such a thing as *a differential coefficient has absolutely no meaning in quaternions*, except in those special cases in which we are dealing with degraded quaternions, such as numbers, Cartesian cöordinates, &c. But a quaternion expression has always a *differential*, which is, simply, what Newton called a *fluxion*.

As with the Laws of Motion, the basis of Dynamics, so with the foundations of the Differential Calculus; we are gradually coming to the conclusion that Newton's system is the best after all.

34.] Suppose ρ to be the vector of a curve in space. Then, generally, ρ may be expressed as the sum of a number of terms, each of which is a multiple of a given vector by a function of some *one* indeterminate; or, as in § 31 (*l*), if P be a point on the curve,

$$\overline{OP} = \rho = \phi(t).$$

And, similarly, if Q be *any other* point on the curve,

$$\overline{OQ} = \rho_1 = \phi(t_1) = \phi(t+\delta t),$$

where δt is *any number whatever.*

The vector-chord $\overline{PQ}$ is therefore, rigorously,

$$\delta\rho = \rho_1 - \rho = \phi(t+\delta t) - \phi t.$$

35.] It is obvious that, in the present case, *because the vectors involved in ϕ are constant, and their numerical multipliers alone vary,* the expression $\phi(t+\delta t)$ is, by Taylor's Theorem equivalent to

$$\phi(t) + \frac{d\phi(t)}{dt}\delta t + \frac{d^2\phi(t)}{dt^2}\frac{(\delta t)^2}{1.2} + \ldots\ldots$$

Hence,

$$\delta\rho = \frac{d\phi(t)}{dt}\delta t + \frac{d^2\phi(t)}{dt^2}\frac{(\delta t)^2}{1.2} + \&c.$$

And we are thus entitled to write, when δt has been made indefinitely small,

$$\text{Limit}\left(\frac{\delta\rho}{\delta t}\right)_{\delta t=0} = \frac{d\rho}{dt} = \frac{d\phi(t)}{dt} = \phi'(t).$$

In such a case as this, then, we are permitted to differentiate, or to form the differential coefficient of, a vector, according to the ordinary rules of the Differential Calculus. But great additional insight into the process is gained by applying Newton's method.

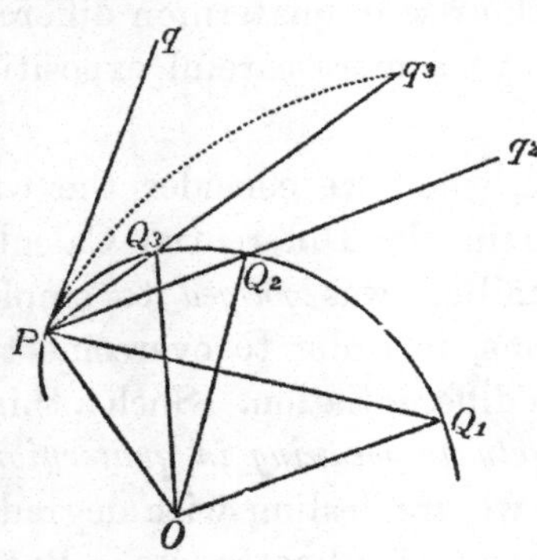

36.] Let $\overline{OP}$ be

$$\rho = \phi(t),$$

and $\overline{OQ_1}$

$$\rho = \phi(t+dt),$$

where dt is any number whatever.

The number t may here be taken as representing *time,* i. e. we may suppose a point to move along the curve in such a way that the value of t for the vector of point P of the curve denotes the interval which has elapsed (since a fixed epoch) when the moving point has reached the extremity of that vector. If, then, dt represent any interval, finite or not, we see that

$$\overline{OQ_1} = \phi(t+dt)$$

will be the vector of the point after the additional interval dt.

But this, in general, gives us little or no information as to the velocity of the point at P. We shall get a better approximation by halving the interval dt, and finding Q_2, where $\overline{OQ_2} = \phi(t + \frac{1}{2}dt)$, as the position of the moving point at that time. Here the vector virtually described in $\frac{1}{2}dt$ is $\overline{PQ_2}$. To find, on this supposition, the vector described in dt, we must double $\overline{PQ_2}$, and we find, as a second approximation to the vector which the moving point would have described in time dt, if it had moved for that period in the direction and with the velocity it had at P,

$$\begin{aligned}\overline{Pq_2} = 2\,\overline{PQ_2} &= 2\,(\overline{OQ_2} - \overline{OP}) \\ &= 2\,\{\phi(t + \tfrac{1}{2}dt) - \phi(t)\}.\end{aligned}$$

The next approximation gives

$$\begin{aligned}\overline{Pq_3} = 3\,\overline{PQ_3} &= 3\,(\overline{OQ_3} - \overline{OP}) \\ &= 3\,\{\phi(t + \tfrac{1}{3}dt) - \phi(t)\}.\end{aligned}$$

And so on, each step evidently leading us nearer the sought truth. Hence, to find the vector which would have been described in time dt had the circumstances of the motion at P remained undisturbed, we must find the value of

$$d\rho = \overline{Pq} = \mathcal{L}_{x=\infty}\, x\left\{\phi\left(t + \frac{1}{x}dt\right) - \phi(t)\right\}.$$

We have seen that in this particular case we may use Taylor's Theorem. We have, therefore,

$$\begin{aligned}d\rho &= \mathcal{L}_{x=\infty}\, x\left\{\phi'(t)\frac{1}{x}dt + \phi''(t)\frac{1}{x^2}\frac{(dt)^2}{1.2} + \&c.\right\} \\ &= \phi'(t)\,dt.\end{aligned}$$

And, if we choose, we may now write

$$\frac{d\rho}{dt} = \phi'(t).$$

37.] But it is to be most particularly remarked that in the whole of this investigation no regard whatever has been paid to the magnitude of dt. The question which we have now answered may be put in the form—*A point describes a given curve in a given manner. At any point of its path its motion suddenly ceases to be accelerated. What space will it describe in a definite interval?* As Hamilton well observes, this is, for a planet or comet, the case of a 'celestial Atwood's machine.'

38.] If we suppose the variable, in terms of which ρ is expressed, to be the arc, s, of the curve measured from some fixed point, we find as before

$$\begin{aligned}d\rho = \phi'(t)\,dt &= \frac{\phi'(t)\,ds}{\frac{ds}{dt}} \\ &= \phi'(s)\,ds.\end{aligned}$$

From the very nature of the question it is obvious that the length of $d\rho$ must in this case be ds. This remark is of importance, as we shall see later; and it may therefore be useful to obtain afresh the above result without any reference to time or velocity.

39.] Following strictly the process of Newton's VIIth Lemma, let us describe on Pq_2 an arc similar to PQ_2, and so on. Then obviously, as the subdivision of ds is carried farther, the new arc (whose length is always ds) more and more nearly coincides with the line which expresses the corresponding approximation to $d\rho$.

40.] As a final example let us take the hyperbola

$$\rho = \alpha t + \frac{\beta}{t}.$$

Here
$$d\rho = \left(\alpha - \frac{\beta}{t^2}\right) dt.$$

This shews that the tangent is parallel to the vector

$$\alpha t - \frac{\beta}{t}.$$

In words, *if the vector (from the centre) of a point in a hyperbola be one diagonal of a parallelogram, two of whose sides coincide with the asymptotes, the other diagonal is parallel to the tangent at the point.*

41.] Let us reverse this question, and *seek the envelope of a line which cuts off from two fixed axes a triangle of constant area.*

If the axes be in the directions of α and β, the intercepts may evidently be written αt and $\frac{\beta}{t}$. Hence the equation of the line is (§ 30)

$$\rho = \alpha t + x\left(\frac{\beta}{t} - \alpha t\right).$$

The condition of envelopment is, obviously, (see Chap. IX.)

$$d\rho = 0.$$

This gives
$$0 = \left\{\alpha - x\left(\frac{\beta}{t^2} + \alpha\right)\right\} dt + \left(\frac{\beta}{t} - \alpha t\right) dx\,*$$

Hence
$$(1-x)\,dt - t\,dx = 0,$$

and
$$-\frac{x}{t^2}\,dt + \frac{dx}{t} = 0.$$

* *We are not here to equate to zero the coefficients of* dt and dx; for we must remember that this equation is of the form

$$0 = p\alpha + q\beta,$$

where p and q are numbers; and that, so long as α and β are actual and non-parallel vectors, the existence of such an equation requires

$$p = 0, \quad q = 0.$$

From these, at once, $x = \frac{1}{2}$, since dx and dt are indeterminate. Thus the equation of the envelope is

$$\rho = \alpha t + \tfrac{1}{2}\left(\frac{\beta}{t} - \alpha t\right)$$

$$= \tfrac{1}{2}\left(\alpha t + \frac{\beta}{t}\right),$$

the hyperbola as before; α, β being portions of its asymptotes.

42.] It may assist the student to a thorough comprehension of the above process, if we put it in a slightly different form. Thus the equation of the enveloping line may be written

$$\rho = \alpha t(1-x) + \beta \frac{x}{t},$$

which gives $$d\rho = 0 = \alpha d\,(t(1-x)) + \beta d\left(\frac{x}{t}\right).$$

Hence, as α is not parallel to β, we must have

$$d\,(t(1-x)) = 0, \qquad d\left(\frac{x}{t}\right) = 0;$$

and these are, when expanded, the equations we obtained in the preceding section.

43.] For farther illustration we give a solution not directly employing the differential calculus. The equations of any two of the enveloping lines are

$$\rho = \alpha t + x\left(\frac{\beta}{t} - \alpha t\right),$$

$$\rho = \alpha t_1 + x_1\left(\frac{\beta}{t_1} - \alpha t_1\right),$$

t and t_1 being given, while x and x_1 are indeterminate.

At the point of intersection of these lines we have (§ 26),

$$\left.\begin{aligned} t(1-x) &= t_1(1-x_1), \\ \frac{x}{t} &= \frac{x_1}{t_1}. \end{aligned}\right\}$$

These give, by eliminating x_1,

$$t(1-x) = t_1\left(1 - \frac{t_1}{t}x\right),$$

or $$x = \frac{t}{t_1 + t}.$$

Hence the vector of the point of intersection is

$$\rho = \frac{\alpha t t_1 + \beta}{t_1 + t},$$

and thus, for the ultimate intersections, where $\mathcal{L}\frac{t_1}{t} = 1$,

$$\rho = \tfrac{1}{2}\left(\alpha t + \frac{\beta}{t}\right) \text{ as before.}$$

COR. (1). If $tt_1 = 1$,

$$\rho = \frac{\alpha + \beta}{t + \frac{1}{t}};$$

or the intersection lies in the diagonal of the parallelogram on α, β.

COR. (2). If $t_1 = mt$, where m is constant,

$$\rho = \frac{mt\alpha + \frac{\beta}{t}}{m+1}.$$

But we have also $x = \frac{1}{m+1}$.

Hence *the locus of a point which divides in a given ratio a line cutting off a given area from two fixed axes, is a hyperbola of which these axes are the asymptotes.*

COR. (3). If we take

$$tt_1(t+t_1) = \text{constant}$$

the locus is a parabola; and so on.

44.] The reader who is fond of Anharmonic Ratios and Trans versals will find in the early chapters of Hamilton's *Elements of Quaternions* an admirable application of the composition of vectors to these subjects. The Theory of Geometrical Nets, in a plane, and in space, is there very fully developed; and the method is shewn to include, as particular cases, the processes of Grassmann's *Ausdehnungslehre* and Möbius' *Barycentrische Calcul.* Some very curious investigations connected with curves and surfaces of the second and third orders are also there founded upon the composition of vectors.

EXAMPLES TO CHAPTER I.

1. The lines which join, towards the same parts, the extremities of two equal and parallel lines are themselves equal and parallel. (*Euclid*, I. xxxiii.)

2. Find the vector of the middle point of the line which joins

the middle points of the diagonals of any quadrilateral, plane or gauche, the vectors of the corners being given; and so prove that this point is the mean point of the quadrilateral.

If two opposite sides be divided proportionally, and two new quadrilaterals be formed by joining the points of division, the mean points of the three quadrilaterals lie in a straight line.

Shew that the mean point may also be found by bisecting the line joining the middle points of a pair of opposite sides.

3. Verify that the property of the coefficients of three vectors whose extremities are in a line (§ 30) is not interfered with by altering the origin.

4. If two triangles ABC, abc, be so situated in space that Aa, Bb, Cc meet in a point, the intersections of AB, ab, of BC, bc, and of CA, ca, lie in a straight line.

5. Prove the converse of 4, i. e. if lines be drawn, one in each of two planes, from any three points in the straight line in which these planes meet, the two triangles thus formed are sections of a common pyramid.

6. If five quadrilaterals be formed by omitting in succession each of the sides of any pentagon, the lines bisecting the diagonals of these quadrilaterals meet in a point. (H. Fox Talbot.)

7. Assuming, as in § 7, that the operator

$$\cos\theta + \sqrt{-1}\sin\theta$$

turns any radius of a given circle through an angle θ in the positive direction of rotation, without altering its length, deduce the ordinary formulæ for $\cos(A+B)$, $\cos(A-B)$, $\sin(A+B)$, and $\sin(A-B)$, in terms of sines and cosines of A and B.

8. If two tangents be drawn to a hyperbola, the line joining the centre with their point of intersection bisects the lines joining the points where the tangents meet the asymptotes: and the tangent at the point where it meets the curves bisects the intercepts of the asymptotes.

9. Any two tangents, limited by the asymptotes, divide each other proportionally.

10. If a chord of a hyperbola be one diagonal of a parallelogram whose sides are parallel to the asymptotes, the other diagonal passes through the centre.

11. Shew that $\rho = x^2\alpha + y^2\beta + (x+y)^2\gamma$

is the equation of a cone of the second degree, and that its section by the plane

$$\rho = \frac{p\alpha + q\beta + r\gamma}{p+q+r}$$

is an ellipse which touches, at their middle points, the sides of the triangle of whose corners α, β, γ are the vectors. (Hamilton, *Elements*, p. 96.)

12. The lines which divide, proportionally, the pairs of opposite sides of a gauche quadrilateral, are the generating lines of a hyper bolie paraboloid. (*Ibid.* p. 97.)

13. Shew that $\rho = x^3\alpha + y^3\beta + z^3\gamma$,

where $x + y + z = 0$,

represents a cone of the third order, and that its section by the plane

$$\rho = \frac{p\alpha + q\beta + r\gamma}{p + q + r}$$

is a cubic curve, of which the lines

$$\rho = \frac{p\alpha + q\beta}{p + q}, \text{ \&c.}$$

are the asymptotes and the three (real) tangents of inflexion. Also that the mean point of the triangle formed by these lines is a conjugate point of the curve. Hence that the vector $\alpha + \beta + \gamma$ is a conjugate ray of the cone. (*Ibid.* p. 96.)

CHAPTER II.

PRODUCTS AND QUOTIENTS OF VECTORS.

45.] We now come to the consideration of points in which the Calculus of Quaternions differs entirely from any previous mathematical method; and here we shall get an idea of what a Quaternion is, and whence it derives its name. These points are fundamentally involved in the novel use of the symbols of multiplication and division. And the simplest introduction to the subject seems to be the consideration of the quotient, or ratio, of two vectors.

46.] If the given vectors be parallel to each other, we have already seen (§ 22) that either may be expressed as a *numerical* multiple of the other; the multiplier being simply the ratio of their lengths, taken positively if they are similarly directed, negatively if they run opposite ways.

47.] If they be not parallel, let $\overline{OA}$ and $\overline{OB}$ be drawn parallel and equal to them from any point O; and the question is reduced to finding the value of the ratio of two vectors drawn from the same point. Let us try to find *upon how many distinct numbers this ratio depends.*

We may suppose $\overline{OA}$ to be changed into $\overline{OB}$ by the following processes.

1st. Increase or diminish the length of $\overline{OA}$ till it becomes equal to that of $\overline{OB}$. For this only *one* number is required, viz. the ratio of the lengths of the two vectors. As Hamilton remarks, this is a positive, or rather a *signless*, number.

2nd. Turn $\overline{OA}$ about O until its direction coincides with that of $\overline{OB}$, and (remembering the effect of the first operation) we see that the two vectors now coincide or become identical. To specify this operation *three* more numbers are required, viz. *two* angles (such as node and inclination in the case of a planet's orbit) to fix the plane in which the rotation takes place, and *one* angle for the amount of this rotation.

Thus it appears that the ratio of two vectors, or the multiplier required to change one vector into another, in general depends upon *four* distinct numbers, whence the name QUATERNION.

The particular case of perpendicularity of the two vectors, where their quotient is a vector perpendicular to their plane, is fully considered below; §§ 64, 65, 72, &c.

48.] It is obvious that the operations just described may be performed, with the same result, in the opposite order, being perfectly independent of each other. Thus it appears that a quaternion, considered as the factor or agent which changes one definite vector into another, may itself be decomposed into two factors of which the order is immaterial.

The *stretching* factor, or that which performs the first operation in § 47, is called the TENSOR, and is denoted by prefixing T to the quaternion considered.

The *turning* factor, or that corresponding to the second operation in § 47, is called the VERSOR, and is denoted by the letter U prefixed to the quaternion.

49.] Thus, if $\overline{OA} = \alpha$, $\overline{OB} = \beta$, and if q be the quaternion which changes α to β, we have

$$\beta = q\alpha,$$

which we may write in the form

$$\frac{\beta}{\alpha} = q, \quad \text{or} \quad \beta\alpha^{-1} - q,$$

if we agree to *define* that

$$\frac{\beta}{\alpha} . \alpha = \beta\alpha^{-1} . \alpha = \beta.$$

Here it is to be particularly noticed that we write q *before* α to signify that α is multiplied by q, not q multiplied by α.

This remark is of extreme importance in quaternions, for, as we shall soon see, the Commutative Law does not generally apply to the factors of a product.

We have also, by §§ 47, 48,

$$q = Tq\,Uq = Uq\,Tq,$$

where, as before, Tq depends merely on the relative lengths of α and β, and Uq depends solely on their directions.

Thus, if α_1 and β_1 be vectors of unit length parallel to α and β respectively,

$$T\frac{\beta_1}{\alpha_1} - 1, \qquad U\frac{\beta_1}{\alpha_1} = U\frac{\beta}{\alpha}.$$

As will soon be shewn, when α is perpendicular to β, the versor of the quotient is quadrantal, i. e. it is a unit-vector.

√50.] We must now carefully notice that the quaternion which is the quotient when β is divided by α in no way depends upon the *absolute* lengths, or directions, of these vectors. Its value will remain unchanged if we substitute for them any other pair of vectors which

(1) have their lengths in the same ratio,

(2) have their common plane the same or parallel,

and (3) make the same angle with each other.

Thus in the annexed figure

$$\frac{\overline{O_1 B_1}}{\overline{O_1 A_1}} = \frac{\overline{OB}}{\overline{OA}}$$

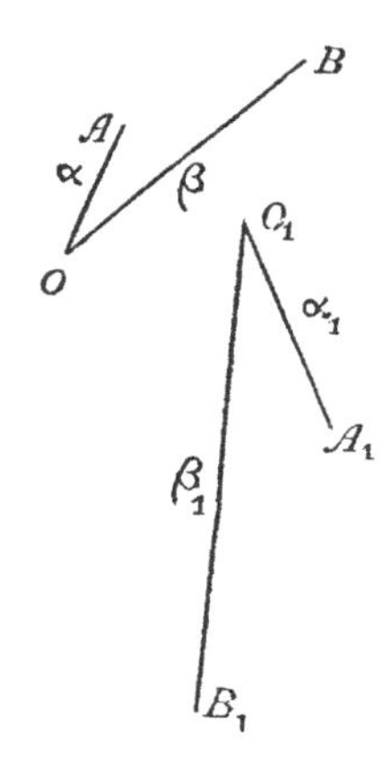

if, and only if,

(1) $\dfrac{O_1 B_1}{O_1 A_1} = \dfrac{OB}{OA}$,

(2) plane AOB parallel to plane $A_1 O_1 B_1$,

(3) $\angle AOB = \angle A_1 O_1 B_1$.

[Equality of angles is understood to include similarity in direction. Thus the rotation about an upward axis is negative (or right-handed) from OA to OB, and also from $O_1 A_1$ to $O_1 B_1$.]

√51.] The *Reciprocal* of a quaternion q is defined by the equation,

$$q\frac{1}{q} = qq^{-1} = 1.$$

Hence if $$\frac{\beta}{\alpha} = q, \text{ or}$$

$$\beta = q\alpha,$$

we must have $$\frac{\alpha}{\beta} = \frac{1}{q} = q^{-1}.$$

For this gives $$\frac{\alpha}{\beta}\cdot\beta = q^{-1}.q\alpha,$$

and each member of the equation is evidently equal to α.

Or, we may reason thus, q changes $\overline{OA}$ to $\overline{OB}$, q^{-1} must therefore change $\overline{OB}$ to $\overline{OA}$, and is therefore expressed by $\frac{\alpha}{\beta}$ (§ 49).

The tensor of the reciprocal of a quaternion is therefore the reciprocal of the tensor; and the versor differs merely by the *reversal* of its representative angle. The versor, it must be remembered, gives the plane and angle of the turning—it has nothing to do with the extension.

52.] The *Conjugate* of a quaternion q, written Kq, has the same tensor, plane, and angle, only the angle is taken the reverse way. Thus, if OA, OB, OA', lie in one plane, and if

$$OA' = OA, \text{ and } \angle A'OB = \angle AOB, \text{ we have}$$

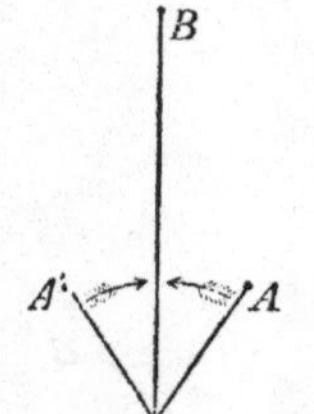

$$\frac{\overline{OB}}{\overline{OA}} = q, \text{ and } \frac{\overline{OB}}{\overline{OA'}} = \text{conjugate of } q = Kq.$$

By last section we see that

$$Kq = (Tq)^2 q^{-1},$$

Hence $$qKq = Kq \, . \, q = (Tq)^2.$$

This proposition is obvious, if we recollect that the tensors of q and Kq are equal, and that the versors are such that either *annuls* the effect of the other. The joint effect of these factors is therefore merely to multiply twice over by the common tensor.

53.] It is evident from the results of § 50 that, if α and β be of equal length, their quaternion quotient becomes a versor (the tensor being unity) and may be represented indifferently by any one of an infinite number of arcs of given length lying on the circumference of a circle, of which the two vectors are radii. This is of considerable importance in the proofs which follow.

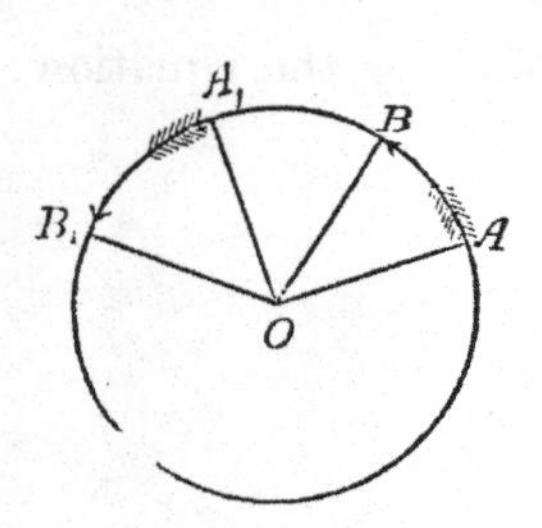

Thus the versor $\frac{\overline{OB}}{\overline{OA}}$ may be represented in magnitude, plane, and direction (§ 50) by the arc AB, which may in this extended sense be written $\overset{\frown}{AB}$.

And, similarly, the versor $\frac{\overline{OB_1}}{\overline{OA_1}}$ is represented by $\overset{\frown}{A_1B_1}$ which is *equal* to (and measured in the same direction as) $\overset{\frown}{AB}$ if

$$\angle A_1OB_1 = \angle AOB,$$

i.e. if the versors are equal, in the quaternion meaning of the word.

4.] By the aid of this process, when a versor is represented as rc of a great circle on the unit-sphere, we can easily prove that *ternion multiplication is not generally commutative.*

'hus let q be the versor $\widehat{AB}$ or $\frac{\overline{OB}}{\overline{OA}}$. ke $BC = AB$, (which, it must be embered, makes the points A, B, C one great circle), then q may also 'epresented by $\frac{\overline{OC}}{\overline{OB}}$.

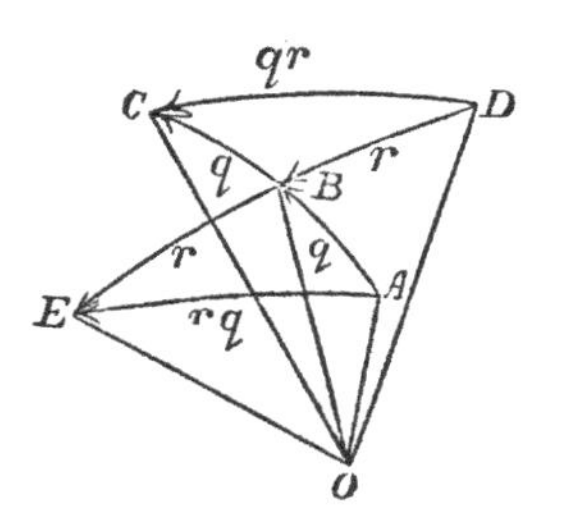

n the same way any other versor r be represented by $\widehat{DB}$ or BE and by $\frac{\overline{OB}}{\overline{OD}}$ or $\frac{\overline{OE}}{\overline{OB}}$.

'he line OB in the figure is definite, and is given by the interion of the planes of the two versors; O being the centre of the -sphere.

ow $\qquad r\overline{OD} = \overline{OB}, \quad$ and $\quad q\overline{OB} = \overline{OC},$

ce $\qquad qr\,OD = \overline{OC},$

$r = \frac{OC}{\overline{OD}}$, and may therefore be represented by the arc $\widehat{DC}$ of eat circle.

ut rq is easily seen to be represented by the arc $\widehat{AE}$.

$$q\overline{OA} = OB, \quad \text{and} \quad r\overline{OB} = \overline{OE},$$

nce $\qquad rq\overline{OA} = \overline{OE}, \quad$ and $\quad rq = \frac{\overline{OE}}{\overline{OA}}.$

s the versors rq and qr, though represented by arcs of equal 'th, are not generally in the same plane and are therefore unl: unless the planes of q and r coincide.

alling OA α, we see that we have assumed, or defined, in the e proof, that $q.r\alpha = qr.\alpha$ and $r.q\alpha = rq.\alpha$ when $q\alpha$, $r\alpha$, $q.r\alpha$, and are all *vectors.*

5.] Obviously CB is Kq, $\widehat{BD}$ is Kr, and CD is $K(qr)$. But $= \widehat{BD}.CB$, which gives us the very important theorem

$$K(qr) = Kr.Kq,$$

the conjugate of the product of two quaternions is the product of r conjugates in inverted order.

6.] The propositions just proved are, of course, true of quaters as well as of versors; for the former involve only an additional numerical factor which has reference to the length merely, and not the direction, of a vector (§ 48).

57.] Seeing thus that the commutative law does not in general hold in the multiplication of quaternions, let us enquire whether the Associative Law holds. That is, if p, q, r be three quaternions, have we

$$p.qr = pq.r\,?$$

This is, of course, obviously true if p, q, r be numerical quantities, or even any of the imaginaries of algebra. But it cannot be considered as a truism for symbols which do not in general give

$$pq = qp.$$

58.] In the first place we remark that p, q, and r may be considered as versors only, and therefore represented by arcs of great circles, for their tensors may obviously (§ 48) be divided out from both sides, being commutative with the versors.

Let $AB = p$, $\widehat{ED} = \widehat{CA} = q$, and $\widehat{FE} = r$.

Join BC and produce the great circle till it meets EF in H, and make $\widehat{KH} = \widehat{FE} = r$, and $\widehat{HG} = CB = pq$ (§ 54).

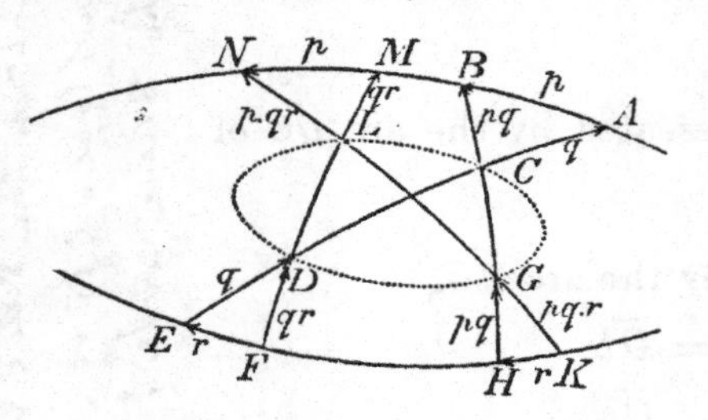

Join GK. Then

$$\widehat{KG} = \widehat{HG}\,.\,\widehat{KH} = pq\,.\,r.$$

Join FD and produce it to meet AB in M. Make

$$\widehat{LM} = \widehat{FD},$$

and $\widehat{MN} = \widehat{AB}$,

and join NL. Then

$$LN = \widehat{MN}\,.\,LM = p.qr.$$

Hence to shew that $p.qr = pq.r$ all that is requisite is to prove that LN, and KG, described as above, are *equal arcs of the same great circle*, since, by the figure, they are evidently measured in the same direction. This is perhaps most easily effected by the help of the fundamental properties of the curves known as *Spherical Conics*. As they are not usually familiar to students, we make a slight digression for the purpose of proving these fundamental properties; after Chasles, by whom and Magnus they were discovered. An independent proof of the associative principle will presently be indicated, and in Chapter VII we shall employ quaternions to give an independent proof of the theorems now to be established.

59.*] Def. *A spherical conic is the curve of intersection of a cone of the second degree with a sphere, the vertex of the cone being the centre of the sphere.*

LEMMA. If a cone have one series of circular sections, it has another series, and any two circles belonging to different series lie on a sphere. This is easily proved as follows.

Describe a sphere, A, cutting the cone in one circular section, C, and in any other point whatever, and let the side OpP of the cone meet A in p, P; P being a point in C. Then $PO \cdot Op$ is constant, and, therefore, since P lies in a plane, p lies on a sphere, a, passing through O. Hence the locus, c, of p is a circle, being the intersection of the two spheres A and a.

Let OqQ be any other side of the cone, q and Q being points in c, C respectively. Then the quadrilateral $qQPp$ is inscribed in a circle (that in which its plane cuts the sphere A) and the exterior angle at p is equal to the interior angle at Q. If OL, OM be the lines in which the plane POQ cuts the *cyclic planes* (planes through O parallel to the two series of circular sections) they are obviously parallel to pq, QP, respectively; and therefore

$$\angle LOp = \angle Opq = \angle OQP = \angle MOQ.$$

Let any third side, OrR, of the cone be drawn, and let the plane OPR cut the cyclic planes in Ol, Om respectively. Then, evidently,

$$\angle lOL = \angle qpr,$$
$$\angle MOm = \angle QPR,$$

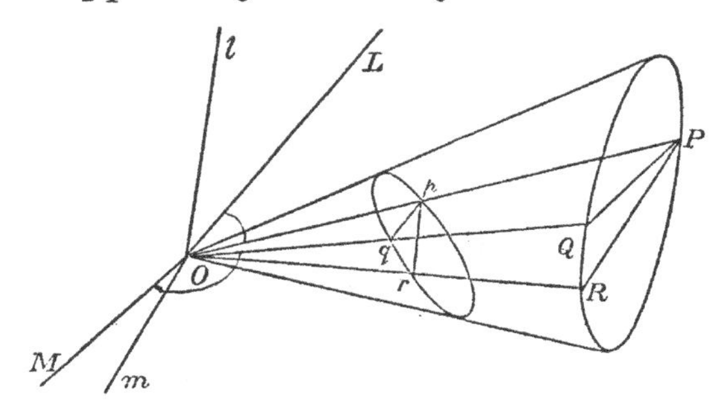

and these angles are independent of the position of the points p and P, if Q and R be fixed points.

In a section of the above diagram by a sphere whose centre is O, lL, Mm are the great circles which represent the cyclic planes, PQR is the spherical conic which represents the cone. The point P represents the line OpP, and so with the others. The propositions above may now be stated thus

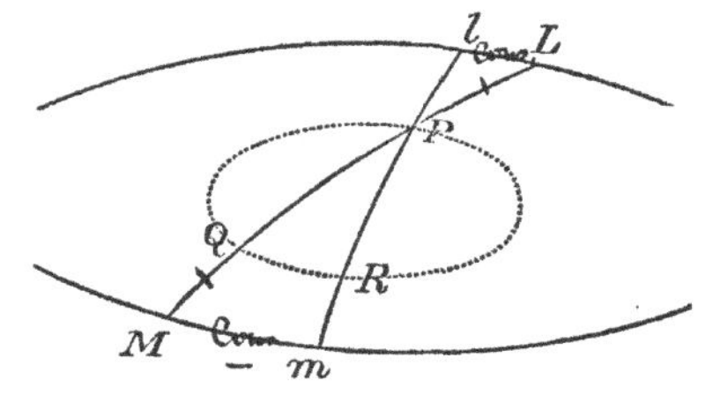

$$\text{Arc } PL = \text{arc } MQ;$$

and, if Q and R be fixed, Mm and lL are constant arcs whatever be the position of P.

60.] The application to § 58 is now obvious. In the figure of that article we have

$$\widehat{FE} = \widehat{KH},\ \widehat{ED} = \widehat{CA},\ \widehat{HG} = \widehat{CB},\ \widehat{LM} = \widehat{FD}.$$

Hence L, C, G, D are points of a spherical conic whose cyclic planes are those of AB, FE. Hence also KG passes through L, and with LM intercepts on AB an arc equal to AB. That is, it passes through N, or KG and LN are arcs of the same great circle· and they are equal, for G and L are points in the spherical conic.

Also, the associative principle holds for any number of quaternion factors. For, obviously,

$$qr.st = qrs.t = \&c., \&c.,$$

since we may consider qr as a single quaternion, and the above proof applies directly.

61.] That quaternion addition, and therefore also subtraction, is commutative, it is easy to shew.

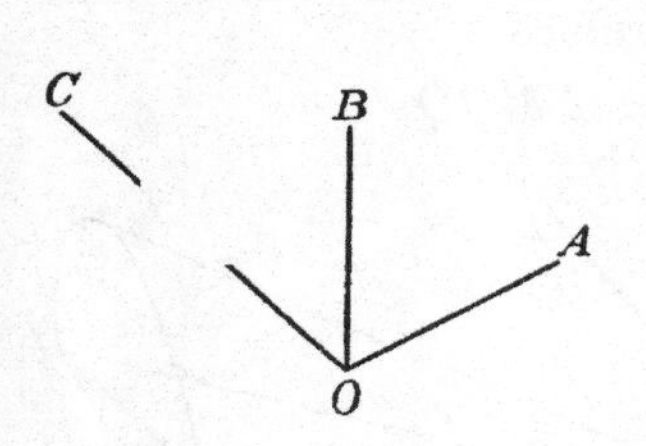

For if the planes of two quaternions, q and r, intersect in the line OA, we may take any vector $\overline{OA}$ in that line, and at once find two others, $\overline{OB}$ and $\overline{OC}$, such that

$$\overline{OB} = q\,\overline{OA},$$

and $$\overline{OC} = r\,\overline{OA}.$$

And $\quad (q+r)\,\overline{OA} = \overline{OB} + \overline{OC} = \overline{OC} + \overline{OB} = (r+q)\,OA,$

since *vector* addition is commutative (§ 27).

Here it is obvious that $(q+r)\,OA$, being the diagonal of the parallelogram on OB, $\overline{OC}$, divides the angle between OB and OC in a ratio depending solely on the ratio of the lengths of these lines, i. e. on the ratio of the tensors of q and r. This will be useful to us in the proof of the distributive law, to which we proceed.

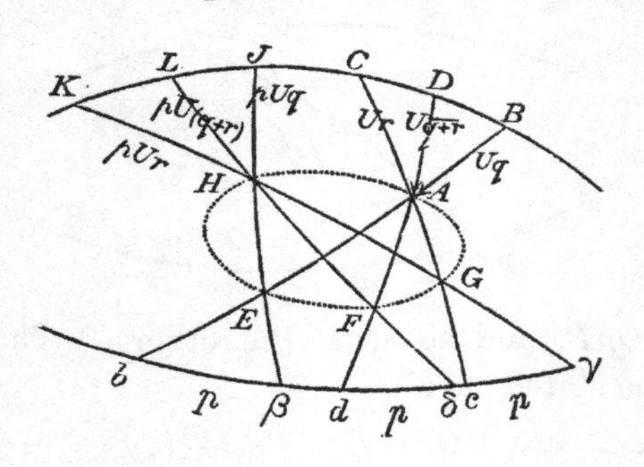

62.] Quaternion multiplication, and therefore division, is distributive. One simple proof of this depends on the possibility, shortly to be proved, of representing *any* quaternion as a linear function of three given rectangular unit-vectors. And when the proposition is thus established, the associative principle may readily be deduced from it.

But we may employ for its proof the properties of Spherical

Conics already employed in demonstrating the truth of the associative principle. For continuity we give an outline of the proof by this process.

Let $\widehat{BA}$, $\widehat{CA}$ represent the versors of q and r, and bc the great circle whose plane is that of p.

Then, if we take as operand the vector $\overline{OA}$, it is obvious that $U(q+r)$ will be represented by some such arc as DA where B, D, C are in one great circle; for $(q+r)\,\overline{OA}$ is in the same plane as $q\,\overline{OA}$ and $r\,\overline{OA}$, and the relative magnitudes of the arcs BD and DC depend solely on the tensors of q and r. Produce BA, DA, CA to meet bc in b, d, c respectively, and make

$$\widehat{Eb} = \widehat{BA}, \quad \widehat{Fd} = \widehat{DA}, \quad \widehat{Gc} = \widehat{CA}.$$

Also make $\widehat{b\beta} = \widehat{d\delta} = \widehat{c\gamma} = p$. Then E, F, G, A lie on a spherical conic of which BC and bc are the cyclic arcs. And, because $\widehat{b\beta} = \widehat{d\delta} = \widehat{c\gamma}$, $\widehat{\beta E}$, $\widehat{\delta F}$, $\widehat{\gamma G}$, when produced, meet in a point H which is also on the spherical conic (§ 59*). Let these arcs meet BC in J, L, K respectively. Then we have

$$\begin{aligned} JH &= E\beta = p\,Uq, \\ \widehat{LH} &= F\delta = p\,U(q+r), \\ KH &= G\gamma = p\,Ur. \end{aligned}$$

Also $$LJ = DB,$$

and $$\widehat{KL} = \widehat{CD}.$$

And, on comparing the portions of the figure bounded respectively by HKJ and by ACB we see that (when considered with reference to their effects as factors multiplying $\overline{OH}$ and $\overline{OA}$ respectively)

$p\,U(q+r)$ bears the same relation to $p\,Uq$ and $p\,Ur$

that $U(q+r)$ bears to Uq and Ur.

But $$T(q+r)\,U(q+r) = q+r = Tq\,Uq + Tr\,Ur.$$

Hence $$T(q+r).p\,U(q+r) = Tq.p\,Uq + Tr.p\,Ur;$$

or, since the tensors are mere numbers and commutative with all other factors, $$p(q+r) = pq + pr.$$

In a similar manner it may be proved that

$$(q+r)p = qp + rp.$$

And then it follows at once that

$$(p+q)(r+s) = pr + ps + qr + qs.$$

63.] By similar processes to those of § 53 we see that versors, and therefore also quaternions, are subject to the index-law

$$q^m.q^n = q^{m+n},$$

at least so long as m and n are positive integers.

The extension of this property to negative and fractional exponents must be deferred until we have defined a negative or fractional power of a quaternion.

64.] We now proceed to the special case of *quadrantal* versors, from whose properties it is easy to deduce all the foregoing results of this chapter. These properties were indeed those whose invention by Hamilton in 1843 led almost intuitively to the establishment of the Quaternion Calculus. We shall content ourselves at present with an assumption, which will be shewn to lead to consistent results; but at the end of the chapter we shall shew that no other assumption is possible, following for this purpose a very curious quasi-metaphysical speculation of Hamilton.

65.] Suppose we have a system of three mutually perpendicular unit-vectors, drawn from one point, which we may call for shortness I, J, K. Suppose also that these are so situated that a positive (i. e. *left-handed*) rotation through a right angle about I as an axis brings J to coincide with K. Then it is obvious that positive quadrantal rotation about J will make K coincide with I; and, about K, will make I coincide with J.

For definiteness we may suppose I to be drawn *eastwards*, J *northwards*, and K *upwards*. Then it is obvious that a positive (left-handed) rotation about the eastward line (I) brings the northward line (J) into a vertically upward position (K); and so of the others.

66.] Now the operator which turns J into K is a quadrantal versor (§ 53); and, as its axis is the vector I, we may call it i.

Thus $$\frac{K}{J} = i, \quad \text{or} \quad K = iJ. \quad \text{.................... (1)}$$

Similarly we may put $$\frac{I}{K} = j, \quad \text{or} \quad I = jK, \quad \text{.................... (2)}$$

and $$\frac{J}{I} = k, \quad \text{or} \quad J = kI. \quad \text{.................... (3)}$$

[It may here be noticed, merely to shew the symmetry of the system we are explaining, that if the three mutually perpendicular vectors I, J, K be made to revolve about a line equally inclined to all, so that I is brought to coincide with J, J will then coincide with K, and K with I: and the above equations will still hold good, only (1) will become (2), (2) will become (3), and (3) will become (1).]

67.] By the results of § 50 we see that

$$\frac{-J}{K} = \frac{K}{J};$$

i. e. a southward unit-vector bears the same ratio to an upward unit-vector that the latter does to a northward one; and therefore we have

$$\frac{-J}{K} = i, \quad \text{or} \quad -J = iK. \quad \ldots\ldots (4)$$

Similarly $$\frac{-K}{I} = j, \quad \text{or} \quad -K = jI; \quad \ldots\ldots (5)$$

and $$\frac{-I}{J} = k, \quad \text{or} \quad I = kJ. \quad \ldots\ldots (6)$$

68.] By (4) and (1) we have

$$J = iK = i(iJ) = i^2 J.$$

Hence $$i^2 = -1. \quad \ldots\ldots (7)$$

And, in the same way, (5) and (2) give

$$j^2 = -1, \quad \ldots\ldots (8)$$

and (6) and (3) $$k^2 = -1. \quad \ldots\ldots (9)$$

Thus, as the directions of I, J, K are perfectly arbitrary, we see that *the square of every quadrantal versor is negative unity.*

Though the following proof is in principle exactly the same as the foregoing, it may perhaps be of use to the student, in shewing him precisely the nature as well as the simplicity of the step we have taken.

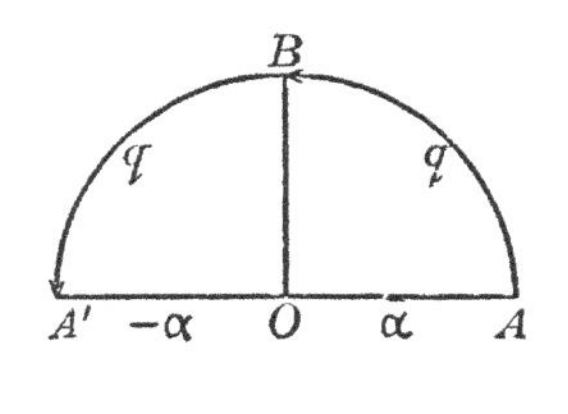

Let ABA' be a semicircle, whose centre is O, and let OB be perpendicular to AOA'.

Then $\dfrac{\overline{OB}}{\overline{OA}}$, $= q$ suppose, is a quadrantal versor, and is evidently equal to $\dfrac{\overline{OA'}}{\overline{OB}}$; §§ 50, 53.

Hence $$q^2 = \frac{\overline{OA'}}{\overline{OB}} \cdot \frac{\overline{OB}}{\overline{OA}} = \frac{\overline{OA'}}{\overline{OA}} = -1.$$

69.] Having thus found that the squares of i, j, k are each equal to negative unity; it only remains that we find the values of their products two and two. For, as we shall see, the result is such as to shew that the value of any other combination whatever of i, j, k (as factors of a product) may be deduced from the values of these squares and products.

Now it is obvious that

$$\frac{K}{-I} = \frac{I}{K} = j;$$

(i. e. the versor which turns a westward unit-vector into an upward one will turn the upward into an eastward unit);

$$\text{or} \quad K = j(-I) = -jI^*. \quad\text{.......................}\quad (10)$$

Now let us operate on the two equal vectors in (10) by the same versor, i, and we have

$$iK = i(-jI) = -ijI.$$

But by (4) and (3)

$$iK = -J = -kI.$$

Comparing these equations, we have

$$-ijI = -kI;$$

or, by § 54 (end), and symmetry gives

$$\left.\begin{aligned} ij &= k, \\ jk &= i, \\ ki &= j. \end{aligned}\right\} \quad\text{.............................}\quad (11)$$

The meaning of these important equations is very simple; and is, in fact, obvious from our construction in § 54 for the multiplication of versors; as we see by the annexed figure, where we must remember that i, j, k are quadrantal versors whose planes are at right angles, so that the figure represents a hemisphere divided into quadrantal triangles.

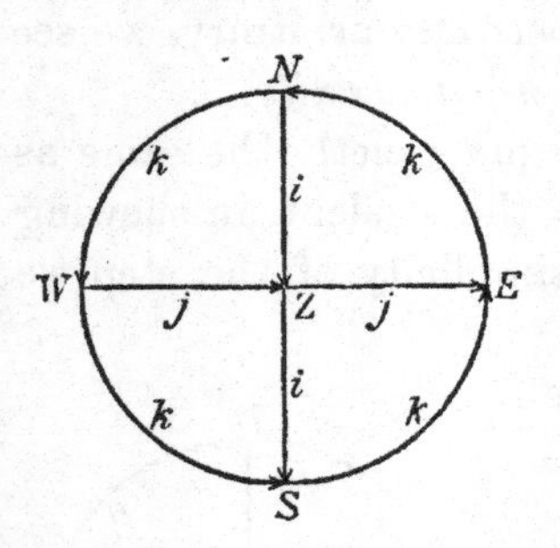

Thus, to shew that $ij = k$, we have, O being the centre of the sphere, N, E, S, W the north, east, south, and west and Z the zenith (as in § 65);

$$j\overline{OW} = \overline{OZ},$$

$$\text{whence} \quad ij\overline{OW} = i\overline{OZ} = \overline{OS} = k\overline{OW}.$$

70.] But, by the same figure,

$$i\overline{ON} = \overline{OZ},$$

$$\text{whence} \quad ji\overline{ON} = j\overline{OZ} = \overline{OE} = -\overline{OW} = -k\overline{ON}.$$

71.] From this it appears that

$$\left.\begin{aligned} ji &= -k, \\ kj &= -i, \\ ik &= -j, \end{aligned}\right\} \quad\text{.............................}\quad (12)$$

and similarly

and thus, by comparing (11),

$$\left.\begin{aligned} ij &= -ji = k, \\ jk &= -kj = i, \\ ki &= -ik = j. \end{aligned}\right\} \quad ((11), (12)).$$

* The negative sign, being a mere numerical factor, is evidently commutative with j; indeed we may, if necessary, easily assure ourselves of the fact that to turn the negative (or reverse) of a vector through a right (or indeed any) angle, is the same thing as to turn the vector through that angle and then reverse it.

These equations, along with

$$i^2 = j^2 = k^2 = -1 \quad ((7), (8), (9)),$$

contain essentially the whole of Quaternions. But it is easy to see that, for the first group, we may substitute the single equation

$$ijk = -1, \qquad (13)$$

since from it, by the help of the values of the squares of i, j, k, all the other expressions may be deduced. We may consider it proved in this way, or deduce it afresh from the figure above, thus

$$k\overline{ON} = \overline{OW},$$
$$jk\overline{ON} = j\overline{OW} = \overline{OZ},$$
$$ijk\overline{ON} = ij\overline{OW} = i\overline{OZ} = \overline{OS} = -\overline{ON}.$$

§ 72.] One most important step remains to be made, to wit the assumption referred to in § 64. We have treated i, j, k simply as quadrantal versors; and I, J, K as unit-vectors at right angles to each other, and coinciding with the axes of rotation of these versors. But if we collate and compare the equations just proved we have

$$\begin{cases} ij = k, & (11) \\ iJ = K, & (1) \end{cases}$$

$$\begin{cases} ji = -k, & (12) \\ jI = -K, & (10) \end{cases}$$

with the other similar groups symmetrically derived from them. Now the meanings we have assigned to i, j, k are quite independent of, and not inconsistent with, those assigned to I, J, K. And it is superfluous to use two sets of characters when one will suffice. Hence it appears that i, j, k may be substituted for I, J, K; in other words, *a unit-vector when employed as a factor may be considered as a quadrantal versor whose plane is perpendicular to the vector*. This is one of the main elements of the singular simplicity of the quaternion calculus.

§ 73.] Thus *the product, and therefore the quotient, of two perpendicular vectors is a third vector perpendicular to both.*

Hence the reciprocal (§ 51) of a vector is a vector which has the *opposite* direction to that of the vector, and its length is the reciprocal of the length of the vector.

The conjugate (§ 52) of a vector is simply the vector reversed.

Hence, by § 52, if α be a vector

$$(T\alpha)^2 = \alpha K\alpha = \alpha(-\alpha) = -\alpha^2.$$

§ 74.] We may now see that *every versor may be represented by a power of a unit-vector.*

For, if α be any vector perpendicular to i (which is *any* definite unit-vector),

$i\alpha, = \beta$, is a vector equal in length to α, but perpendicular to both i and α;

$$i^2\alpha = -\alpha,$$
$$i^3\alpha = -i\alpha = -\beta,$$
$$i^4\alpha = -i\beta - -i^2\alpha = \alpha.$$

Thus, by successive applications of i, α is turned round i as an axis through successive right angles. Hence it is natural to *define* i^m *as a versor which turns any vector perpendicular to i through m right angles in the positive direction of rotation about i as an axis.* Here m may have any real value whatever, whole or fractional, for it is easily seen that analogy leads us to interpret a negative value of m as corresponding to rotation in the negative direction.

75.] From this again it follows that *any quaternion may be expressed as a power of a vector.* For the tensor and versor elements of the vector may be so chosen that, when raised to the same power, the one may be the tensor and the other the versor of the given quaternion. The vector must be, of course, perpendicular to the plane of the quaternion.

76.] And we now see, as an immediate result of the last two sections, that the index-law holds with regard to powers of a quaternion (§ 63).

77.] So far as we have yet considered it, a quaternion has been regarded as the *product* of a tensor and a versor: we are now to consider it as a *sum*. The easiest method of so analysing it seems to be the following.

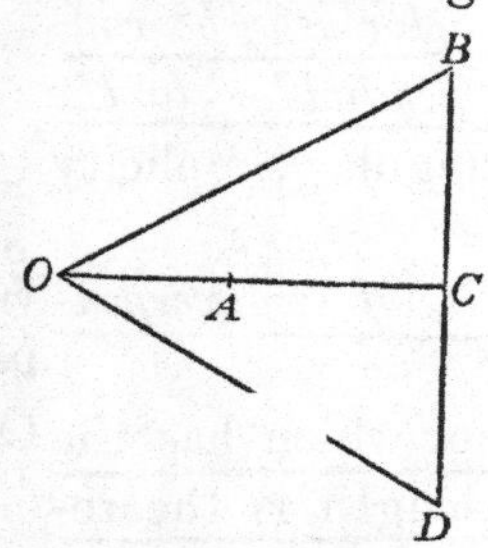

Let $\frac{\overline{OB}}{\overline{OA}}$ represent any quaternion. Draw BC perpendicular to OA, produced if necessary.

Then, § 19, $\overline{OB} = \overline{OC} + \overline{CB}$.

But, § 22, $\overline{OC} = x\overline{OA}$,

where x is a number, whose sign is the same as that of the cosine of $\angle AOB$.

Also, § 73, since CB is perpendicular to OA,

$$\overline{CB} = \gamma\overline{OA},$$

where γ is a vector perpendicular to OA and CB, i.e. to the plane of the quaternion.

Hence $$\frac{\overline{OB}}{\overline{OA}} = \frac{x\overline{OA} + \gamma\overline{OA}}{\overline{OA}} = x + \gamma.$$

Thus a quaternion, in general, may be decomposed into the sum of two parts, one numerical, the other a vector. Hamilton calls them the SCALAR, and the VECTOR, and denotes them respectively by the letters S and V prefixed to the expression for the quaternion.

78.] Hence $q = Sq + Vq$, and if in the above example

$$\frac{\overline{OB}}{\overline{OA}} = q,$$

then $$\overline{OB} = \overline{OC} + \overline{CB} = Sq.\overline{OA} + Vq.\overline{OA}\,*.$$

The equation above gives

$$\overline{OC} = Sq.\overline{OA},$$
$$\overline{CB} = Vq.\overline{OA}.$$

79.] If, in the figure of last section, we produce BC to D, so as to double its length, and join OD, we have, by § 52,

$$\frac{\overline{OD}}{\overline{OA}} = Kq = SKq + VKq;$$

$$\therefore \quad \overline{OD} = \overline{OC} + \overline{CD} = SKq.OA + VKq.\overline{OA}.$$

Hence $$\overline{OC} = SKq.\overline{OA},$$

and $$\overline{CD} = VKq.\overline{OA}.$$

Comparing this value of $\overline{OC}$ with that in last section, we find

$$SKq = Sq, \quad \ldots\ldots\ldots\ldots\ldots\ldots \quad (1)$$

or *the scalar of the conjugate of a quaternion is equal to the scalar of the quaternion.*

Again, $\overline{CD} = -\overline{CB}$ by the figure, and the substitution of their values gives $$VKq = -Vq, \quad \ldots\ldots\ldots\ldots\ldots\ldots \quad (2)$$

or *the vector of the conjugate of a quaternion is the vector of the quaternion reversed.*

We may remark that the results of this section are simple consequences of the fact that the symbols S, V, K are commutative †

Thus $$SKq = KSq = Sq,$$

since the conjugate of a number is the number itself; and

$$VKq = KVq = -Vq \quad (§\ 73).$$

* The points are inserted to shew that S and V apply only to q, and not to $q\overline{OA}$.

† It is curious to compare the properties of these quaternion symbols with those of the Elective Symbols of Logic, as given in BOOLE's wonderful treatise on the *Laws of Thought;* and to think that the same grand science of mathematical analysis, by processes remarkably similar to each other, reveals to us truths in the science of *position* far beyond the powers of the geometer, and truths of deductive reasoning to which unaided thought could never have led the logician.

Again, it is obvious that

$$\Sigma Sq = S\Sigma q, \quad \Sigma Vq = V\Sigma q,$$

and thence

$$\Sigma Kq = K\Sigma q.$$

80.] Since any vector whatever may be represented by

$$xi + yj + zk$$

where x, y, z are numbers (or Scalars), and i, j, k may be any three non-coplanar vectors, §§ 23, 25—though they are usually understood as representing a rectangular system of unit-vectors—and since any scalar may be denoted by w; we may write, for any quaternion q, the expression

$$q = w + xi + yj + zk \quad (§\ 78).$$

Here we have the essential dependence on four distinct numbers, from which the quaternion derives its name, exhibited in the most simple form.

And now we see at once that an equation such as

$$q' = q,$$

where

$$q' = w' + x'i + y'j + z'k,$$

involves, of course, the *four* equations

$$w' = w, \quad x' = x, \quad y' = y, \quad z' = z.$$

81.] We proceed to indicate another mode of proof of the distributive law of multiplication.

We have already defined, or assumed (§ 61), that

$$\frac{\beta}{\alpha} + \frac{\gamma}{\alpha} = \frac{\beta + \gamma}{\alpha},$$

or

$$\beta\alpha^{-1} + \gamma\alpha^{-1} - (\beta + \gamma)\,\alpha^{-1},$$

and have thus been able to understand what is meant by adding two quaternions.

But, writing α for α^{-1}, we see that this involves the equality

$$(\beta + \gamma)\alpha = \beta\alpha + \gamma\alpha;$$

from which, by taking the conjugates of both sides, we derive

$$\alpha'(\beta' + \gamma') = \alpha'\beta' + \alpha'\gamma' \quad (§\ 55).$$

And a combination of these results (putting $\beta + \gamma$ for α' in the latter, for instance) gives

$$(\beta + \gamma)(\beta' + \gamma') = (\beta + \gamma)\beta' + (\beta + \gamma)\gamma'$$
$$= \beta\beta' + \gamma\beta' + \beta\gamma' + \gamma\gamma' \text{ by the former.}$$

Hence *the distributive principle is true in the multiplication of vectors.*

It only remains to shew that it is true as to the scalar and vector parts of a quaternion, and then we shall easily attain the general proof.

Now, if a be any scalar, α any vector, and q any quaternion,

$$(a+\alpha)\,q = aq + \alpha q.$$

For, if β be the vector in which the plane of q is intersected by a plane perpendicular to α, we can find other two vectors, γ and δ, in these planes such that

$$\alpha = \frac{\gamma}{\beta}, \qquad q = \frac{\beta}{\delta}.$$

And, of course, a may be written $\frac{a\beta}{\beta}$; so that

$$(a+\alpha)q = \frac{a\beta+\gamma}{\beta}\cdot\frac{\beta}{\delta} = \frac{a\beta+\gamma}{\delta}$$

$$= a\frac{\beta}{\delta} + \frac{\gamma}{\delta} = a\frac{\beta}{\delta} + \frac{\gamma}{\beta}\,\frac{\beta}{\delta}$$

$$= aq + \alpha q.$$

And the conjugate may be written

$$q'(a'+\alpha') = q'a' + q'\alpha' \quad (\S\ 55).$$

Hence, generally,

$$(a+\alpha)(b+\beta) = ab + a\beta + b\alpha + \alpha\beta;$$

or, breaking up a and b each into the sum of two scalars, and α, β each into the sum of two vectors,

$$(a_1+a_2+\alpha_1+\alpha_2)(b_1+b_2+\beta_1+\beta_2)$$
$$= (a_1+a_2)(b_1+b_2) + (a_1+a_2)(\beta_1+\beta_2) + (b_1+b_2)(\alpha_1+\alpha_2)$$
$$+ (\alpha_1+\alpha_2)(\beta_1+\beta_2)$$

(by what precedes, all the factors on the right are distributive, so that we may easily put it in the form)

$$= (a_1+\alpha_1)(b_1+\beta_1) + (a_1+\alpha_1)(b_2+\beta_2) + (a_2+\alpha_2)(b_1+\beta_1)$$
$$+ (a_2+\alpha_2)(b_2+\beta_2).$$

Putting $\quad a_1+\alpha_1 = p, \quad a_2+\alpha_2 = q, \quad b_1+\beta_1 = r, \quad b_2+\beta_2 = s,$

we have $$(p+q)(r+s) = pr + ps + qr + qs.$$

82.] For variety, we shall now for a time forsake the geometrical mode of proof we have hitherto adopted, and deduce some of our next steps from the analytical expression for a quaternion given in § 80, and the properties of a rectangular system of unit-vectors as in § 71.

We will commence by proving the result of § 77 anew.

83.] Let $$\alpha = x\,i + yj + z\,k,$$
$$\beta = x'i + y'j + z'k.$$

Then, because by § 71 every product or quotient of i, j, k is reducible to one of them or to a number, we are entitled to assume

$$q = \frac{\beta}{a} = \omega + \xi i + \eta j + \zeta k,$$

where ω, ξ, η, ζ are numbers. This is the proposition of § 80.

84.] But it may be interesting to find ω, ξ, η, ζ in terms of x, y, z, x', y', z'.

We have $$\beta = q a,$$

or $$x'i + y'j + z'k = (\omega + \xi i + \eta j + \zeta k)(xi + yj + zk)$$
$$= -(\xi x + \eta y + \zeta z) + (\omega x + \eta z - \zeta y)i + (\omega y + \zeta x - \xi z)j + (\omega z + \xi y - \eta x)k,$$

as we easily see by the expressions for the powers and products of i, j, k, given in § 71. But the student must pay particular attention to the *order* of the factors, else he is certain to make mistakes.

This (§ 80) resolves itself into the four equations

$$\begin{aligned} 0 &= \quad\;\; \xi x + \eta y + \zeta z, \\ x' &= \omega x \quad\;\; + \eta z - \zeta y, \\ y' &= \omega y - \xi z \quad\;\; + \zeta x, \\ z' &= \omega z + \xi y - \eta x. \end{aligned}$$

The three last equations give

$$xx' + yy' + zz' = \omega(x^2 + y^2 + z^2),$$

which determines ω.

Also we have, from the same three, by the help of the first

$$\xi x' + \eta y' + \zeta z' = 0;$$

which, combined with the first, gives

$$\frac{\xi}{yz' - zy'} = \frac{\eta}{zx' - xz'} = \frac{\zeta}{xy' - yx'},$$

and the common value of these three fractions is then easily seen to be

$$\frac{1}{x^2 + y^2 + z^2}.$$

It is easy enough to interpret these expressions by means of ordinary cöordinate geometry: but a much simpler process will be furnished by quaternions themselves in the next chapter, and, in giving it, we shall refer back to this section.

85.] The associative law of multiplication is now to be proved by means of the distributive (§ 81). We leave the proof to the student. He has merely to multiply together the factors

$$w + xi + yj + zk, \quad w' + x'i + y'j + z'k, \quad \text{and} \quad w'' + x''i + y''j + z''k,$$

as follows:—

First, multiply the third factor by the second, and then multiply the product by the first; next, multiply the second factor by the first and employ the product to multiply the third: always remembering that the multiplier in any product is placed *before* the multiplicand. He will find the scalar parts and the coefficients of i, j, k, in these products, respectively equal, each to each.

86.] With the same expressions for α, β, as in section 83, we have

$$\alpha\beta=(xi+yj+zk)(x'i+y'j+z'k)$$
$$=-(xx'+yy'+zz')+(yz'-zy')i+(zx'-xz')j+(xy'-yx')k.$$

But we have also

$$\beta\alpha=-(xx'+yy'+zz')-(yz'-zy')i-(zx'-xz')j-(xy'-yx')k.$$

The only difference is in the *sign* of the vector parts.

Hence $$S\alpha\beta=S\beta\alpha, \quad (1)$$
$$V\alpha\beta=-V\beta\alpha, \quad (2)$$
also $$\alpha\beta+\beta\alpha=2S\alpha\beta, \quad (3)$$
$$\alpha\beta-\beta\alpha=2V\alpha\beta, \quad (4)$$
and, finally, by § 79, $$\alpha\beta=K\beta\alpha. \quad (5)$$

87.] If $\alpha=\beta$ we have of course (§ 25)

$$x=x', \quad y=y', \quad z=z',$$

and the formulae of last section become

$$\alpha\beta=\beta\alpha=\alpha^2=-(x^2+y^2+z^2);$$

which was anticipated in § 73, where we proved the formula

$$(T\alpha)^2=-\alpha^2,$$

and also, to a certain extent, in § 25.

88.] Now let q and r be any quaternions, then

$$S.qr=S.(Sq+Vq)(Sr+Vr),$$
$$=S.(Sq\,Sr+Sr.Vq+Sq.Vr+Vq\,Vr),$$
$$=Sq\,Sr+SVq\,Vr,$$

since the two middle terms are vectors.

Similarly, $$S.rq=Sr\,Sq+SVr\,Vq.$$

Hence, since by (1) of § 86 we have

$$SVq\,Vr=SVr\,Vq,$$

we see that $$S.qr=S.rq, \quad (1)$$

a formula of considerable importance.

It may easily be extended to any number of quaternions, because, r being arbitrary, we may put for it rs. Thus we have

$$S.qrs=S.rsq,$$
$$=S.sqr$$

by a second application of the process. In words, we have the theorem—*the scalar of the product of any number of given quaternions depends only upon the cyclical order in which they are arranged.*

89.] An important case is that of three factors, each a vector. The formula then becomes

$$S.\alpha\beta\gamma = S.\beta\gamma\alpha = S.\gamma\alpha\beta.$$

But
$$\begin{aligned} S.\alpha\beta\gamma &= S\alpha(S\beta\gamma + V\beta\gamma) \\ &= S\alpha V\beta\gamma, \quad \text{since } \alpha S\beta\gamma \text{ is a vector,} \\ &= -S\alpha V\gamma\beta, \quad \text{by (2) of § 86,} \\ &= -S\alpha(S\gamma\beta + V\gamma\beta) \\ &= -S.\alpha\gamma\beta. \end{aligned}$$

Hence *the scalar of the product of three vectors changes sign when the cyclical order is altered.*

Other curious propositions connected with this will be given later, as we wish to devote this chapter to the production of the fundamental formulae in as compact a form as possible.

90.] By (4) of § 86,

$$2V\beta\gamma = \beta\gamma - \gamma\beta.$$

Hence
$$2V\alpha V\beta\gamma = V\alpha(\beta\gamma - \gamma\beta)$$

(by multiplying both by α, and taking the vector parts of each side)

$$= V(\alpha\beta\gamma + \beta\alpha\gamma - \beta\alpha\gamma - \alpha\gamma\beta)$$

(by introducing the null term $\beta\alpha\gamma - \beta\alpha\gamma$).

That is

$$\begin{aligned} 2V\alpha V\beta\gamma &= V.(\alpha\beta + \beta\alpha)\gamma - V(\beta S\alpha\gamma + \beta V\alpha\gamma + S\alpha\gamma.\beta + V\alpha\gamma.\beta) \\ &= V(2S\alpha\beta)\gamma - 2V\beta S\alpha\gamma \end{aligned}$$

(if we notice that $V.V\alpha\gamma.\beta = -V\beta V\alpha\gamma$, by (2) of § 86).

Hence
$$V\alpha V\beta\gamma = \gamma S\alpha\beta - \beta S\gamma\alpha, \quad \ldots\ldots\ldots\ldots\ldots\ldots\ldots \quad (1)$$

a formula of constant occurrence.

Adding $\alpha S\beta\gamma$ to both sides we get another most valuable formula

$$V.\alpha\beta\gamma = \alpha S\beta\gamma - \beta S\gamma\alpha + \gamma S\alpha\beta; \quad \ldots\ldots\ldots\ldots\ldots\ldots \quad (2)$$

and the form of this shews that we may interchange γ and α without altering the right-hand member. This gives

$$V.\alpha\beta\gamma = V.\gamma\beta\alpha,$$

a formula which may be greatly extended.

91.] We have also

$$\begin{aligned} VV\alpha\beta V\gamma\delta &= -VV\gamma\delta V\alpha\beta \quad \text{by (2) of § 86:} \\ &= \delta S\gamma V\alpha\beta - \gamma S\delta V\alpha\beta = \delta S.\alpha\beta\gamma - \gamma S.\alpha\beta\delta, \\ &= -\beta S\alpha V\gamma\delta + \alpha S\beta V\gamma\delta = -\beta S.\alpha\gamma\delta + \alpha S.\beta\gamma\delta, \end{aligned}$$

all of these being arrived at by the help of § 90 (1) and of § 89; and by treating alternately $V\alpha\beta$ and $V\gamma\delta$ as *simple* vectors.

Equating two of these values, we have

$$\delta S.\alpha\beta\gamma = \alpha S.\beta\gamma\delta + \beta S.\gamma\alpha\delta + \gamma S.\alpha\beta\delta, \quad \ldots\ldots\ldots\ldots \quad (3)$$

a very useful formula, expressing any vector whatever in terms of three given vectors.

✓ 92.] That such an expression is possible we knew already by § 23. For variety we may seek another expression of a similar character, by a process which differs entirely from that employed in last section.

α, β, γ being any three vectors, we may derive from them three others $V\alpha\beta$, $V\beta\gamma$, $V\gamma\alpha$; and, as these will not generally be coplanar, any other vector δ may be expressed as the sum of the three, each multiplied by some scalar (§ 23). It is required to find this expression for δ.

Let $$\delta = xV\alpha\beta + yV\beta\gamma + zV\gamma\alpha.$$

Then $$S\gamma\delta = xS.\gamma\alpha\beta = xS.\alpha\beta\gamma,$$

the terms in y and z going out, because

$$S\gamma V\beta\gamma = S.\gamma\beta\gamma = S\beta\gamma^2 = \gamma^2 S\beta = 0,$$

for γ^2 is (§ 73) a number.

Similarly $$S\beta\delta = zS.\beta\gamma\alpha = zS.\alpha\beta\gamma,$$

and $$S\alpha\delta = yS.\alpha\beta\gamma.$$

Very Important

Thus $$\delta S.\alpha\beta\gamma = V\alpha\beta S\gamma\delta + V\beta\gamma S\alpha\delta + V\gamma\alpha S\beta\delta. \quad \text{(4)}$$

✓ 93.] We conclude the chapter by shewing (as promised in § 64) that the assumption that the product of two parallel vectors is a number, and the product of two perpendicular vectors a third vector perpendicular to both, is not only useful and convenient, but absolutely inevitable, if our system is to deal indifferently with all directions in space. We abridge Hamilton's reasoning.

Suppose that there is no direction in space pre-eminent, and that the product of two vectors is something which has quantity, so as to vary in amount if the factors are changed, and to have its sign changed if that of one of them is reversed; if the vectors be parallel, their product cannot be, in whole or in part, a vector *inclined* to them, for there is nothing to determine the direction in which it must lie. It cannot be a vector *parallel* to them; for by changing the sign of both factors the product is unchanged, whereas, as the whole system has been reversed, the product vector ought to have been reversed. Hence it must be a number. Again, the product of two perpendicular vectors cannot be wholly or partly a number, because on inverting one of them the sign of that number ought to change; but inverting one of them is simply equivalent to a rotation through two right angles about the other, and (from the symmetry of space) ought to leave the number unchanged. Hence the product of two perpendicular vectors must be a vector, and a simple extension of the same reasoning shews that it must be perpendicular to each of the factors. It is easy to carry this farther, but enough has been said to shew the character of the reasoning.

EXAMPLES TO CHAPTER II.

1. It is obvious from the properties of polar triangles that any mode of representing versors by the *sides* of a triangle must have an equivalent statement in which they are represented by *angles* in the polar triangle.

   Shew directly that the product of two versors represented by two angles of a spherical triangle is a third versor represented by the *supplement* of the remaining angle of the triangle; and determine the rule which connects the *directions* in which these angles are to be measured.

2. Hence derive another proof that we have not generally

$$pq = qp.$$

3. Hence shew that the proof of the associative principle, § 57, may be made to depend upon the fact that if from any point of the sphere tangent arcs be drawn to a spherical conic, and also arcs to the foci, the inclination of either tangent arc to one of the focal arcs is equal to that of the other tangent arc to the other focal arc.

4. Prove the formulae

$$2S.\alpha\beta\gamma = \alpha\beta\gamma - \gamma\beta\alpha,$$
$$2V.\alpha\beta\gamma = \alpha\beta\gamma + \gamma\beta\alpha.$$

5. Shew that, whatever odd number of vectors be represented by α, β, γ, &c., we have always

$$V.\alpha\beta\gamma\delta\epsilon = V.\epsilon\delta\gamma\beta\alpha,$$
$$V.\alpha\beta\gamma\delta\epsilon\zeta\eta = V.\eta\zeta\epsilon\delta\gamma\beta\alpha, \text{ \&c.}$$

6. Shew that

$$S.V\alpha\beta V\beta\gamma V\gamma\alpha = -(S.\alpha\beta\gamma)^2,$$
$$V.V\alpha\beta V\beta\gamma V\gamma\alpha = V\alpha\beta(\gamma^2 S\alpha\beta - S\beta\gamma S\gamma\alpha) + \ldots\ldots,$$

and
$$V.(V\alpha\beta V.V\beta\gamma V\gamma\alpha) = (\beta S\alpha\gamma - \alpha S\beta\gamma)S.\alpha\beta\gamma.$$

7. If α, β, γ be any vectors at right angles to each other, shew that

$$(\alpha^3 + \beta^3 + \gamma^3)S.\alpha\beta\gamma = \alpha^4 V\beta\gamma + \beta^4 V\gamma\alpha + \gamma^4 V\alpha\beta.$$

8. If α, β, γ be non-coplanar vectors, find the relations among the six scalars, x, y, z and ξ, η, ζ, which are implied in the equation

$$x\alpha + y\beta + z\gamma = \xi V\beta\gamma + \eta V\gamma\alpha + \zeta V\alpha\beta.$$

9. If α, β, γ be any three non-coplanar vectors, express any fourth vector, δ, as a linear function of each of the following sets of three derived vectors,

$$V.\gamma\alpha\beta, \quad V.\alpha\beta\gamma, \quad V.\beta\gamma\alpha,$$

and $$V.V\alpha\beta V\beta\gamma V\gamma\alpha, \quad V.V\beta\gamma V\gamma\alpha V\alpha\beta, \quad V.V\gamma\alpha V\alpha\beta V\beta\gamma.$$

10. Eliminate ρ from the equations

$$S\alpha\rho = a, \quad S\beta\rho = b, \quad S\gamma\rho = c, \quad S\delta\rho = d,$$

where α, β, γ, δ are vectors, and a, b, c, d scalars.

11. In any quadrilateral, plane or gauche, the sum of the squares of the diagonals is double the sum of the squares of the lines joining the middle points of opposite sides.

CHAPTER III.

INTERPRETATIONS AND TRANSFORMATIONS OF QUATERNION EXPRESSIONS.

94.] Among the most useful characteristics of the Calculus of Quaternions, the ease of interpreting its formulae geometrically, and the extraordinary variety of transformations of which the simplest expressions are susceptible, deserve a prominent place. We devote this Chapter to some of the more simple of these, together with a few of somewhat more complex character but of constant occurrence in geometrical and physical investigations. Others will appear in every succeeding Chapter. It is here, perhaps, that the student is likely to feel most strongly the peculiar difficulties of the new Calculus. But on that very account he should endeavour to master them, for the variety of forms which any one formula may assume, though puzzling to the beginner, is of the most extraordinary advantage to the advanced student, not alone as aiding him in the solution of complex questions, but as affording an invaluable mental discipline.

95.] If we refer again to the figure of § 77 we see that

$$OC = OB \cos AOB,$$
$$CB = OB \sin AOB.$$

Hence, if $\overline{OA} = \alpha$, $\overline{OB} = \beta$, and $\angle AOB = \theta$, we have

$$OB = T\beta, \qquad OA = T\alpha,$$
$$OC = T\beta \cos\theta, \qquad CB = T\beta \sin\theta.$$

Hence $$S\frac{\beta}{\alpha} = \frac{OC}{OA} = \frac{T\beta}{T\alpha}\cos\theta.$$

Similarly $$TV\frac{\beta}{\alpha} = \frac{CB}{OA} = \frac{T\beta}{T\alpha}\sin\theta.$$

Hence, if ϵ be a unit-vector perpendicular to α and β, or

$$\epsilon = \frac{U\overline{CB}}{U\overline{OA}} = UV\frac{\beta}{\alpha},$$

we have
$$V\frac{\beta}{\alpha} = \frac{T\beta}{T\alpha}\sin\theta . \epsilon.$$

96.] In the same way we may shew that

$$S\alpha\beta = -T\alpha\, T\beta \cos\theta,$$

$$TV\alpha\beta = T\alpha\, T\beta \sin\theta,$$

and
$$V\alpha\beta = T\alpha\, T\beta \sin\theta . \eta$$

where
$$\eta = UV\alpha\beta = UV\frac{\beta}{\alpha}.$$

Thus *the scalar of the product of two vectors is the continued product of their tensors and of the cosine of the supplement of the contained angle.*

The tensor of the vector of the product of two vectors is the continued product of their tensors and the sine of the contained angle; and the versor of the same is a unit-vector perpendicular to both, and such that the rotation about it from the first vector (i. e. *the multiplier*) *to the second is left-handed or positive.*

Hence $TV\alpha\beta$ *is double the area of the triangle two of whose sides are* α, β.

97.]

(*a*.) In any triangle ABC we have

$$\overline{AC} = \overline{AB} + \overline{BC}.$$

Hence
$$\overline{AC}^2 = S\overline{AC}\,\overline{AC} = S.\overline{AC}(\overline{AB} + \overline{BC}).$$

With the usual notation for a plane triangle the interpretation of this formula is

$$b^2 = -bc\cos A - ab\cos C,$$

or
$$b = a\cos C + c\cos A.$$

(*b*.) Again we have, obviously,

$$V\overline{AB}\,\overline{AC} = V\overline{AB}(\overline{AB} + \overline{BC})$$
$$= V\overline{AB}\,\overline{BC},$$

or
$$cb\sin A = ca\sin B,$$

whence
$$\frac{\sin A}{a} = \frac{\sin B}{b} = \frac{\sin C}{c}.$$

These are truths, but not truisms, as we might have been led to fancy from the excessive simplicity of the process employed.

98.] From § 96 it follows that, if α and β be both actual (i. e. real and non-evanescent) vectors, the equation

$$S\alpha\beta = 0$$

shews that $\cos\theta = 0$, or that *α is perpendicular to β*. And, in fact, we know already that the product of two perpendicular vectors is a vector.

Again, if

$$V\alpha\beta = 0,$$

we must have $\sin\theta = 0$, or *α is parallel to β*. We know already that the product of two parallel vectors is a scalar.

Hence we see that

$$S\alpha\beta = 0$$

is equivalent to

$$\alpha = V\gamma\beta,$$

where γ is an undetermined vector; and that

$$V\alpha\beta = 0$$

is equivalent to

$$\alpha = x\beta,$$

where x is an undetermined scalar.

99.] If we write, as in § 83,

$$\alpha = ix + jy + kz,$$
$$\beta = ix' + jy' + kz',$$

we have, at once, by § 86,

$$S\alpha\beta = -xx' - yy' - zz'$$
$$= -rr'\left(\frac{x}{r}\frac{x'}{r'} + \frac{y}{r}\frac{y'}{r'} + \frac{z}{r}\frac{z'}{r'}\right)$$

where $r = \sqrt{x^2 + y^2 + z^2}$, $r' = \sqrt{x'^2 + y'^2 + z'^2}$.

Also $$V\alpha\beta = rr'\left\{\frac{yz' - zy'}{rr'}\,i + \frac{zx' - xz'}{rr'}\,j + \frac{xy' - yx'}{rr'}\,k\right\}.$$

These express in Cartesian cöordinates the propositions we have just proved. In commencing the subject it may perhaps assist the student to see these more familiar forms for the quaternion expressions; and he will doubtless be induced by their appearance to prosecute the subject, since he cannot fail even at this stage to see how much more simple the quaternion expressions are than those to which he has been accustomed.

100.] The expression

$$S.\alpha\beta\gamma$$

may be written

$$S(V\alpha\beta)\gamma,$$

because the quaternion $\alpha\beta\gamma$ may be broken up into

$$(S\alpha\beta)\gamma + (V\alpha\beta)\gamma$$

of which the first term is a vector.

But, by § 96,

$$S(V\alpha\beta)\gamma = T\alpha\, T\beta \sin\theta\, S\eta\gamma.$$

Here $T\eta = 1$, let ϕ be the angle between η and γ, then finally

$$S.\alpha\beta\gamma = -T\alpha\, T\beta\, T\gamma \sin\theta \cos\phi.$$

But as η is perpendicular to α and β, $T\gamma\cos\phi$ is the length of the perpendicular from the extremity of γ upon the plane of α, β. And as the product of the other three factors is (§ 96) the area of the parallelogram two of whose sides are α, β, we see that the magnitude of $S.\alpha\beta\gamma$, independent of its sign, is *the volume of the parallelepiped of which three coördinate edges are* α, β, γ; or six times the volume of the pyramid which has α, β, γ for edges.

101.] Hence the equation

$$S.\alpha\beta\gamma = 0,$$

if we suppose α, β, γ to be actual vectors, shews either that

$$\sin\theta = 0,$$

or

$$\cos\phi = 0,$$

i. e. *two of the three vectors are parallel*, or *all three are parallel to one plane.*

This is consistent with previous results, for if $\gamma = p\beta$ we have

$$S.\alpha\beta\gamma = pS.\alpha\beta^2 = 0;$$

and, if γ be coplanar with α, β, we have $\gamma = p\alpha + q\beta$, and

$$S.\alpha\beta\gamma = S.\alpha\beta(p\alpha+q\beta) = 0.$$

102.] This property of the expression $S.\alpha\beta\gamma$ prepares us to find that it is a determinant. And, in fact, if we take α, β as in § 83, and in addition

$$\gamma = ix''+jy''+kz'',$$

we have at once

$$S.\alpha\beta\gamma = -x''(yz'-zy')-y''(zx'-xz')-z''(xy'-yx'),$$

$$= -\begin{vmatrix} x & y & z \\ x' & y' & z' \\ x'' & y'' & z'' \end{vmatrix}.$$

The determinant changes sign if we make any two rows change places. This is the proposition we met with before (§ 89) in the form

$$S.\alpha\beta\gamma = -S.\beta\alpha\gamma = S.\beta\gamma\alpha, \text{ \&c.}$$

If we take three new vectors

$$\alpha_1 = ix+jx'+kx'',$$

$$\beta_1 = iy+jy'+ky'',$$

$$\gamma_1 = iz+jz'+kz'',$$

we thus see that they are coplanar if α, β, γ are so. That is, if

$$S.\alpha\beta\gamma = 0,$$

then

$$S.\alpha_1\beta_1\gamma_1 = 0.$$

103.] We have, by § 52,

$$(Tq)^2 = qKq = (Sq+Vq)(Sq-Vq) \quad (\S\ 79),$$
$$= (Sq)^2-(Vq)^2 \text{ by algebra,}$$
$$= (Sq)^2+(TVq)^2 \quad (\S\ 73).$$

If $q = \alpha\beta$, we have $Kq = \beta\alpha$, and the formula becomes

$$\alpha\beta.\beta\alpha = \alpha^2\beta^2 = (S\alpha\beta)^2-(V\alpha\beta)^2.$$

In Cartesian cöordinates this is

$$(x^2+y^2+z^2)(x'^2+y'^2+z'^2)$$
$$= (xx'+yy'+zz')^2+(yz'-zy')^2+(zx'-xz')^2+(xy'-yx')^2.$$

More generally we have

$$(T(qr))^2 = qr\,K(qr)$$
$$= qr\,Kr\,Kq \;(\S\ 55) = (Tq)^2\,(Tr)^2 \quad (\S\ 52).$$

If we write

$$q = w+\alpha = w+ix+jy+kz,$$
$$r = w'+\beta = w'+ix'+jy'+kz';$$

this becomes

$$(w^2+x^2+y^2+z^2)(w'^2+x'^2+y'^2+z'^2)$$
$$= (ww'-xx'-yy'-zz')^2+(wx'+w'x+yz'-zy')^2$$
$$+(wy'+w'y+zx'-xz')^2+(wz'+w'z+xy'-yx')^2,$$

a formula of algebra due to Euler.

104.] We have, of course, by multiplication,

$$(\alpha+\beta)^2 = \alpha^2+\alpha\beta+\beta\alpha+\beta^2 = \alpha^2+2S\alpha\beta+\beta^2 \quad (\S\ 86\ (3)).$$

Translating into the usual notation of plane trigonometry, this becomes

$$c^2 = a^2-2ab\cos C+b^2,$$

the common formula.

Again, $V(\alpha+\beta)(\alpha-\beta) = -V\alpha\beta+V\beta\alpha = -2V\alpha\beta$ (§ 86 (2)).

Taking tensors of both sides we have the theorem, *the parallelogram whose sides are parallel and equal to the diagonals of a given parallelogram, has double its area* (§ 96).

Also $$S(\alpha+\beta)(\alpha-\beta) = \alpha^2-\beta^2,$$

and vanishes only when $\alpha^2 = \beta^2$, or $T\alpha = T\beta$; that is, *the diagonals of a parallelogram are at right angles to one another, when, and only when, it is a rhombus.*

Later it will be shewn that this contains a proof that the angle in a semicircle is a right angle.

105.] The expression $\rho = \alpha\beta\alpha^{-1}$

obviously denotes a vector whose tensor is equal to that of β.

But we have $$S.\beta\alpha\rho = 0,$$

so that ρ is in the plane of α, β.

Also we have $$S\alpha\rho = S\alpha\beta,$$

so that β and ρ make equal angles with α, evidently on opposite sides of it. Thus if α be the perpendicular to a reflecting surface and β the path of an incident ray, ρ will be the path of the reflected ray.

Another mode of obtaining these results is to expand the above expression, thus, § 90 (2),

$$\begin{aligned} \rho &= 2\alpha^{-1}S\alpha\beta - \beta \\ &= 2\alpha^{-1}S\alpha\beta - \alpha^{-1}(S\alpha\beta + V\alpha\beta) \\ &- \alpha^{-1}(S\alpha\beta - V\alpha\beta), \end{aligned}$$

so that in the figure of § 77 we see that if $\overline{OA} = \alpha$, and $\overline{OB} = \beta$, we have $\overline{OD} = \rho = \alpha\beta\alpha^{-1}$.

Or, again, we may get the result at once by transforming the equation to $U\frac{\rho}{\alpha} - U\frac{\alpha}{\beta}$

106.] For any three coplanar vectors the expression

$$\rho = \alpha\beta\gamma$$

is (§ 101) a vector. It is interesting to determine what this vector is. The reader will easily see that if a circle be described about the triangle, two of whose sides are (in order) α and β, and if from the extremity of β a line parallel to γ be drawn again cutting the circle, the vector joining the point of intersection with the origin of α is the direction of the vector $\alpha\beta\gamma$. For we may write it in the form

$$\rho = \alpha\beta^2\beta^{-1}\gamma = -(T\beta)^2\alpha\beta^{-1}\gamma = -(T\beta)^2\frac{\alpha}{\beta}\gamma,$$

which shews that the *versor* $\left(\frac{\alpha}{\beta}\right)$ which turns β into a direction parallel to α, turns γ into a direction parallel to ρ. And this expresses the long-known property of opposite angles of a quadrilateral inscribed in a circle.

Hence if α, β, γ be the sides of a triangle taken in order, the tangents to the circumscribing circle at the angles of the triangle are parallel respectively to

$$\alpha\beta\gamma, \quad \beta\gamma\alpha, \quad \text{and} \quad \gamma\alpha\beta.$$

Suppose two of these to be parallel, i. e. let

$$\alpha\beta\gamma = x\beta\gamma\alpha = x\alpha\gamma\beta \quad (\S\ 90),$$

since the expression is a vector. Hence

$$\beta\gamma = x\gamma\beta,$$

which requires either

$$x = 1, \quad V\gamma\beta = 0 \quad \text{or} \quad \gamma \parallel \beta,$$

a case not contemplated in the problem,

$$\text{or} \qquad x = -1, \qquad S\beta\gamma = 0,$$

i. e. the triangle is right-angled. And geometry shews us at once that this is correct.

Again, if the triangle be isosceles, the tangent at the vertex is parallel to the base. Here we have

$$x\beta = \alpha\beta\gamma,$$

or $$x(\alpha+\gamma) = \alpha(\alpha+\gamma)\gamma;$$

whence $x = \gamma^2 = \alpha^2$, or $T\gamma = T\alpha$, as required.

As an elegant extension of this proposition the reader may prove that the vector of the continued product $\alpha\beta\gamma\delta$ of the vector-sides of a quadrilateral inscribed in a sphere is parallel to the radius drawn to the corner (α, δ).

107.] To exemplify the variety of possible transformations even of simple expressions, we will take two cases which are of frequent occurrence in applications to geometry.

Thus $$T(\rho+\alpha) = T(\rho-\alpha),$$

[which expresses that if

$$\overline{OA} = \alpha, \quad \overline{OA'} = -\alpha, \quad \text{and} \quad \overline{OP} = \rho,$$

we have $$AP = A'P,$$

and thus that P is any point equidistant from two fixed points,] may be written $$(\rho+\alpha)^2 = (\rho-\alpha)^2,$$

or $$\rho^2 + 2S\alpha\rho + \alpha^2 = \rho^2 - 2S\alpha\rho + \alpha^2 \text{ (§ 104)},$$

whence $$S\alpha\rho = 0.$$

This may be changed to

$$\alpha\rho + \rho\alpha = 0,$$

or $$\alpha\rho + K\alpha\rho = 0,$$

$$SU\frac{\rho}{\alpha} - 0,$$

or finally, $$TVU\frac{\rho}{\alpha} - 1,$$

all of which express properties of a plane.

Again, $$T\rho = T\alpha$$

may be written $$T\frac{\rho}{\alpha} = 1,$$

$$\left(S\frac{\rho}{\alpha}\right)^2 - \left(V\frac{\rho}{\alpha}\right)^2 = 1,$$

$$(\rho+\alpha)^2 - 2S\alpha(\rho+\alpha) = 0,$$

$$\rho = (\rho+\alpha)^{-1}\alpha(\rho+\alpha),$$

$$S(\rho+\alpha)(\rho-\alpha) = 0, \text{ or finally,}$$

$$T.(\rho+\alpha)(\rho-\alpha) = 2TV\alpha\rho.$$

All of these express properties of a sphere. They will be interpreted when we come to geometrical applications.

108.] We have seen in § 95 that a quaternion may be divided into its scalar and vector parts as follows :—

$$\frac{\beta}{\alpha} = S\frac{\beta}{\alpha} + V\frac{\beta}{\alpha} = \frac{T\beta}{T\alpha}(\cos\theta + \epsilon\sin\theta)\,;$$

where θ is the angle between the directions of α and β, and $\epsilon = UV\frac{\beta}{\alpha}$ is the unit-vector perpendicular to the plane of α and β so situated that positive (i. e. left-handed) rotation about it turns α towards β.

Similarly we have (§ 96)

$$\begin{aligned} \alpha\beta &= S\alpha\beta + V\alpha\beta \\ &= T\alpha T\beta(-\cos\theta + \epsilon\sin\theta), \end{aligned}$$

θ and ϵ having the same signification as before.

109.] Hence, considering the versor parts alone, we have

$$U\frac{\beta}{\alpha} = \cos\theta + \epsilon\sin\theta.$$

Similarly $$U\frac{\gamma}{\beta} = \cos\phi + \epsilon\sin\phi\,;$$

ϕ being the positive angle between the directions of γ and β, and ϵ the same vector as before, if α, β, γ be coplanar.

Also we have

$$U\frac{\gamma}{\alpha} = \cos(\theta+\phi) + \epsilon\sin(\theta+\phi).$$

But we have always

$$\frac{\gamma}{\beta}\cdot\frac{\beta}{\alpha} = \frac{\gamma}{\alpha}, \text{ and therefore}$$

$$U\frac{\gamma}{\beta}\cdot U\frac{\beta}{\alpha} = U\frac{\gamma}{\alpha},$$

or $$\begin{aligned} \cos(\phi+\theta) + \epsilon\sin(\phi+\theta) &= (\cos\phi + \epsilon\sin\phi)(\cos\theta + \epsilon\sin\theta) \\ &= \cos\phi\cos\theta - \sin\phi\sin\theta + \epsilon(\sin\phi\cos\theta + \cos\phi\sin\theta), \end{aligned}$$

from which we have at once the fundamental formulae for the cosine and sine of the sum of two arcs, by equating separately the scalar and vector parts of these quaternions.

And we see, as an immediate consequence of the expressions above, that

$$\cos m\theta + \epsilon\sin m\theta = (\cos\theta + \epsilon\sin\theta)^m$$

if m be a positive whole number. For the left-hand side is a versor which turns through the angle $m\theta$ at once, while the right-hand side is a versor which effects the same object by m successive turnings each through an angle θ. See § 8.

110.] To extend this proposition to fractional indices we have only to write $\frac{\theta}{n}$ for θ, when we obtain the results as in ordinary trigonometry.

From De Moivre's Theorem, thus proved, we may of course deduce the rest of Analytical Trigonometry. And as we have already deduced, as interpretations of self-evident quaternion transformations (§§ 97, 104), the fundamental formulae for the solution of plane triangles, we will now pass to the consideration of spherical trigonometry, a subject specially adapted for treatment by quaternions; but to which we cannot afford more than a very few sections. (More on this subject will be found in Chap. X, in connexion with the Kinematics of rotation.) The reader is referred to Hamilton's works for the treatment of this subject by quaternion exponentials.

111.] Let α, β, γ be unit-vectors drawn from the centre to the corners A, B, C of a triangle on the unit-sphere. Then it is evident that, with the usual notation, we have (§ 96),

$$S\alpha\beta = -\cos c, \qquad S\beta\gamma = -\cos a, \qquad S\gamma\alpha = -\cos b,$$
$$TV\alpha\beta = \sin c, \qquad TV\beta\gamma = \sin a, \qquad TV\gamma\alpha = \sin b.$$

Also $UV\alpha\beta$, $UV\beta\gamma$, $UV\gamma\alpha$ are evidently the vectors of the corners of the polar triangle.

Hence
$$S.UV\alpha\beta UV\beta\gamma = \cos B, \text{ \&c.},$$
$$TV.UV\alpha\beta UV\beta\gamma = \sin B, \text{ \&c.}$$

Now (§ 90 (1)) we have
$$SV\alpha\beta V\beta\gamma = S.\alpha V.\beta V\beta\gamma$$
$$= -S\alpha\beta S\beta\gamma + \beta^2 S\alpha\gamma.$$

Remembering that we have
$$SV\alpha\beta V\beta\gamma = TV\alpha\beta TV\beta\gamma S.UV\alpha\beta UV\beta\gamma,$$
we see that the formula just written is equivalent to
$$\sin a \sin c \cos B = -\cos a \cos c + \cos b,$$
or
$$\cos b = \cos a \cos c + \sin a \sin c \cos B.$$

112.] Again, $V.V\alpha\beta V\beta\gamma = -\beta S\alpha\beta\gamma$,

which gives
$$TV.V\alpha\beta V\beta\gamma = S.\alpha\beta\gamma = S.\alpha V\beta\gamma = S.\beta V\gamma\alpha = S.\gamma V\alpha\beta,$$
or
$$\sin a \sin c \sin B = \sin a \sin p_a = \sin b \sin p_b = \sin c \sin p_c;$$
where p_a is the arc drawn from A perpendicular to BC, &c.

Hence
$$\sin p_a = \sin c \sin B,$$
$$\sin p_b = \frac{\sin a \sin c}{\sin b} \sin B,$$
$$\sin p_c = \sin a \sin B.$$

113.] Combining the results of the last two sections, we have
$$V\alpha\beta . V\beta\gamma = \sin a \sin c \cos B - \beta \sin a \sin c \sin B$$
$$= \sin a \sin c (\cos B - \beta \sin B).$$
Hence
$$U. V\alpha\beta V\beta\gamma = (\cos B - \beta \sin B),$$
and
$$U. V\gamma\beta V\beta\alpha = (\cos B + \beta \sin B).$$
These are therefore versors which turn the system negatively or positively about OB through the angle B.

As another instance, we have
$$\tan B = \frac{\sin B}{\cos B}$$
$$= \frac{TV. V\alpha\beta V\beta\gamma}{S. V\alpha\beta V\beta\gamma}$$
$$= -\beta^{-1} \frac{V. V\alpha\beta V\beta\gamma}{S. V\alpha\beta V\beta\gamma}$$
$$= -\frac{S.\alpha\beta\gamma}{S\alpha\gamma + S\alpha\beta S\beta\gamma} = \&c.$$
The interpretation of each of these forms gives a different theorem in spherical trigonometry.

Again, let us square the equal quantities
$$V.\alpha\beta\gamma \quad \text{and} \quad \alpha S\beta\gamma - \beta S\alpha\gamma + \gamma S\alpha\beta,$$
supposing α, β, γ to be any unit-vectors whatever. We have
$$-(V.\alpha\beta\gamma)^2 = S^2\beta\gamma + S^2\gamma\alpha + S^2\alpha\beta + 2S\beta\gamma S\gamma\alpha S\alpha\beta.$$
But the left-hand member may be written as
$$T^2.\alpha\beta\gamma - S^2.\alpha\beta\gamma,$$
whence
$$1 - S^2.\alpha\beta\gamma = S^2\beta\gamma + S^2\gamma\alpha + S^2\alpha\beta + 2S\beta\gamma S\gamma\alpha S\alpha\beta,$$
or
$$1 - \cos^2 a - \cos^2 b - \cos^2 c + 2 \cos a \cos b \cos c$$
$$- \sin^2 a \sin^2 p_a = \&c.$$
$$- \sin^2 a \sin^2 b \sin^2 C = \&c.,$$
all of which are well-known formulae.

Such results may be multiplied indefinitely by any one who has mastered the elements of quaternions.

114.] A curious proposition, due to Hamilton, gives us a quaternion expression for the *spherical excess* in any triangle. The following proof, which is very nearly the same as one of his, though by no means the simplest that can be given, is chosen here because it incidentally gives a good deal of other information. We leave the quaternion proof as an exercise.

Let the unit-vectors drawn from the centre of the sphere to A, B, C, respectively, be α, β, γ. It is required to express, as an arc and as an angle on the sphere, the quaternion

$$\beta\alpha^{-1}\gamma,$$

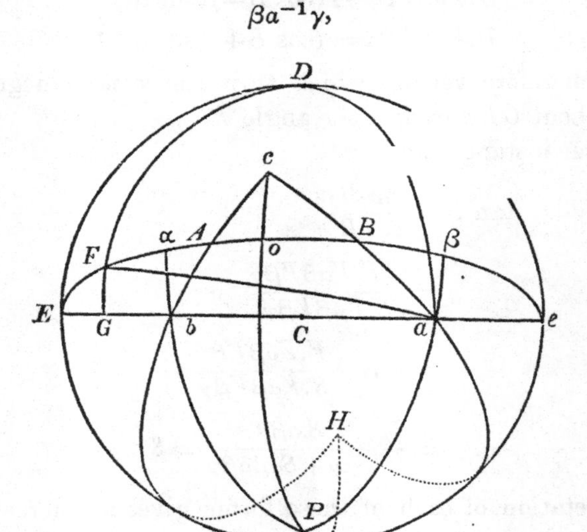

The figure represents an orthographic projection made on a plane perpendicular to γ. Hence C is the centre of the circle DEe. Let the great circle through A, B meet DEe in E, e, and let DE be a quadrant. Thus DE represents γ (§ 72). Also make $\widehat{EF} = AB = \beta\alpha^{-1}$. Then, evidently,

$$DF = \beta\alpha^{-1}\gamma,$$

which gives the arcual representation required.

Let DF cut Ee in G. Make $Ca = EG$, and join D, a, and a, F. Obviously, as D is the pole of Ee, Da is a quadrant; and since $EG = Ca$, $Ga = EC$, a quadrant also. Hence a is the pole of DG, and therefore the quaternion may be represented by the angle DaF.

Make $Cb = Ca$, and draw the arcs $Pa\beta$, Pba from P, the pole of AB. Comparing the triangles Eba and $ea\beta$, we see that $Ea = e\beta$. But, since P is the pole of AB, $F\beta a$ is a right angle: and therefore as Fa is a quadrant, so is $F\beta$. Thus AB is the complement of Ea or βe, and therefore

$$a\beta = 2AB.$$

Join bA and produce it to c so that $Ac = bA$; join c, P, cutting AB in o. Also join c, B, and B, a.

Since P is the pole of AB, the angles at o are right angles; and therefore, by the equal triangles baA, coA, we have

$$aA = Ao.$$

But $$a\beta = 2\,AB,$$

whence $$oB = B\beta,$$

and therefore the triangles coB and $Ba\beta$ are equal, and c, B, a lie on the same great circle.

Produce cA and cB to meet in H (on the opposite side of the sphere). H and c are diametrically opposite, and therefore cP, produced, passes through H.

Now $Pa = Pb = PH$, for they differ from quadrants by the equal arcs $a\beta$, ba, oc. Hence these arcs divide the triangle Hab into three isosceles triangles.

But $$\angle PHb + \angle PHa = \angle aHb = \angle bca.$$

Also $$\angle Pab = \pi - \angle cab - \angle PaH,$$

$$\angle Pba = \angle Pab = \pi - \angle cba - \angle PbH.$$

Adding, $$2\angle Pab = 2\pi - \angle cab - \angle cba - \angle bca$$
$$= \pi - (\text{spherical excess of } abc).$$

But, as $\angle Fa\beta$ and $\angle Dae$ are right angles, we have

$$\text{angle of } \beta a^{-1}\gamma = \angle FaD = \angle \beta ae = \angle Pab$$
$$= \frac{\pi}{2} - \tfrac{1}{2}(\text{spherical excess of } abc).$$

[Numerous singular geometrical theorems, easily proved *ab initio* by quaternions, follow from this: e.g. The arc AB, which bisects two sides of a spherical triangle abc, intersects the base at the distance of a quadrant from its middle point. All spherical triangles, with a common side, and having their other sides bisected by the same great circle (i.e. having their vertices in a small circle parallel to this great circle) have equal areas, &c., &c.]

115.] Let $\overline{Oa} = \alpha'$, $\overline{Ob} = \beta'$, $\overline{Oc} = \gamma'$, and we have

$$\left(\frac{\alpha'}{\beta'}\right)^{\frac{1}{2}}\left(\frac{\beta'}{\gamma'}\right)^{\frac{1}{2}}\left(\frac{\gamma'}{\alpha'}\right)^{\frac{1}{2}} = Ca.cA.Bc$$
$$= \frown{Ca}.\frown{BA}$$
$$= \frown{EG}.\frown{FE} = \frown{FG}.$$

But FG is the complement of DF. Hence the *angle of the quaternion*

$$\left(\frac{\alpha'}{\beta'}\right)^{\frac{1}{2}}\left(\frac{\beta'}{\gamma'}\right)^{\frac{1}{2}}\left(\frac{\gamma'}{\alpha'}\right)^{\frac{1}{2}}$$

is half the spherical excess of the triangle whose angular points are at the extremities of the unit-vectors α', β', γ'.

[In seeking a purely quaternion proof of the preceding propositions, the student may commence by shewing that for any three unit-vectors we have

$$\frac{\beta\gamma\alpha}{\alpha\beta\gamma} = (\beta\alpha^{-1}\gamma)^2.$$

The angle of the first of these quaternions can be easily assigned; and the equation shews how to find that of $\beta\alpha^{-1}\gamma$. But a still simpler method of proof is easily derived from the composition of rotations.]

116.] A *scalar* equation in ρ, the vector of an undetermined point, is generally the equation of a *surface;* since we may substitute for ρ the expression

$$\rho = x\alpha,$$

where x is an unknown scalar, and α any assumed unit-vector. The result is an equation to determine x. Thus one or more points are found on the vector $x\alpha$ whose coördinates satisfy the equation; and the locus is a surface whose degree is determined by that of the equation which gives the values of x.

But a *vector* equation in ρ, as we have seen, generally leads to three scalar equations, from which the three rectangular or other components of the sought vector are to be derived. Such a vector equation, then, usually belongs to a definite number of *points* in space. But in certain cases these may form a *line*, and even a *surface*, the vector equation losing as it were one or two of the three scalar equations to which it is usually equivalent.

Thus while the equation

$$\alpha\rho = \beta$$

gives at once

$$\rho = \alpha^{-1}\beta,$$

which is the vector of a definite point (since we have evidently

$$S\alpha\beta = 0);$$

the closely allied equation

$$V\alpha\rho = \beta$$

is easily seen to involve

$$S\alpha\beta = 0,$$

and to be satisfied by

$$\rho = \alpha\ {}^{1}\beta + x\alpha,$$

whatever be x. Hence the vector of any point whatever in the line drawn parallel to α from the extremity of $\alpha^{-1}\beta$ satisfies the given equation.

117.] Again,

$$V\alpha\rho . V\rho\beta = (V\alpha\beta)^2$$

is equivalent to but two scalar equations. For it shews that $V\alpha\rho$

and $V\beta\rho$ are parallel, i. e. ρ lies in the same plane as α and β, and can therefore be written (§ 24)

$$\rho = x\alpha + y\beta,$$

where x and y are scalars as yet undetermined.

We have now
$$V\alpha\rho = yV\alpha\beta,$$
$$V\rho\beta = xV\alpha\beta,$$

which, by the given equation, lead to

$$xy = 1, \quad \text{or} \quad y = \frac{1}{x}, \quad \text{or finally}$$

$$\rho = x\alpha + \frac{1}{x}\beta;$$

which (§ 40) is the equation of a hyperbola whose asymptotes are in the directions of α and β.

118.] Again, the equation

$$V.V\alpha\beta V\alpha\rho = 0,$$

though apparently equivalent to three scalar equations, is really equivalent to one only. In fact we see by § 91 that it may be written

$$-\alpha S.\alpha\beta\rho = 0,$$

whence, if α be not zero, we have

$$S.\alpha\beta\rho = 0,$$

and thus (§ 101) the only condition is that ρ is coplanar with α, β. Hence the equation represents the plane in which α and β lie.

119.] Some very curious results are obtained when we extend these processes of interpretation to functions of a *quaternion*

$$q = w + \rho$$

instead of functions of a mere *vector* ρ.

A scalar equation containing such a quaternion, along with quaternion constants, gives, as in last section, the equation of a surface, if we assign a definite value to w. Hence for successive values of w, we have successive surfaces belonging to a system; and thus when w is indeterminate the equation represents not a *surface*, as before, but a *volume*, in the sense that the vector of any point within that volume satisfies the equation.

Thus the equation
$$(Tq)^2 = a^2,$$
or
$$w^2 - \rho^2 = a^2,$$
or
$$(T\rho)^2 = a^2 - w^2,$$

represents, for any assigned value of w, not greater than a, a sphere whose radius is $\sqrt{a^2 - w^2}$. Hence the equation is satisfied by the vector of any point whatever in the *volume* of a sphere of radius a, whose centre is origin.

Again, by the same kind of investigation,

$$(T(q-\beta))^2 = a^2,$$

where $q = w+\rho$, is easily seen to represent the volume of a sphere of radius a described about the extremity of β as centre.

Also $S(q^2) = -a^2$ is the equation of infinite space less the space contained in a sphere of radius a about the origin.

Similar consequences as to the interpretation of vector equations in quaternions may be readily deduced by the reader.

120.] The following transformation is enuntiated without proof by Hamilton (*Lectures*, p. 587, and *Elements*, p. 299).

$$r^{-1}(r^2q^2)^{\frac{1}{2}}q^{-1} = U(rq+KrKq).$$

To prove it, let $\quad r^{-1}(r^2q^2)^{\frac{1}{2}}q^{-1} = t,\quad$ then

$$Tt = 1,\quad \text{and therefore}$$

$$Kt = t^{-1};$$

But $\quad (r^2q^2)^{\frac{1}{2}} = rtq,$

or $\quad r^2q^2 = rtqrtq,$

or $\quad rq = tqrt.$

Hence $\quad KqKr = t^{-1}KrKqt^{-1},$

or $\quad KrKq = tKqKrt.$

Thus we have $\quad U(rq+KrKq) = tU(qr+KqKr)t,$

or, if we put $\quad s = U(qr+KqKr),$

$$Ks = +tst.$$

Hence $\quad sKs = (Ts)^2 = 1 = +stst,$

which, if we take the positive sign, requires

$$st = +1,$$

or $\quad t = +s^{-1} - +UKs,$

which is the required transformation.

[It is to be noticed that there are other results which might have been arrived at by using the negative sign above; some in volving an arbitrary unit-vector, others involving the imaginary of ordinary algebra.]

121.] As a final example, we take a transformation of Hamilton's, of great importance in the theory of surfaces of the second order.

Transform the expression

$$(S\alpha\rho)^2+(S\beta\rho)^2+(S\gamma\rho)^2$$

in which α, β, γ are any three mutually rectangular vectors, into the form

$$\left(\frac{T(\iota\rho+\rho\kappa)}{\kappa^2-\iota^2}\right)^2,$$

which involves only two vector-constants, ι, κ.

$$\begin{aligned}\{T(\iota\rho+\rho\kappa)\}^2 &= (\iota\rho+\rho\kappa)(\rho\iota+\kappa\rho) \quad (\S\S\ 52,\ 55)\\ &-(\iota^2+\kappa^2)\rho^2+(\iota\rho\kappa\rho+\rho\kappa\rho\iota)\\ &=(\iota^2+\kappa^2)\rho^2+2\,S.\iota\rho\kappa\rho\\ &=(\iota-\kappa)^2\rho^2+4\,S\iota\rho S\kappa\rho.\end{aligned}$$

Hence $\quad (S\alpha\rho)^2+(S\beta\rho)^2+(S\gamma\rho)^2=\dfrac{(\iota-\kappa)^2}{(\kappa^2-\iota^2)^2}\rho^2+4\dfrac{S\iota\rho S\kappa\rho}{(\kappa^2-\iota^2)^2}.$

But $\quad \alpha^{-2}(S\alpha\rho)^2+\beta^{-2}(S\beta\rho)^2+\gamma^{-2}(S\gamma\rho)^2=\rho^2 \quad (\S\S\ 25,\ 73).$

Multiply by β^2 and subtract, we get

$$\left(1-\frac{\beta^2}{\alpha^2}\right)(S\alpha\rho)^2-\left(\frac{\beta^2}{\gamma^2}-1\right)(S\gamma\rho)^2=\left\{\frac{(\iota-\kappa)^2}{(\kappa^2-\iota^2)^2}-\beta^2\right\}\rho^2+4\frac{S\iota\rho S\kappa\rho}{(\kappa^2-\iota^2)^2}.$$

The left side breaks up into two real factors if β^2 be intermediate in value to α^2 and γ^2: and that the right side may do so the term in ρ^2 must vanish. This condition gives

$$\beta^2=\frac{(\iota-\kappa)^2}{(\kappa^2-\iota^2)^2}; \text{ and the identity becomes}$$

$$S\left(\alpha\sqrt{(1-\frac{\beta^2}{\alpha^2})}+\gamma\sqrt{(\frac{\beta^2}{\gamma^2}-1)}\right)\rho S\left(\alpha\sqrt{(1-\frac{\beta^2}{\alpha^2})}-\gamma\sqrt{(\frac{\beta^2}{\gamma^2}-1)}\right)\rho=4\frac{S\iota\rho S\kappa\rho}{(\kappa^2-\iota^2)^2}.$$

Hence we must have

$$\frac{2\iota}{\kappa^2-\iota^2}=p\left(\alpha\sqrt{(1-\frac{\beta^2}{\alpha^2})}+\gamma\sqrt{(\frac{\beta^2}{\gamma^2}-1)}\right),$$

$$\frac{2\kappa}{\kappa^2-\iota^2}=\frac{1}{p}\left(\alpha\sqrt{(1-\frac{\beta^2}{\alpha^2})}-\gamma\sqrt{(\frac{\beta^2}{\gamma^2}-1)}\right),$$

where p is an undetermined scalar.

To determine p, substitute in the expression for β^2, and we find

$$\begin{aligned}4\beta^2=\frac{4(\iota-\kappa)^2}{(\kappa^2-\iota^2)^2}&=\left(p-\frac{1}{p}\right)^2(\alpha^2-\beta^2)+\left(p+\frac{1}{p}\right)^2(\beta^2-\gamma^2)\\ &=\left(p^2+\frac{1}{p^2}\right)(\alpha^2-\gamma^2)-2(\alpha^2+\gamma^2)+4\beta^2.\end{aligned}$$

Thus the transformation succeeds if

$$p^2+\frac{1}{p^2}=\frac{2(a^2+\gamma^2)}{a^2-\gamma^2},$$

which gives $$p+\frac{1}{p}=+2\sqrt{\frac{a^2}{a^2-\gamma^2}},$$

$$p-\frac{1}{p}-+2\sqrt{\frac{\gamma^2}{a^2-\gamma 2}}.$$

Hence $$\frac{4(\kappa^2-\iota^2)}{(\kappa^2-\iota^2)^2}=\left(\frac{1}{p^2}-p^2\right)(a^2-\gamma^2)=+4\sqrt{a^2\gamma^2},$$

or $$(\kappa^2-\iota^2)^{-1}-+TaT\gamma.$$

Again, $$p=\frac{Ta+T\gamma}{\sqrt{\gamma^2-a^2}},\quad \frac{1}{p}=\frac{Ta-T\gamma}{\sqrt{\gamma^2-a^2}},$$

and therefore

$$2\iota=\frac{Ta+T\gamma}{Ta\,T\gamma}\left(\sqrt{\frac{\beta^2-a^2}{\gamma^2-a^2}}\,Ua+\sqrt{\frac{\gamma^2-\beta^2}{\gamma^2-a^2}}\,U\gamma\right),$$

$$2\kappa=\frac{Ta-T\gamma}{Ta\,T\gamma}\left(\sqrt{\frac{\beta^2-a^2}{\gamma^2-a^2}}\,Ua-\sqrt{\frac{\gamma^2-\beta^2}{\gamma^2-a^2}}\,U\gamma\right).$$

Thus we have proved the possibility of the transformation, and determined the transforming vectors ι, κ.

122.] By differentiating the equation

$$(Sa\rho)^2+(S\beta\rho)^2+(S\gamma\rho)^2=\left(\frac{T(\iota\rho+\rho\kappa)}{(\kappa^2-\iota^2)}\right)^2$$

we obtain, as will be seen in Chapter IV, the following,

$$Sa\rho Sa\rho'+S\beta\rho S\beta\rho'+S\gamma\rho S\gamma\rho'=\frac{S.(\iota\rho+\rho\kappa)(\kappa\rho'+\rho'\iota)}{(\kappa^2-\iota^2)^2},$$

where ρ' also may be any vector whatever.

This is another very important formula of transformation; and it will be a good exercise for the student to prove its truth by processes analogous to those in last section. We may merely observe, what indeed is obvious, that by putting $\rho'=\rho$ it becomes the formula of last section. And we see that we may write, with the recent values of ι and κ in terms of a, β, γ, the identity

$$aSa\rho+\beta S\beta\rho+\gamma S\gamma\rho=\frac{(\iota^2+\kappa^2)\rho+2V.\iota\rho\kappa}{(\kappa^2-\iota^2)^2}$$

$$=\frac{(\iota-\kappa)^2\rho+2(\iota S\kappa\rho+\kappa S\iota\rho)}{(\kappa^2-\iota^2)^2}.$$

123.] In various quaternion investigations, especially in such as involve *imaginary* intersections of curves and surfaces, the old imaginary of algebra of course appears. But it is to be particularly noticed that this expression is analogous to a scalar and not to a vector, and that like real scalars it is commutative in multiplication with all other factors. Thus it appears, by the same proof as in algebra, that any quaternion expression which contains this imaginary can always be broken up into the sum of two parts, one real, the other multiplied by the first power of $\sqrt{-1}$. Such an expression, viz.

$$q = q' + \sqrt{-1}\,q'',$$

where q' and q'' are real quaternions, is called a BIQUATERNION. Some little care is requisite in the management of these expressions, but there is no new difficulty. The points to be observed are: first, that any biquaternion can be divided into a real and an imaginary part, the latter being the product of $\sqrt{-1}$ by a real quaternion; second, that this $\sqrt{-1}$ is commutative with all other quantities in multiplication; third, that if two biquaternions be equal, as

$$q' + \sqrt{-1}\,q'' = r' + \sqrt{-1}\,r'',$$

we have, as in algebra, $q' = r', \quad q'' = r''$;

so that an equation between biquaternions involves in general *eight* equations between scalars. Compare § 80.

124.] We have, obviously, since $\sqrt{-1}$ is a scalar,

$$S(q' + \sqrt{-1}\,q'') = Sq' + \sqrt{-1}\,Sq'',$$
$$V(q' + \sqrt{-1}\,q'') = Vq' + \sqrt{-1}\,Vq''$$

Hence (§ 103)

$$\{T(q' + \sqrt{-1}\,q'')\}^2$$
$$= (Sq' + \sqrt{-1}Sq'' + Vq' + \sqrt{-1}Vq'')(Sq' + \sqrt{-1}Sq'' - Vq' - \sqrt{-1}Vq'')$$
$$= (Sq' + \sqrt{-1}Sq'')^2 - (Vq' + \sqrt{-1}\,Vq'')^2,$$
$$= (Tq')^2 - (Tq'')^2 + 2\sqrt{-1}S.q'Kq''.$$

The only remark which need be made on such formulæ is this, that *the tensor of a biquaternion may vanish while both of the component quaternions are finite.*

Thus, if $\qquad Tq' = Tq'',$

and $\qquad S.q'Kq'' = 0,$

the above formula gives

$$T(q' + \sqrt{-1}\,q'') = 0.$$

The condition $\qquad S.q'Kq'' = 0$

may be written

$$Kq'' = q'^{-1}a, \quad \text{or} \quad q'' = -aKq'^{-1} = -\frac{aq'}{(Tq')^2},$$

where a is any vector whatever.

Hence $$Tq' = Tq'' = TKq'' = \frac{Ta}{Tq'},$$
and therefore
$$Tq'(Uq' - \sqrt{-1}\,Ua.Uq') = (1 - \sqrt{-1}\,Ua)q'$$
is the general form of a biquaternion whose tensor is zero.

125.] More generally we have, q, r, q', r' being any four real and non-evanescent quaternions,
$$(q + \sqrt{-1}\,q')(r + \sqrt{-1}\,r') = qr - q'r' + \sqrt{-1}(qr' + q'r).$$
That this product may vanish we must have
$$qr = q'r',$$
and
$$qr' = -q'r.$$
Eliminating r' we have $\quad qq'^{-1}qr = -q'r,$
which gives $\quad (q'^{-1}q)^2 = -1,$
i. e. $\quad q = q'a$
where a is some unit-vector.

And the two equations now agree in giving
$$-r = ar',$$
so that we have the biquaternion factors in the form
$$q'(a + \sqrt{-1}) \quad \text{and} \quad -(a - \sqrt{-1})r';$$
and their product is
$$q'(a + \sqrt{-1})(a - \sqrt{-1})r',$$
which, of course, vanishes.

[A somewhat simpler investigation of the same proposition may be obtained by writing the biquaternions as
$$q'(q'^{-1}q + \sqrt{-1}) \quad \text{and} \quad (rr'^{-1} + \sqrt{-1})r',$$
or
$$q'(q'' + \sqrt{-1}) \quad \text{and} \quad (r'' + \sqrt{-1})r',$$
and shewing that
$$q'' = -r'' = a, \text{ where } Ta = 1.]$$

From this it appears that if the product of two *bivectors*
$$\rho + \sigma\sqrt{-1} \quad \text{and} \quad \rho' + \sigma'\sqrt{-1}$$
is zero, we must have
$$\sigma^{-1}\rho - -\rho'\sigma'^{-1} - Ua,$$
where a may be any vector whatever. But this result is still more easily obtained by means of a direct process.

126.] It may be well to observe here (as we intend to avail ourselves of them in the succeeding Chapters) that certain abbreviated forms of expression may be used when they are not liable to confuse, or lead to error. Thus we may write

$$T^2 q \text{ for } (Tq)^2,$$

just as we write

$$\cos^2\theta \text{ for } (\cos\theta)^2,$$

although the true meanings of these expressions are

$$T(Ta) \text{ and } \cos(\cos\theta).$$

The former is justifiable, as $T(Ta) = Ta$, and therefore $T^2 a$ is not required to signify the second tensor (or tensor of the tensor) of a. But the trigonometrical usage is quite indefensible.

Similarly we may write

$$S^2 q \text{ for } (Sq)^2, \text{ \&c.},$$

but it may be advisable not to use

$$Sq^2$$

as the equivalent of either of those just written; inasmuch as it might be confounded with the (generally) different quantity

$$S.q^2 \text{ or } S(q^2),$$

although this is rarely written without the point or the brackets.

127.] The beginner may expect to be a little puzzled with the aspect of this notation at first; but, as he learns more of the subject, he will soon see clearly the distinction between such an expression as

$$S.V\alpha\beta V\beta\gamma,$$

where we may omit at pleasure either the point or the first V without altering the value, and the very different one

$$S\alpha\beta.V\beta\gamma,$$

which admits of no such changes, without altering its value.

All these simplifications of notation are, in fact, merely examples of the transformations of quaternion expressions to which part of this Chapter has been devoted. Thus, to take a very simple example, we easily see that

$$\begin{aligned} S.V\alpha\beta V\beta\gamma = SV\alpha\beta V\beta\gamma = S.\alpha\beta V\beta\gamma = S\alpha V.\beta V\beta\gamma = -S\alpha V.(V\beta\gamma)\beta \\ = S\alpha V.(V\gamma\beta)\beta = S.\alpha V(\gamma\beta)\beta = S.V(\gamma\beta)\beta\alpha = SV\gamma\beta V\beta\alpha \\ = S.\gamma\beta V\beta\alpha = \text{\&c., \&c.} \end{aligned}$$

The above group does not nearly exhaust the list of even the simpler ways of expressing the given quantity. We recommend it to the careful study of the reader. He will find it advisable, at first, to use stops and brackets pretty freely; but will gradually learn to dispense with those which are not absolutely necessary to prevent ambiguity.

EXAMPLES TO CHAPTER III.

1. Investigate, by quaternions, the requisite formulæ for changing from any one set of cöordinate axes to another; and derive from your general result, and also from special investigations, the usual expressions for the following cases:—

(*a.*) Rectangular axes turned about z through any angle.

(*b.*) Rectangular axes turned into any new position by rotation about a line equally inclined to the three.

(*c.*) Rectangular turned to oblique, one of the new axes lying in each of the former cöordinate planes.

2. If $T\rho = T\alpha = T\beta = 1$, and $S.\alpha\beta\rho = 0$, shew by direct transformations that

$$S.U(\rho-\alpha)U(\rho-\beta) = +\sqrt{\tfrac{1}{2}(1-S\alpha\beta)}.$$

Interpret this theorem geometrically.

3. If $S\alpha\beta = 0$, $T\alpha = T\beta = 1$, shew that

$$(1+\alpha^m)\beta = 2\cos\frac{m\pi}{4}\alpha^{\frac{m}{2}}\beta = 2\,S\alpha^{\frac{m}{2}}.\alpha^{\frac{m}{2}}\beta.$$

4. Put in its simplest form the equation

$$\rho S.V\alpha\beta V\beta\gamma V\gamma\alpha = aV.V\gamma\alpha V\alpha\beta + bV.V\alpha\beta V\beta\gamma + cV.V\beta\gamma V\gamma\alpha;$$

and shew that $\qquad a = S.\beta\gamma\rho$, &c.

5. Prove the following theorems, and exhibit them as properties of determinants:—

(*a.*) $S.(\alpha+\beta)(\beta+\gamma)(\gamma+\alpha) = 2\,S.\alpha\beta\gamma,$

(*b.*) $S.V\alpha\beta V\beta\gamma V\gamma\alpha = -(S.\alpha\beta\gamma)^2,$

(*c.*) $S.V(\alpha+\beta)(\beta+\gamma)V(\beta+\gamma)(\gamma+\alpha)V(\gamma+\alpha)(\alpha+\beta) = -4(S.\alpha\beta\gamma)^2,$

(*d.*) $S.V(V\alpha\beta V\beta\gamma)V(V\beta\gamma V\gamma\alpha)V(V\gamma\alpha V\alpha\beta) = -(S.\alpha\beta\gamma)^4,$

(*e.*) $S.\delta\epsilon\zeta = -16(S.\alpha\beta\gamma)^4,$ where

$$\delta = V(V(\alpha+\beta)(\beta+\gamma)V(\beta+\gamma)(\gamma+\alpha)),$$
$$\epsilon = V(V(\beta+\gamma)(\gamma+\alpha)V(\gamma+\alpha)(\alpha+\beta)),$$
$$\zeta = V(V(\gamma+\alpha)(\alpha+\beta)V(\alpha+\beta)(\beta+\gamma)).$$

6. Prove the common formula for the product of two determinants of the third order in the form

$$S.\alpha\beta\gamma S.\alpha_1\beta_1\gamma_1 = -\begin{vmatrix} S\alpha\alpha_1 & S\beta\alpha_1 & S\gamma\alpha_1 \\ S\alpha\beta_1 & S\beta\beta_1 & S\gamma\beta_1 \\ S\alpha\gamma_1 & S\beta\gamma_1 & S\gamma\gamma_1 \end{vmatrix}.$$

7. If, in § 102, α, β, γ be three mutually perpendicular vectors, can anything be predicted as to α_1, β_1, γ_1? If α, β, γ be rectangular *unit* vectors, what of α_1, β_1, γ_1?

8. If $\alpha, \beta, \gamma, \alpha', \beta', \gamma'$ be two sets of rectangular unit-vectors, shew that $\quad S\alpha\alpha' = S\gamma\beta' S\beta\gamma' - S\beta\beta' S\gamma\gamma'$, &c., &c.

9. The lines bisecting pairs of opposite sides of a quadrilateral are perpendicular to each other when the diagonals of the quadrilateral are equal.

10. Shew that

(*a*.) $S.q^2 = 2S^2q - T^2q$,

(*b*.) $S.q^3 = S^3q - 3SqT^2Vq$,

(*c*.) $\alpha^2\beta^2\gamma^2 + S^2.\alpha\beta\gamma = V^2.\alpha\beta\gamma$,

(*d*.) $S(V.\alpha\beta\gamma V.\beta\gamma\alpha V.\gamma\alpha\beta) = 4S\alpha\beta S\beta\gamma S\gamma\alpha S.\alpha\beta\gamma$,

(*e*.) $V.q^3 = (3S^2q - T^2Vq)Vq$,

(*f*.) $qUVq^{-1} = -Sq.UVq + TVq$;

and interpret each as a formula in plane or spherical trigonometry.

11. If q be an undetermined quaternion, what loci are represented by

(*a*.) $(q\alpha^{-1})^2 = -a^2$,

(*b*.) $(q\alpha^{-1})^4 = a^4$,

(*c*.) $S.(q - \alpha)^2 = a^2$,

where a is any given scalar and α any given vector?

12. If q be any quaternion, shew that the equation

$$Q^2 = q^2$$

is satisfied, not alone by $Q = \pm q$ but also, by

$$Q = \pm\sqrt{-1}(Sq.UVq - TVq).$$

(Hamilton, *Lectures*, p. 673.)

13. Wherein consists the difference between the two equations

$$T^2\frac{\rho}{\alpha} = 1, \quad \text{and} \quad \left(\frac{\rho}{\alpha}\right)^2 = -1\,?$$

What is the full interpretation of each, α being a given, and ρ an undetermined, vector?

14. Find the full consequences of each of the following groups of equations, both as regards the unknown vector ρ and the given vectors α, β, γ:—

(*a*.) $S.\alpha\beta\rho = 0$, $S.\beta\gamma\rho = 0$;

(*b*.) $S\alpha\rho = 0$, $S.\alpha\beta\rho = 0$, $S\beta\rho = 0$;

(*c*.) $S\alpha\rho = 0$, $S.\alpha\beta\rho = 0$, $S.\alpha\beta\gamma\rho = 0$.

15. From §§ 74, 109, shew that, if ϵ be any unit-vector, and m any scalar, $\quad \epsilon^m = \cos\frac{m\pi}{2} + \epsilon\sin\frac{m\pi}{2}$.

Hence shew that if α, β, γ be radii drawn to the corners of a triangle on the unit-sphere, whose spherical excess is m right angles,

$$\frac{\alpha+\beta}{\beta+\gamma}\cdot\frac{\gamma+\alpha}{\alpha+\beta}\cdot\frac{\beta+\gamma}{\gamma+\alpha}=\alpha^m.$$

Also that, if A, B, C be the angles of the triangle, we have

$$\gamma^{\frac{2C}{\pi}}\beta^{\frac{2B}{\pi}}\alpha^{\frac{2A}{\pi}}=-1.$$

16. Shew that for *any* three vectors α, β, γ we have

$$(U\alpha\beta)^2+(U\beta\gamma)^2+(U\alpha\gamma)^2+(U.\alpha\beta\gamma)^2+4\,U\alpha\gamma.SU\alpha\beta SU\beta\gamma=\quad 2.$$

(Hamilton, *Elements*, p. 388.)

17. If a_1, a_2, a_3, x be any four scalars, and ρ_1, ρ_2, ρ_3 any three vectors, shew that

$$(S.\rho_1\rho_2\rho_3)^2+(\Sigma.a_1V\rho_2\rho_3)^2+x^2(\Sigma V\rho_1\rho_2)^2-x^2(\Sigma.a_1(\rho_2-\rho_3))^2$$
$$+2\Pi(x^2+S\rho_1\rho_2+a_1a_2)=2\Pi(x^2+\rho^2)+2\Pi a^2$$
$$+\Sigma\{(x^2+a_1^2+\rho_1^2)((V\rho_2\rho_3)^2+2a_2a_3(x^2+S\rho_2\rho_3)-x^2(\rho_2-\rho_3)^2)\};$$

where $\qquad \Pi a^2=a_1^2a_2^2a_3^2.$

Verify this formula by a simple process in the particular case

$$a_1=a_2=a_3=x=0.$$

(*Ibid.*)

CHAPTER IV.

DIFFERENTIATION OF QUATERNIONS.

128.] In Chapter I, we have already considered as a special case the differentiation of a *vector* function of a scalar independent variable: and it is easy to see at once that a similar process is applicable to a *quaternion* function of a scalar independent variable. The differential, or differential coefficient, thus found, is in general another function of the same scalar variable; and can therefore be differentiated anew by a second, third, &c. application of the same process. And precisely similar remarks apply to partial differentiation of a quaternion function of any number of *scalar* independent variables. In fact, this process is identical with ordinary differentiation.

129.] But when we come to differentiate a function of a vector, or of a quaternion, some caution is requisite; there is, in general, nothing which can be called a differential coefficient; and in fact we require (as already hinted in § 33) to employ a definition of a differential, somewhat different from the ordinary one but, coinciding with it when applied to functions of mere scalar variables.

130.] If $r = F(q)$ be a function of a quaternion q,

$$dr = dFq = \mathcal{L}_{\infty} n \left\{F\left(q + \frac{dq}{n}\right) - F(q)\right\},$$

where n is a scalar which is ultimately to be made infinite, is *defined* to be the differential of r or Fq.

Here dq may be *any quaternion whatever*, and the right-hand member may be written

$$f(q, dq),$$

where f is a new function, depending on the form of F; homogeneous and of the *first* degree in dq; but not, in general, capable of being put in the form

$$\mathrm{f}(q)\, dq.$$

131.] To make more clear these last remarks, we may observe that the function

$$f(q, dq),$$

thus derived as the differential of $F(q)$, is *distributive* with respect to dq. That is

$$f(q, r+s) = f(q, r)+f(q, s),$$

r and s being any quaternions.

For $$f(q, r+s) = \mathcal{L}_\infty n\left(F\left(q+\frac{r+s}{n}\right)-F(q)\right)$$

$$= \mathcal{L}_\infty n\left\{F\left(q+\frac{r}{n}+\frac{s}{n}\right)-F\left(q+\frac{s}{n}\right)+F\left(q+\frac{s}{n}\right)-Fq\right\}$$

$$= \mathcal{L}_\infty n\left\{F\left(q+\frac{s}{n}+\frac{r}{n}\right)-F\left(q+\frac{s}{n}\right)\right\}+\mathcal{L}_\infty n\left\{F\left(q+\frac{s}{n}\right)-Fq\right\}$$

$$= f(q, r)+f(q, s).$$

And, as a particular case, it is obvious that if x be any scalar

$$f(q, xr) = xf(q, r).$$

132.] And if we define in the same way

$$dF(q, r, s \ldots\ldots)$$

as being the value of

$$\mathcal{L}_\infty n\left\{F\left(q+\frac{dq}{n}, r+\frac{dr}{n}, s+\frac{ds}{n}, \ldots\ldots\right)-F(q, r, s, \ldots\ldots)\right\},$$

where $q, r, s, \ldots dq, dr, ds,$ are any quaternions whatever; we shall obviously arrive at a result which may be written

$$f(q, r, s, \ldots dq, dr, ds, \ldots\ldots),$$

where f is homogeneous and linear in the system of quaternions $dq, dr, ds, \ldots$ .. and distributive with respect to each of them. Thus, in differentiating any power, product, &c. of one or more quaternions, each factor is to be differentiated as if it alone were variable; and the terms corresponding to these are to be added for the complete differential. This differs from the ordinary process of scalar differentiation solely in the fact that, on account of the non-commutative property of quaternion multiplication, each factor must in general be differentiated *in situ*. Thus

$$d(qr) = dq.r+qdr, \text{ but \textit{not} generally } = rdq+qdr.$$

133.] As Examples we take chiefly those which lead to results which will be of constant use to us in succeeding Chapters. Some of the work will be given at full length as an exercise in quaternion transformations.

$$(1) \quad (T\rho)^2 = -\rho^2$$

The differential of the left-hand side is simply, since $T\rho$ is a scalar,

$$2\,T\rho\,dT\rho.$$

That of ρ^2 is $\quad \mathcal{L}_\infty n\left(\left(\rho+\frac{d\rho}{n}\right)^2-\rho^2\right)$

$$= \mathcal{L}_\infty n\left(\frac{2}{n}S\rho d\rho+\frac{(d\rho)^2}{n^2}\right) \text{ (§ 104)}$$
$$= 2S\rho d\rho.$$

Hence $\quad T\rho\, dT\rho = -S\rho d\rho,$

or $\quad dT\rho = -S.U\rho\, d\rho = S\frac{d\rho}{U\rho},$

or $\quad \frac{dT\rho}{T\rho} = S\frac{d\rho}{\rho}.$

(2) Again, $\quad \rho = T\rho\, U\rho$

$$d\rho = dT\rho.U\rho + T\rho\, dU\rho,$$

whence $\quad \frac{d\rho}{\rho} = \frac{dT\rho}{T\rho}+\frac{dU\rho}{U\rho}$

$$= S\frac{d\rho}{\rho}+\frac{dU\rho}{U\rho} \text{ by (1).}$$

Hence $\quad \frac{dU\rho}{U\rho} = V\frac{d\rho}{\rho}.$

This may be transformed into $V\frac{d\rho.\rho}{\rho^2}$ or $\frac{V\rho d\rho}{T\rho^2}$, &c.

(3) $\quad (Tq)^2 = qKq$

$$2TqdTq = d(qKq) = \mathcal{L}_\infty n\left[\left(q+\frac{dq}{n}\right)K\left(q+\frac{dq}{n}\right)-qKq\right],$$
$$= \mathcal{L}_\infty n\left(\frac{qKdq+dqKq}{n}+\frac{1}{n^2}dqKdq\right),$$
$$= qKdq+dqKq,$$
$$= qKdq+K(qKdq) \text{ (§ 55)},$$
$$= 2S.qKdq = 2S.Kqdq.$$

Hence $\quad dTq = S.UKqdq = S.Uq^{-1}dq$

since $\quad Tq = TKq, \quad$ and $\quad UKq = Uq^{-1}.$

If $q=\rho$, a vector, $Kq = K\rho = -\rho$, and the formula becomes

$$dT\rho = -S.U\rho d\rho, \text{ as in (1).}$$

Again, $\quad \frac{dTq}{Tq} = S\frac{dq}{q}.$

But $\quad dq = TqdUq + UqdTq,$

which gives $\quad \frac{dq}{q} = \frac{dTq}{Tq}+\frac{dUq}{Uq};$

whence, as $\quad S\frac{dq}{q} = \frac{dTq}{Tq}$

we have $\quad V\frac{dq}{q} = \frac{dUq}{Uq}.$

(4) $$d(q^2) = \mathfrak{L}_\infty n\left(\left(q + \frac{dq}{n}\right)^2 - q^2\right)$$
$$= qdq + dq.q$$
$$= 2S.qdq + 2Sq.Vdq + 2Sdq.Vq.$$

If q be a vector, as ρ, Sq and Sdq vanish, and we have
$$d(\rho^2) = 2S\rho d\rho, \text{ as in (1).}$$

(5) Let $q = r^{\frac{1}{2}}$.

This gives $dr^{\frac{1}{2}} = dq$. But
$$dr = d(q^2) = qdq + dq.q.$$

This, multiplied *by* q and *into* Kq, gives
$$qdr = q^2dq + qdq.q,$$
and $$drKq = dq.Tq^2 + qdq.Kq.$$

Adding, we have
$$qdr + dr.Kq = (q^2 + Tq^2 + 2Sq.q)\,dq\,;$$
whence dq, i.e. $dr^{\frac{1}{2}}$, is at once found in terms of dr. This process is given by Hamilton, *Lectures*, p. 628.

(6) $$qq^{-1} = 1,$$
$$qdq^{-1} + dq.q^{-1} = 0\,;$$
$$\therefore\quad dq^{-1} = -q^{-1}dq.q^{-1}.$$

If q is a vector, $= \rho$ suppose,
$$dq^{-1} = -\rho^{-1}d\rho.\rho^{-1}$$
$$= \frac{d\rho}{\rho^2} - \frac{2}{\rho}S\frac{d\rho}{\rho}$$
$$= \left(\frac{d\rho}{\rho} - 2S\frac{d\rho}{\rho}\right)\frac{1}{\rho}$$
$$= -K\left(\frac{d\rho}{\rho}\right)\frac{1}{\rho}$$

(7) $$q = Sq + Vq,$$
$$dq = dSq + dVq.$$
But $$dq = Sdq + Vdq.$$

Comparing, we have
$$dSq = Sdq, \qquad dVq = Vdq.$$

Since $Kq = Sq - Vq$, we find by a similar process
$$dKq = Kdq.$$

134.] Successive differentiation of course presents no new difficulty.

Thus, we have seen that
$$d(q^2) = dq.q + qdq.$$

Differentiating again, we have

$$d^2(q^2) = d^2q.q + 2(dq)^2 + qd^2q,$$

and so on for higher orders.

If q be a vector, as ρ, we have, § 133 (1),

$$d(\rho^2) = 2S\rho\, d\rho.$$

Hence $d^2(\rho^2) = 2(d\rho)^2 + 2S\rho\, d^2\rho$, and so on.

Similarly $$d^2 U\rho = -d\left(\frac{U\rho}{T\rho^2} V\rho\, d\rho\right).$$

But $$d\frac{1}{T\rho^2} = -\frac{2dT\rho}{T\rho^3} = \frac{2S\rho\, d\rho}{T\rho^4},$$

and $$d.V\rho d\rho = V.\rho d^2\rho.$$

Hence $$-d^2 U\rho = -\frac{U\rho}{T\rho^4}(V\rho d\rho)^2 + \frac{U\rho V\rho d^2\rho}{T\rho^2} + \frac{2U\rho V\rho d\rho S\rho d\rho}{T\rho^4}$$

$$= -\frac{U\rho}{T\rho^4}((V\rho d\rho)^2 + \rho^2 V\rho d^2\rho - 2V\rho d\rho S\rho d\rho)\ *.$$

135.] If the first differential of q be considered as a *constant* quaternion, we have, of course,

$$d^2q = 0, \qquad d^3q = 0, \text{ \&c.},$$

and the preceding formulæ become considerably simplified.

Hamilton has shewn that in this case *Taylor's Theorem* admits of an easy extension to quaternions. That is, we may write

$$f(q + x\,dq) = f(q) + xdf(q) + \frac{x^2}{1.2} d^2 f(q) +$$

if $d^2q = 0$; subject, of course, to particular exceptions and limita tions as in the ordinary applications to functions of scalar variables. Thus, let

$$\begin{aligned} f(q) &= q^3, \text{ and we have} \\ df(q) &= q^2 dq + qdq.q + dq.q^2, \\ d^2 f(q) &= 2dq.qdq + 2q(dq)^2 + 2(dq)^2 q, \\ d^3 f(q) &= 6(dq)^3, \end{aligned}$$

and it is easy to verify by multiplication that we have rigorously

$$(q + xdq)^3 = q^3 + x(q^2dq + qdq.q + dq.q^2) + x^2(dq.qdq + q(dq)^2 + (dq)^2q) + x^3(dq)^3;$$

which is the value given by the application of the above form of Taylor's Theorem.

As we shall not have occasion to employ this theorem, and as the demonstrations which have been found are all too laborious for an elementary treatise, we refer the reader to Hamilton's works, where he will find several of them.

* This may be farther simplified; but it may be well to caution the student that we cannot, for such a purpose, write the above expression as

$$-\frac{U\rho}{T\rho^4} V.\rho\{d\rho V\rho d\rho + d^2\rho.\rho^2 - 2d\rho S\rho d\rho\}.$$

136.] To differentiate a function of a function of a quaternion we proceed as with scalar variables, attending to the peculiarities already pointed out.

137.] A case of considerable importance in geometrical applications of quaternions is the differentiation of a scalar function of ρ, the vector of any point in space.

Let
$$F(\rho) = C,$$
where F is a scalar function and C an arbitrary constant, be the equation of a series of surfaces. Its differential,
$$f(\rho, d\rho) = 0,$$
is, of course, a scalar function: and, being homogeneous and linear in $d\rho$, § 130, may be thus written,
$$S\nu d\rho = 0,$$
where ν is a vector, in general a function of ρ.

This vector, ν, is easily seen to have the direction of the *normal* to the given surface at the extremity of ρ; being, in fact, perpendicular to every tangent line $d\rho$, §§ 36, 98. Its length, when F is a surface of the second degree, is as the reciprocal of the distance of the tangent-plane from the origin. And we will shew, later, that if
$$\rho = ix + jy + kz,$$
then
$$\nu = \left(i\frac{d}{dx} + j\frac{d}{dy} + k\frac{d}{dz}\right)F(\rho).$$

EXAMPLES TO CHAPTER IV.

1. Shew that

(*a.*) $d.SUq = S.UqV\dfrac{dq}{q} = -S\dfrac{dq}{qUVq}TVUq,$

(*b.*) $d.VUq = V.Uq^{-1}V(dq.q^{-1}),$

(*c.*) $d.TVUq = S\dfrac{dUq}{UVq} = S\dfrac{dq}{qUVq}SUq,$

(*d.*) $d.a^x = \dfrac{\pi}{2}a^{x+1}dx,$

(*e.*) $d^2.Tq = \{S^2.dqq^{-1} - S.(dqq^{-1})^2\}Tq = -TqV^2\dfrac{dq}{q}.$

2. If $F\rho = \Sigma.S\alpha\rho S\beta\rho + \frac{1}{2}g\rho^2$

give $dF\rho = S\nu d\rho,$

shew that $\nu = \Sigma V.\alpha\rho\beta + (g + \Sigma S\alpha\beta)\rho.$

CHAPTER V

THE SOLUTION OF EQUATIONS OF THE FIRST DEGREE.

138.] We have seen that the differentiation of any function whatever of a quaternion, q, leads to an equation of the form

$$dr = f(q, dq),$$

where f is linear and homogeneous in dq. To complete the process of differentiation, we must have the means of solving this equation so as to be able to exhibit directly the value of dq.

This general equation is not of so much practical importance as the particular case in which dq is a vector; and, besides, as we proceed to shew, the solution of the general question may easily be made to depend upon that of the particular case; so that we shall commence with the latter.

The most general expression for the function f is easily seen to be

$$dr = f(q, dq) = \Sigma V.a\,dq\,b + S.c\,dq,$$

where a, b, and c may be any quaternion functions of q whatever. Every possible term of a linear and homogeneous function is reducible to this form, as the reader may easily see by writing down all the forms he can devise.

Taking the scalars of both sides, we have

$$Sdr = S.c\,dq = Sdq\,Sc + S.Vdq\,Vc.$$

But we have also, by taking the vector parts

$$Vdr = \Sigma V.a\,dq\,b = Sdq.\Sigma Vab + \Sigma V.a(Vdq)b.$$

Eliminating Sdq between the equations for Sdr and Vdr it is obvious that a linear and vector expression in Vdq will remain. Such an expression, so far as it contains Vdq, may always be reduced to the form of a sum of terms of the type $aS.\beta Vdq$, by the help of formulæ like those in §§ 90, 91. Solving this, we have Vdq, and Sdq is then found from the preceding equation.

139.] The problem may now be stated thus.

Find the value of ρ from the equation

$$\alpha S\beta\rho + \alpha_1 S\beta_1\rho + \ldots = \Sigma . \alpha S\beta\rho = \gamma,$$

where α, β, α_1, β_1, ... γ are given vectors. [It will be shewn later that the most general form requires but three terms, i. e. *six* vector constants α, β, α_1, β_1, α_2, β_2 in all.]

If we write, with Hamilton,

$$\phi\rho = \Sigma . \alpha S\beta\rho,$$

the given equation may be written

$$\phi\rho = \gamma,$$

or $$\rho = \phi^{-1}\gamma,$$

and the object of our investigation is to find the value of the inverse function ϕ^{-1}.

140.] We have seen that any vector whatever may be expressed in terms of any three non-coplanar vectors. Hence, we should expect *à priori* that a vector such as $\phi\phi\phi\rho$, or $\phi^3\rho$, for instance, should be capable of expression in terms of ρ, $\phi\rho$, and $\phi^2\rho$. [This is, of course, on the supposition that ρ, $\phi\rho$, and $\phi^2\rho$ are not generally coplanar. But it may easily be seen to extend to this case also. For if these vectors be generally coplanar, so are $\phi\rho$, $\phi^2\rho$, and $\phi^3\rho$, since they may be written σ, $\phi\sigma$, and $\phi^2\sigma$. And thus, of course, $\phi^3\rho$ can be expressed as above. If in a particular case, we should have, for some *definite* vector ρ, $\phi\rho = g\rho$ where g is a scalar, we shall obviously have $\phi^2\rho = g^2\rho$ and $\phi^3\rho = g^3\rho$, so that the equation will still subsist. And a similar explanation holds for the particular case when, for some *definite* value of ρ, the three vectors ρ, $\phi\rho$, $\phi^2\rho$ are coplanar. For then we have an equation of the form

$$\phi^2\rho = A\rho + B\phi\rho,$$

which gives
$$\phi^3\rho = A\phi\rho + B\phi^2\rho$$
$$= AB\rho + (A + B^2)\phi\rho.$$

So that $\phi^3\rho$ is in the same plane.]

If, then, we write

$$-\phi^3\rho = x\rho + y\phi\rho + z\phi^2\rho, \qquad (1)$$

it is evident that x, y, z are quantities independent of the vector ρ, and we can determine them at once by processes such as those in §§ 91, 92.

If any three vectors, as i, j, k, be substituted for ρ, they will in general enable us to assign the values of the three coefficients on the right side of the equation, and the solution is complete. For by putting $\phi^{-1}\rho$ for ρ and transposing, the equation becomes

$$x\phi^{-1} = y\rho + z\phi\rho + \phi^2\rho;$$

that is, the unknown inverse function is expressed in terms of direct operations. If x vanish, while y remains finite, we substitute $\phi^{\ 2}\rho$ for ρ, and have

$$-y\phi^{-1}\rho = z\rho + \phi\rho,$$

and if x and y both vanish

$$z\phi^{-1}\rho = \rho.$$

141.] To illustrate this process by a simple example we shall take the very important case in which ϕ belongs to a *central* surface of the second order; suppose an ellipsoid; in which case it will be shewn (in Chap. VIII.) that we may write

$$\phi\rho = -a^2 iSi\rho - b^2 jSj\rho - c^2 kSk\rho.$$

Here we have

$$\begin{aligned} \phi i &= a^2 i, & \phi^2 i &= a^4 i, & \phi^3 i &= a^6 i, \\ \phi j &= b^2 j, & \phi^2 j &= b^4 j, & \phi^3 j &= b^6 j, \\ \phi k &= c^2 k, & \phi^2 k &= c^4 k, & \phi^3 k &= c^6 k. \end{aligned}$$

Hence, putting separately i, j, k for ρ in the equation (1) of last section, we have

$$\begin{aligned} -a^6 &= x + ya^2 + za^4, \\ -b^6 &= x + yb^2 + zb^4, \\ -c^6 &= x + yc^2 + zc^4. \end{aligned}$$

Hence a^2, b^2, c^2 are the roots of the cubic

$$\xi^3 + z\xi^2 + y\xi + x = 0,$$

which involves the conditions

$$\begin{aligned} z &= -(a^2 + b^2 + c^2), \\ y &= a^2b^2 + b^2c^2 + c^2a^2, \\ x &= -a^2b^2c^2. \end{aligned}$$

Thus, with the above value of ϕ, we have

$$\phi^3\rho = a^2b^2c^2\rho - (a^2b^2 + b^2c^2 + c^2a^2)\phi\rho + (a^2 + b^2 + c^2)\phi^2\rho.$$

142.] Putting $\phi^{-1}\sigma$ in place of ρ (which is *any* vector whatever) and changing the order of the terms, we have the desired inversion of the function ϕ in the form

$$a^2b^2c^2\phi^{-1}\sigma = (a^2b^2 + b^2c^2 + c^2a^2)\sigma - (a^2 + b^2 + c^2)\phi\sigma + \phi^2\sigma,$$

where the inverse function is expressed in terms of the direct func tion. For this particular case the solution we have given is com plete, and satisfactory; and it has the advantage of preparing the reader to expect a similar form of solution in more complex cases.

143.] It may also be useful as a preparation for what follows, if we put the equation of § 141 in the form

$$\begin{aligned} 0 = \Phi(\rho) &= \phi^3\rho-(a^2+b^2+c^2)\,\phi^2\rho+(a^2b^2+b^2c^2+c^2a^2)\,\phi\rho-a^2b^2c^2\,\rho \\ &= \{\phi^3-(a^2+b^2+c^2)\,\phi^2+(a^2b^2+b^2c^2+c^2a^2)\,\phi-a^2b^2c^2\}\,\rho \\ &= \{(\phi-a^2)(\phi-b^2)(\phi-c^2)\}\,\rho. \quad \ldots\ldots\ldots\ldots\ldots\ldots\ldots\ldots (2) \end{aligned}$$

This last transformation is permitted because ϕ is commutative with scalars like a^2, i. e. $\phi(a^2\rho) = a^2\phi\rho$.

Here we remark that (by § 140) the equation

$$V.\rho\phi\rho = 0, \quad \text{or} \quad \phi\rho = g\rho,$$

where g is some undetermined scalar, is satisfied, not merely by every vector of null-length, but by the definite system of three rectangular vectors Ai, Bj, Ck whatever be their tensors, the corresponding particular values of g being a^2, b^2, c^2.

144.] We now give Hamilton's admirable investigation.

The most general form of a linear and vector function of a vector may of course be written as

$$\phi\rho = \Sigma V.q\rho r,$$

where q and r are any constant quaternions, either or both of which may degrade to a scalar or a vector.

Hence, operating by $S.\sigma$ where σ is any vector whatever,

$$S\sigma\phi\rho = \Sigma S\sigma V.q\rho r = \Sigma S\rho V.r\sigma q = S\rho\phi'\sigma, \quad \ldots\ldots\ldots\ldots (3)$$

if we agree to write $\phi'\sigma = \Sigma V.r\sigma q$,

and remember the proposition of § 88. The functions ϕ and ϕ' are thus *conjugate* to one another, and on this property the whole investigation depends.

145.] Let λ, μ be any two vectors, such that

$$\phi\rho = V\lambda\mu.$$

Operating by $S.\lambda$ and $S.\mu$ we have

$$S\lambda\phi\rho = 0, \qquad S\mu\phi\rho = 0.$$

But, introducing the conjugate function ϕ', these become

$$S\rho\phi'\lambda = 0, \qquad S\rho\phi'\mu = 0,$$

and give ρ in the form $m\rho = V\phi'\lambda\phi'\mu$,

where m is a scalar which, as we shall presently see, is independent of λ, μ, and ρ.

But our original assumption gives

$$\rho = \phi^{-1}V\lambda\mu;$$

hence we have $m\phi^{-1}V\lambda\mu = V\phi'\lambda\phi'\mu, \quad \ldots\ldots\ldots\ldots\ldots\ldots (4)$

and the problem of inverting ϕ is solved.

146.] It remains to find the value of the constant m, and to express the vector
$$V\phi'\lambda\phi'\mu$$
as a function of $V\lambda\mu$.

Operate on (4) by $S.\phi'\nu$, where ν is any vector not coplanar with λ and μ, and we get

$$mS.\phi'\nu\phi^{-1}V\lambda\mu = mS.\nu\phi\phi^{-1}V\lambda\mu \quad \text{(by (3) of § 144)}$$
$$= mS.\lambda\mu\nu = S.\phi'\lambda\phi'\mu\phi'\nu, \text{ or}$$
$$m = \frac{S.\phi'\lambda\phi'\mu\phi'\nu}{S.\lambda\mu\nu}. \qquad (5)$$

[That this quantity is independent of the particular vectors λ, μ, ν is evident from the fact that if

$\lambda' = p\lambda + q\mu + r\nu$, $\mu' = p_1\lambda + q_1\mu + r_1\nu$, and $\nu' = p_2\lambda + q_2\mu + r_2\nu$

be any other three vectors (which is possible since λ, μ, ν are not coplanar), we have

$$\phi'\lambda' = p\phi'\lambda + q\phi'\mu + r\phi'\nu, \text{ \&c., \&c.};$$

from which we deduce

$$S.\phi'\lambda'\phi'\mu'\phi'\nu' = \begin{vmatrix} p & q & r \\ p_1 & q_1 & r_1 \\ p_2 & q_2 & r_2 \end{vmatrix} S.\phi'\lambda\phi'\mu\phi'\nu,$$

and
$$S.\lambda'\mu'\nu' = \begin{vmatrix} p & q & r \\ p_1 & q_1 & r_1 \\ p_2 & q_2 & r_2 \end{vmatrix} S.\lambda\mu\nu,$$

so that the numerator and denominator of the fraction which expresses m are altered in the same ratio. Each of these quantities is in fact an *Invariant*, and the numerical multiplier is the same for both when we pass from any one set of three vectors to another. A still simpler proof is obtained at once by writing $\lambda + p\mu$ for λ in (5), and noticing that neither numerator nor denominator is altered.]

147.] Let us now change ϕ to $\phi + g$, where g is any scalar. It is evident that ϕ' becomes $\phi' + g$, and our equation (4) becomes

$$m_g(\phi+g)^{-1}V\lambda\mu = V(\phi'+g)\lambda(\phi'+g)\mu;$$
$$= V\phi'\lambda\phi'\mu + gV(\phi'\lambda\mu + \lambda\phi'\mu) + g^2V\lambda\mu,$$
$$= (m\phi^{-1} + g\chi + g^2)V\lambda\mu \text{ suppose.}$$

In the above equation

$$m_g = \frac{S.(\phi'+g)\lambda(\phi'+g)\mu(\phi'+g)\nu}{S.\lambda\mu\nu}$$
$$= m + m_1 g + m_2 g^2 + g^3$$

is what m becomes when ϕ is changed into $\phi+g$; m_1 and m_2 being two new scalar constants whose values are

$$m_1 = \frac{S.(\lambda\phi'\mu\phi'\nu + \phi'\lambda\mu\phi'\nu + \phi'\lambda\phi'\mu\nu)}{S.\lambda\mu\nu},$$

$$m_2 = \frac{S.(\lambda\mu\phi'\nu + \phi'\lambda\mu\nu + \lambda\phi'\mu\nu)}{S.\lambda\mu\nu}.$$

If, in these expressions, we put $\lambda + p\mu$ for λ, we find that the terms in p vanish identically; so that they also are invariants. Substituting for m_g, and equating the coefficients of the various powers of g after operating on both sides by $\phi+g$, we have two identities and the following two equations,

$$m_2 = \phi + \chi,$$
$$m_1 = \phi\chi + m\phi^{-1}$$

[The first determines χ, and shews that we were justified in treat ing $V(\phi'\lambda\mu + \lambda\phi'\mu)$ as a linear and vector function of $V.\lambda\mu$. The result might have been also obtained thus,

$$S.\lambda\chi V\lambda\mu = S.\lambda\phi'\lambda\mu = -S.\lambda\mu\phi'\lambda = -S.\lambda\phi V\lambda\mu,$$
$$S.\mu\chi V\lambda\mu = S.\mu\lambda\phi'\mu = -S.\mu\phi V\lambda\mu,$$
$$S.\nu\chi V\lambda\mu = S.(\nu\phi'\lambda\mu + \nu\lambda\phi'\mu)$$
$$= m_2 S\lambda\mu\nu - S.\lambda\mu\phi'\nu$$
$$= S.\nu\,(m_2 V\lambda\mu - \phi V\lambda\mu);$$

and all three (the utmost generality) are satisfied by

$$\chi = m_2 - \phi.]$$

148.] Eliminating χ from these equations we find

$$m_1 = \phi\,(m_2 - \phi) + m\phi^{-1},$$
$$\text{or}\quad m\phi^{-1} = m_1 - m_2\,\phi + \phi^2,$$

which contains the complete solution of linear and vector equations.

149.] More to satisfy the student of the validity of the above investigation, about whose logic he may at first feel some difficulties, than to obtain easy solutions, we take a few very simple examples to begin with: we treat them with all desirable prolixity, and we append for comparison easy solutions obtained by methods specially adapted to each case.

150.] *Example I.*

Let $\phi\rho = V.\alpha\rho\beta = \gamma$.

Then $\quad \phi'\rho = V.\beta\rho\alpha = \phi\rho$.

Hence $\quad m = \dfrac{1}{S.\lambda\mu\nu}\, S\,(V.\alpha\lambda\beta V.\alpha\mu\beta V.\alpha\nu\beta)$.

Now λ, μ, ν are any three non-coplanar vectors; and we may therefore put for them α, β, γ *if the latter be non-coplanar.*

With this proviso

$$m = \frac{1}{S.\alpha\beta\gamma} S(\alpha^2\beta.\alpha\beta^2.V.\alpha\gamma\beta)$$
$$= \alpha^2\beta^2 S\alpha\beta,$$
$$m_1 = \frac{1}{S.\alpha\beta\gamma} S(\alpha^2\beta.\alpha\beta^2.\gamma + \alpha.\alpha\beta^2.V.\alpha\gamma\beta + \alpha^2\beta.\beta.V.\alpha\gamma\beta)$$
$$= -\alpha^2\beta^2,$$
$$m_2 = \frac{1}{S.\alpha\beta\gamma} S(\alpha^2\beta.\beta.\gamma + \alpha.\alpha\beta^2.\gamma + \alpha\beta V.\alpha\gamma\beta)$$
$$= -S\alpha\beta.$$

Hence

$$\alpha^2\beta^2 S\alpha\beta.\phi^{-1}\gamma = \alpha^2\beta^2 S\alpha\beta.\rho = -\alpha^2\beta^2\gamma + S\alpha\beta V.\alpha\gamma\beta + V.\alpha(V.\alpha\gamma\beta)\beta,$$

which is one form of solution.

By expanding the vectors of products we may easily reduce it to the form
$$\alpha^2\beta^2 S\alpha\beta.\rho = -\alpha^2\beta^2\gamma + \alpha\beta^2 S\alpha\gamma + \beta\alpha^2 S\beta\gamma,$$
or
$$\rho = \frac{\alpha^{-1}S\alpha\gamma + \beta^{-1}S\beta\gamma - \gamma}{S\alpha\beta}.$$

151.] To verify this solution, we have

$$V.\alpha\rho\beta = \frac{1}{S\alpha\beta}(\beta S\alpha\gamma + \alpha S\beta\gamma - V.\alpha\gamma\beta) = \gamma,$$

which is the given equation.

152.] An easier mode of arriving at the same solution, in this simple case, is as follows :—

Operating by $S.\alpha$ and $S.\beta$ on the given equation

$$V.\alpha\rho\beta = \gamma,$$

we obtain

$$\alpha^2 S\beta\rho = S\alpha\gamma,$$
$$\beta^2 S\alpha\rho = S\beta\gamma;$$

and therefore

$$\alpha S\beta\rho = \alpha^{-1}S\alpha\gamma,$$
$$\beta S\alpha\rho = \beta^{-1}S\beta\gamma.$$

But the given equation may be written

$$\alpha S\beta\rho - \rho S\alpha\beta + \beta S\alpha\rho = \gamma.$$

Substituting and transposing we get

$$\rho S\alpha\beta = \alpha^{-1}S\alpha\gamma + \beta^{-1}S\beta\gamma - \gamma,$$

which agrees with the result of § 150.

153.] If α, β, γ be coplanar, the above mode of solution is appli cable, but the result may be deduced much more simply.

For (§ 101) $S.\alpha\beta\gamma = 0$, and the equation then gives $S.\alpha\beta\rho = 0$, so that ρ is also coplanar with α, β, γ.

Hence the equation may be written

$$\alpha\rho\beta = \gamma,$$

and at once

$$\rho = \alpha^{-1}\gamma\beta^{-1};$$

and this, being a vector, may be written

$$-\alpha^{-1}S\beta^{-1}\gamma + \beta^{-1}S\alpha^{-1}\gamma - \gamma S\alpha^{-1}\beta^{-1}.$$

This formula is *equivalent* to that just given, but not equal to it term by term. [The student will find it a good exercise to prove *directly* that, if α, β, γ are coplanar, we have

$$\frac{1}{S\alpha\beta}(\alpha^{-1}S\alpha\gamma + \beta^{-1}S\beta\gamma - \gamma) = \alpha^{-1}S\beta^{-1}\gamma + \beta^{-1}S\alpha^{-1}\gamma - \gamma S\alpha^{-1}\beta^{-1}.]$$

The conclusion that

$$S.\alpha\beta\rho = 0,$$

in this case, is not necessarily true if

$$S\alpha\beta = 0.$$

But then the original equation becomes

$$\alpha S\beta\rho + \beta S\alpha\rho = \gamma,$$

which is consistent with

$$S.\alpha\beta\gamma = 0.$$

This equation gives

$$\gamma(\alpha^2\beta^2 - S^2\alpha\beta) = \alpha\begin{vmatrix} S\alpha\gamma & S\alpha\beta \\ S\beta\gamma & \beta^2 \end{vmatrix} + \beta\begin{vmatrix} S\beta\gamma & S\alpha\beta \\ S\alpha\gamma & \alpha^2 \end{vmatrix},$$

by comparison of which with the given equation we find

$$S\alpha\rho \quad \text{and} \quad S\beta\rho.$$

The value of ρ remains therefore with one indeterminate scalar.

154.] *Example II.*

Let $\phi\rho = V.\alpha\beta\rho = \gamma$.

Suppose α, β, γ not to be coplanar, and employ them as λ, μ, ν to calculate the coefficients in the equation for ϕ^{-1} We have

$$S.\sigma\phi\rho = S.\sigma\alpha\beta\rho = S.\rho V.\sigma\alpha\beta = S.\rho\phi'\sigma.$$

Hence

$$\phi'\rho = V.\rho\alpha\beta = V.\beta\alpha\rho.$$

We have now

$$m = \frac{1}{S.\alpha\beta\gamma}S(\beta\alpha^2.\beta\alpha\beta.V.\beta\alpha\gamma) = \frac{\alpha^2\beta^2}{S.\alpha\beta\gamma}S.\alpha\beta V.\beta\alpha\gamma$$

$$= \alpha^2\beta^2 S\alpha\beta,$$

$$m_1 = \frac{1}{S.\alpha\beta\gamma}S(\alpha.\beta\alpha\beta.V.\beta\alpha\gamma + \beta\alpha^2.\beta.V.\beta\alpha\gamma + \beta\alpha^2.\beta\alpha\beta.\gamma)$$

$$= 2(S\alpha\beta)^2 + \alpha^2\beta^2,$$

$$m_2 = \frac{1}{S.\alpha\beta\gamma}S(\alpha.\beta.V.\beta\alpha\gamma + \alpha.\beta\alpha\beta.\gamma + \beta\alpha^2.\beta.\gamma)$$

$$= 3S\alpha\beta.$$

Hence

$$\alpha^2\beta^2 S\alpha\beta.\phi^{-1}\gamma = \alpha^2\beta^2 S\alpha\beta.\rho$$
$$- (2(S\alpha\beta)^2 + \alpha^2\beta^2)\gamma - 3S\alpha\beta V.\alpha\beta\gamma + V.\alpha\beta V.\alpha\beta\gamma,$$

which, by expanding the vectors of products, takes easily the simpler form $\alpha^2\beta^2 S\alpha\beta.\rho = \alpha^2\beta^2\gamma - \alpha\beta^2 S\alpha\gamma + 2\beta S\alpha\beta S\alpha\gamma - \beta\alpha^2 S\beta\gamma.$

155.] To verify this, operate by $V.\alpha\beta$ on both sides, and we have

$$\alpha^2\beta^2 S\alpha\beta V.\alpha\beta\rho = \alpha^2\beta^2 V.\alpha\beta\gamma - V.\alpha\beta\alpha\beta^2 S\alpha\gamma + 2\alpha\beta^2 S\alpha\beta S\alpha\gamma - \alpha\alpha^2\beta^2 S\beta\gamma$$
$$- \alpha^2\beta^2(\alpha S\beta\gamma - \beta S\alpha\gamma + \gamma S\alpha\beta) - (2\alpha S\alpha\beta - \beta\alpha^2)\beta^2 S\alpha\gamma$$
$$+ 2\alpha\beta^2 S\alpha\beta S\alpha\gamma - \alpha\alpha^2\beta^2 S\beta\gamma$$
$$- \alpha^2\beta^2 S\alpha\beta.\gamma,$$

or $$V.\alpha\beta\rho = \gamma.$$

156.] To solve the same equation without employing the general method, we may proceed as follows:—

$$\gamma = V.\alpha\beta\rho = \rho S\alpha\beta + V.V(\alpha\beta)\rho.$$

Operating by $S.V\alpha\beta$ we have

$$S.\alpha\beta\gamma = S.\alpha\beta\rho S\alpha\beta.$$

Divide this by $S\alpha\beta$, and add it to the given equation. We thus obtain

$$\gamma + \frac{S.\alpha\beta}{S\alpha\beta\gamma} = \rho S\alpha\beta + V.V(\alpha\beta)\rho + S.V(\alpha\beta)\rho,$$
$$= (S\alpha\beta + V\alpha\beta)\rho,$$
$$= \alpha\beta\rho.$$

Hence $$\rho = \beta^{-1}\alpha^{-1}\left(\gamma + \frac{S.\alpha\beta\gamma}{S\alpha\beta}\right),$$

a form of solution somewhat simpler than that before obtained.

To shew that they agree, however, let us multiply by $\alpha^2\beta^2 S\alpha\beta$, and we get $$\alpha^2\beta^2 S\alpha\beta.\rho = \beta\alpha\gamma S\alpha\beta + \beta\alpha S.\alpha\beta\gamma.$$

In this form we see at once that the right-hand side is a vector, since its scalar is evidently zero (§ 89). Hence we may write

$$\alpha^2\beta^2 S\alpha\beta.\rho = V.\beta\alpha\gamma S\alpha\beta - V\alpha\beta S.\alpha\beta\gamma.$$

But by (3) of § 91,

$$\gamma S.\alpha\beta V\alpha\beta + \alpha S.\beta(V\alpha\beta)\gamma + \beta S.V(\alpha\beta)\alpha\gamma + V\alpha\beta S.\alpha\beta\gamma = 0.$$

Add this to the right-hand side, and we have

$$\alpha^2\beta^2 S\alpha\beta.\rho = \gamma((S\alpha\beta)^2 - S.\alpha\beta V\alpha\beta) - \alpha(S\alpha\beta S\beta\gamma - S.\beta(V\alpha\beta)\gamma)$$
$$+ \beta(S\alpha\beta S\alpha\gamma + S.V(\alpha\beta)\alpha\gamma).$$

But $$(S\alpha\beta)^2 - S.\alpha\beta V\alpha\beta = (S\alpha\beta)^2 - (V\alpha\beta)^2 = \alpha^2\beta^2,$$
$$S\alpha\beta S\beta\gamma - S.\beta(V\alpha\beta)\gamma = S\alpha\beta S\beta\gamma - S\beta\alpha S\beta\gamma + \beta^2 S\alpha\gamma = \beta^2 S\alpha\gamma$$
$$S\alpha\beta S\alpha\gamma + S.V(\alpha\beta)\alpha\gamma = S\alpha\beta S\alpha\gamma + S\alpha\beta S\alpha\gamma - \alpha^2 S\beta\gamma$$
$$= 2S\alpha\beta S\alpha\gamma - \alpha^2 S\beta\gamma;$$

and the substitution of these values renders our equation identical with that of § 154.

[If α, β, γ be coplanar, the simplified forms of the expression for ρ lead to the equation

$$S\alpha\beta.\beta^{-1}\alpha^{-1}\gamma = \gamma - \alpha^{-1}S\alpha\gamma + 2\beta S\alpha^{-1}\beta^{-1}S\alpha\gamma - \beta^{-1}S\beta\gamma,$$

which, as before, we leave as an exercise to the student.]

157.] *Example III.* The solution of the equation

$$V\epsilon\rho = \gamma$$

leads to the vanishing of some of the quantities m. Before, however, treating it by the general method, we shall deduce its solution from that of

$$V.\alpha\beta\rho = \gamma$$

already given. Our reason for so doing is that we thus have an opportunity of shewing the nature of some of the cases in which one or more of m, m_1, m_2 vanish; and also of introducing an example of the use of vanishing fractions in quaternions. Far simpler solu tions will be given in the following sections.

The solution of the last-written equation is, § 154,

$$\alpha^2\beta^2 S\alpha\beta.\rho = \alpha^2\beta^2\gamma - \alpha\beta^2 S\alpha\gamma - \beta\alpha^2 S\beta\gamma + 2\beta S\alpha\beta S\alpha\gamma.$$

If we now put

$$\alpha\beta = e + \epsilon$$

where e is a scalar, the solution of the first-written equation will evidently be derived from that of the second by making e gradually tend to zero.

We have, for this purpose, the following necessary transformations:

$$\alpha^2\beta^2 = \alpha\beta K.\alpha\beta = (e+\epsilon)(e-\epsilon) = e^2 - \epsilon^2$$

$$\begin{aligned} \alpha\beta^2 S\alpha\gamma + \beta\alpha^2 S\beta\gamma &= \alpha\beta.\beta S\alpha\gamma + \beta\alpha.\alpha S\beta\gamma, \\ &= (e+\epsilon)\,\beta S\alpha\gamma + (e-\epsilon)\,\alpha S\beta\gamma, \\ &= e\,(\beta S\alpha\gamma + \alpha S\beta\gamma) + eV.\gamma V\alpha\beta, \\ &= e\,(\beta S\alpha\gamma + \alpha S\beta\gamma) + \epsilon V\gamma\epsilon. \end{aligned}$$

Hence the solution becomes

$$\begin{aligned} (e^2-\epsilon^2)e\rho &= (e^2-\epsilon^2)\gamma - e\,(\beta S\alpha\gamma + \alpha S\beta\gamma) - \epsilon V\gamma\epsilon + 2e\beta S\alpha\gamma, \\ &= (e^2-\epsilon^2)\gamma + eV.\gamma V\alpha\beta - \epsilon V\gamma\epsilon, \\ &= (e^2-\epsilon^2)\gamma + eV\gamma\epsilon + \gamma\epsilon^2 - \epsilon S\gamma\epsilon, \\ &= e^2\gamma + eV\gamma\epsilon - \epsilon S\gamma\epsilon. \end{aligned}$$

Dividing by e, and then putting $e = 0$, we have

$$-\epsilon^2\rho = V\gamma\epsilon - \epsilon \mathfrak{L}_0\left(\frac{S\gamma\epsilon}{e}\right)$$

Now, by the form of the given equation, we see that

$$S\gamma\epsilon = 0.$$

Hence the limit is indeterminate, and we may put for it x, where x is *any* scalar. Our solution is, therefore,

$$\rho = -V\frac{\gamma}{\epsilon} + x\epsilon^{-1};$$

or, as it may be written, since $S\gamma\epsilon = 0$,

$$\rho = \epsilon^{-1}(\gamma + x).$$

The verification is obvious—for we have

$$\epsilon\rho = \gamma + x.$$

158.] This suggests a very simple mode of solution. For we see that the given equation leaves $S\epsilon\rho$ indeterminate. Assume, therefore,

$$S\epsilon\rho = x$$

and add to the given equation. We obtain

$$\epsilon\rho = x + \gamma,$$

$$\text{or} \quad \rho = \epsilon^{-1}(\gamma + x),$$

if, and only if, ρ satisfies the equation

$$V\epsilon\rho = \gamma.$$

159.] To apply the general method, we may take ϵ, γ and $\epsilon\gamma$ (which is a vector) for λ, μ, ν.

We find $\phi'\rho = V\rho\epsilon$.

Hence

$$m = 0,$$

$$m_1 = -\frac{1}{\epsilon^2\gamma^2} S.(\epsilon.\epsilon\gamma.\epsilon^2\gamma) = -\epsilon^2,$$

$$m_2 = 0.$$

Hence

$$-\epsilon^2\phi + \phi^3 = 0$$

$$\text{or} \quad \phi^{-1} - \frac{1}{\epsilon^2}\phi + \phi^{-2}0.$$

That is,

$$\rho = \frac{1}{\epsilon^2}V\epsilon\gamma + x\epsilon,$$

$$- \epsilon^{-1}\gamma + x\epsilon, \text{ as before.}$$

Our warrant for putting $x\epsilon$, as the equivalent of $\phi^{-2}0$ is this:—

The equation

$$\phi^2\sigma = 0$$

may be written

$$V.\epsilon V\epsilon\sigma = 0 = \sigma\epsilon^2 - \epsilon S\epsilon\sigma.$$

Hence, unless $\sigma = 0$, we have $\sigma \parallel \epsilon = x\epsilon$.

160.] *Example IV.* As a final example let us take the most general form of ϕ, which, as will be soon proved, may be expressed as follows:—

$$\phi\rho = \alpha S\beta\rho + \alpha_1 S\beta_1\rho + \alpha_2 S\beta_2\rho = \gamma.$$

Here

$$\phi'\rho = \beta S\alpha\rho + \beta_1 S\alpha_1\rho + \beta_2 S\alpha_2\rho,$$

and, consequently, taking α, α_1, α_2, which are in this case non-coplanar vectors, for λ, μ, ν, we have

$$m = \frac{1}{S.\alpha\alpha_1\alpha_2} S.(\beta S\alpha\alpha + \beta_1 S\alpha_1\alpha + \beta_2 S\alpha_2\alpha)(\beta S\alpha\alpha_1 + \beta_1 S\alpha_1\alpha_1 + \ldots)(\beta S\alpha\alpha_2 + \ldots\ldots)$$

$$= \frac{S.\beta\beta_1\beta_2}{S.\alpha\alpha_1\alpha_2} \begin{vmatrix} S\alpha\alpha & S\alpha_1\alpha & S\alpha_2\alpha \\ S\alpha\alpha_1 & S\alpha_1\alpha_1 & S\alpha_2\alpha_1 \\ S\alpha\alpha_2 & S\alpha_1\alpha_2 & S\alpha_2\alpha_2 \end{vmatrix}$$

$$= \frac{S.\beta\beta_1\beta_2}{S.\alpha\alpha_1\alpha_2}(AS\alpha\alpha + A_1 S\alpha_1\alpha + A_2 S\alpha_2\alpha),$$

where
$$A = S\alpha_1\alpha_1 S\alpha_2\alpha_2 - S\alpha_2\alpha_1 S\alpha_1\alpha_2$$
$$= -S.V\alpha_1\alpha_2 V\alpha_1\alpha_2$$
$$A_1 = S\alpha_2\alpha_1 S\alpha\alpha_2 - S\alpha\alpha_1 S\alpha_2\alpha_2$$
$$= -S.V\alpha_2\alpha V\alpha_1\alpha_2$$
$$A_2 = S\alpha\alpha_1 S\alpha_1\alpha_2 - S\alpha_1\alpha_1 S\alpha\alpha_2$$
$$= -S.V\alpha\alpha_1 V\alpha_1\alpha_2.$$

Hence the value of the determinant is

$$-(S\alpha\alpha S.V\alpha_1\alpha_2 V\alpha_1\alpha_2 + S\alpha_1\alpha S.V\alpha_2\alpha V\alpha_1\alpha_2 + S\alpha_2\alpha S.V\alpha\alpha_1 V\alpha_1\alpha_2)$$
$$= -S.\alpha(V\alpha_1\alpha_2 S.\alpha\alpha_1\alpha_2) \text{ \{by § 92 (3)\}} = -(S.\alpha\alpha_1\alpha_2)^2.$$

The interpretation of this result in spherical trigonometry is very interesting. (See Ex. (6) p. 68.)

By it we see that

$$m = -S.\alpha\alpha_1\alpha_2 S.\beta\beta_1\beta_2.$$

Similarly,

$$m_1 = \frac{1}{S.\alpha\alpha_1\alpha_2} S.[\alpha(\beta S\alpha\alpha_1 + \beta_1 S\alpha_1\alpha_1 + \beta_2 S\alpha_2\alpha_1)(\beta S\alpha\alpha_2 + \beta_1 S\alpha_1\alpha_2 + \beta_2 S\alpha_2\alpha_2) + \&c.]$$
$$= \frac{1}{S.\alpha\alpha_1\alpha_2}(S.\alpha\beta\beta_1(S\alpha\alpha_1 S\alpha_1\alpha_2 - S\alpha_1\alpha_1 S\alpha\alpha_2) + \ldots\ldots)$$
$$= \frac{1}{S.\alpha\alpha_1\alpha_2}(S.\alpha\beta\beta_1 S.\alpha V.\alpha_1 V\alpha_2\alpha_1 + \ldots\ldots)$$
$$= -\frac{1}{S.\alpha\alpha_1\alpha_2}[S.\alpha(V\beta\beta_1 S.V\alpha\alpha_1 V\alpha_1\alpha_2 + V\beta_2\beta S.V\alpha_2\alpha V\alpha_1\alpha_2 + V\beta_1\beta_2 S.V\alpha_1\alpha_2 V\alpha_1\alpha_2)$$
$$+ S.\alpha_1(V\beta\beta_1 S.V\alpha\alpha_1 V\alpha_2\alpha + \ldots\ldots)$$
$$+ S.\alpha_2(V\beta\beta_1 S.V\alpha\alpha_1 V\alpha\alpha_1 + \ldots\ldots)];$$

or, taking the terms by columns instead of by rows,

$$= -\frac{1}{S.\alpha\alpha_1\alpha_2}[S.V\beta\beta_1(\alpha S.V\alpha\alpha_1 V\alpha_1\alpha_2 + \alpha_1 S.V\alpha\alpha_1 V\alpha_2\alpha + \alpha_2 S.V\alpha\alpha_1 V\alpha\alpha_1)$$
$$+ \quad . \quad . \quad . \quad . \quad . \quad . \quad . \quad + \quad . \quad . \quad . \quad . \quad . \quad . \quad],$$
$$= -\frac{1}{S.\alpha\alpha_1\alpha_2}[S.V\beta\beta_1(V\alpha\alpha_1 S.\alpha\alpha_1\alpha_2) \quad + \quad . \quad],$$
$$= -S(V\alpha\alpha_1 V\beta\beta_1 + V\alpha_1\alpha_2 V\beta_1\beta_2 + V\alpha_2\alpha V\beta_2\beta).$$

Again,

$$m_2 = \frac{1}{S.\alpha\alpha_1\alpha_2} S[\alpha\alpha_1(\beta S\alpha\alpha_2 + \beta_1 S\alpha_1\alpha_2 + \ldots) + \alpha_2\alpha(\beta S\alpha\alpha_1 + \ldots) + \alpha_1\alpha_2(\beta S\alpha\alpha + \ldots)]$$

or, grouping as before,

$$= \frac{1}{S.\alpha\alpha_1\alpha_2} S[\beta(V\alpha\alpha_1 S\alpha\alpha_2 + V\alpha_2\alpha S\alpha\alpha_1 + V\alpha_1\alpha_2 S\alpha\alpha) + \ldots],$$
$$= \frac{1}{S.\alpha\alpha_1\alpha_2} S[\beta(\alpha S.\alpha\alpha_1\alpha_2) + \ldots\ldots] \quad (\text{§ } 92\ (4)),$$
$$= S(\alpha\beta + \alpha_1\beta_1 + \alpha_2\beta_2).$$

And the solution is, therefore,

$$\phi^{-1}\gamma S.\alpha\alpha_1\alpha_2 S.\beta\beta_1\beta_2 = \rho S.\alpha\alpha_1\alpha_2 S.\beta\beta_1\beta_2$$
$$= \gamma\Sigma S.V\alpha\alpha_1 V\beta\beta_1 + \phi\gamma\Sigma S\alpha\beta - \phi^2\gamma.$$

[It will be excellent practice for the student to work out in detail the blank portions of the above investigation, and also to prove directly that the value of ρ we have just found satisfies the given equation.]

161.] But it is not necessary to go through such a long process to get the solution—though it will be advantageous to the student to read it carefully—for if we operate on the equation by $S.\alpha_1\alpha_2$, $S.\alpha_2\alpha$, and $S.\alpha\alpha_1$ we get

$$S.\alpha_1\alpha_2\alpha S\beta\rho = S.\alpha_1\alpha_2\gamma,$$
$$S.\alpha_2\alpha\alpha_1 S\beta_1\rho = S.\alpha_2\alpha\gamma,$$
$$S.\alpha\alpha_1\alpha_2 S\beta_2\rho = S.\alpha\alpha_1\gamma.$$

From these, by § 92 (4), we have at once

$$\rho S.\alpha\alpha_1\alpha_2 S.\beta\beta_1\beta_2 = V\beta\beta_1 S.\alpha\alpha_1\gamma + V\beta_1\beta_2 S.\alpha_1\alpha_2\gamma + V\beta_2\beta S.\alpha_2\alpha\gamma.$$

The student will find it a useful exercise to prove that this is equivalent to the solution in § 160.

To verify the present solution we have

$$(\alpha S\beta\rho + \alpha_1 S\beta_1\rho + \alpha_2 S\beta_2\rho)\, S.\alpha\alpha_1\alpha_2 S.\beta\beta_1\beta_2$$
$$= \alpha S.\beta\beta_1\beta_2 S.\alpha_1\alpha_2\gamma + \alpha_1 S.\beta_1\beta_2\beta S.\alpha_2\alpha\gamma + \alpha_2 S.\beta_2\beta\beta_1 S.\alpha\alpha_1\gamma$$
$$= S.\beta\beta_1\beta_2\,(\gamma S.\alpha\alpha_1\alpha_2), \text{ by § 91 (3).}$$

162.] It is evident, from these examples, that for special cases we can usually find modes of solution of the linear and vector equation which are simpler in application than the general process of § 148. The real value of that process however consists partly in its enabling us to express inverse functions of ϕ, such as $(\phi + g)^{-1}$ for instance, in terms of direct operations, a property which will be of great use to us later; partly in its leading us to the fundamental cubic

$$\phi^3 - m_2\phi^2 + m_1\phi - m = 0,$$

which is an immediate deduction from the equation of § 148, and whose interpretation is of the utmost importance with reference to the axes of surfaces of the second order, principal axes of inertia, the analysis of strains in a distorted solid, and various similar enquiries.

163.] When the function ϕ *is its own conjugate*, that is, when

$$S\rho\phi\sigma = S\sigma\phi\rho$$

for all values of ρ and σ, the vectors for which

$$(\phi - g)\,\rho = 0$$

form in general a real and definite rectangular system. This, of course, may in particular cases degrade into one definite vector, and *any* pair of others perpendicular to it; and cases may occur in which the equation is satisfied for *every* vector.

Suppose the roots of $m_g = 0$ (§ 147) to be real and different, then

$$\left.\begin{array}{l}\phi\rho_1 = g_1\rho_1\\ \phi\rho_2 = g_2\rho_2\\ \phi\rho_3 = g_3\rho_3\end{array}\right\}$$ where ρ_1, ρ_2, ρ_3 are three definite vectors determined by the constants involved in ϕ.

Hence $\quad g_1 g_2 S\rho_1\rho_2 = S.\phi\rho_1\phi\rho_2$

$$= S.\rho_1\phi^2\rho_2, \text{ or } = S.\rho_2\phi^2\rho_1,$$

because ϕ is its own conjugate.

But $$\phi^2\rho_2 = g_2^2\rho_2,$$
$$\phi^2\rho_1 = g_1^2\rho_1,$$

and therefore $\quad g_1 g_2 S\rho_1\rho_2 = g_2^2 S\rho_1\rho_2 = g_1^2 S\rho_1\rho_2;$

which, as g_1 and g_2 are by hypothesis different, requires

$$S\rho_1\rho_2 = 0.$$

Similarly $\quad S\rho_2\rho_3 = 0, \qquad S\rho_3\rho_1 = 0.$

If two roots be equal, as g_2, g_3, we still have, by the above proof, $S\rho_1\rho_2 = 0$ and $S\rho_1\rho_3 = 0$. But there is nothing farther to determine ρ_2 and ρ_3, which are therefore *any* vectors perpendicular to ρ_1.

If all three roots be equal, *every* real vector satisfies the equation $(\phi - g)\rho = 0$.

164.] Next, as to the *reality* of the three directions in this case.

Suppose $g_2 + h_2\sqrt{-1}$ to be a root, and let $\rho_2 + \sigma_2\sqrt{-1}$ be the corresponding value of ρ, where g_2 and h_2 are real numbers, ρ_2 and σ_2 real vectors, and $\sqrt{-1}$ the old imaginary of algebra.

Then $\quad \phi(\rho_2 + \sigma_2\sqrt{-1}) = (g_2 + h_2\sqrt{-1})(\rho_2 + \sigma_2\sqrt{-1}),$

and this divides itself, as in algebra, into the two equations

$$\phi\rho_2 = g_2\rho_2 - h_2\sigma_2,$$
$$\phi\sigma_2 = h_2\rho_2 + g_2\sigma_2.$$

Operating on these by $S.\sigma_2$, $S.\rho_2$ respectively, and subtracting the results, remembering our condition as to the nature of ϕ

$$S\sigma_2\phi\rho_2 = S\rho_2\phi\sigma_2,$$

we have $$h_2(\sigma_2^2 + \rho_2^2) = 0.$$

But, as σ_2 and ρ_2 are both real vectors, the sum of their squares cannot vanish. Hence h_2 vanishes, and with it the impossible part of the root.

165.] When ϕ is self-conjugate, we have shewn that the equation

$$g^3 - m_2 g^2 + m_1 g - m = 0$$

has three real roots, in general different from one another.

Hence the cubic in ϕ may be written

$$(\phi - g_1)(\phi - g_2)(\phi - g_3) = 0,$$

and in this form we can easily see the meaning of the cubic. For, let ρ_1, ρ_2, ρ_3 be three vectors such that

$$(\phi-g_1)\rho_1 = 0, \quad (\phi-g_2)\rho_2 = 0, \quad (\phi-g_3)\rho_3 = 0.$$

Then any vector ρ may be expressed by the equation

$$\rho S.\rho_1\rho_2\rho_3 = \rho_1 S.\rho_2\rho_3\rho + \rho_2 S.\rho_3\rho_1\rho + \rho_3 S.\rho_1\rho_2\rho \quad (\S\ 91),$$

and we see that when the complex operation, denoted by the left-hand member of the above symbolic equation, is performed on ρ, the first of the three factors makes the term in ρ_1 vanish, the second and third those in ρ_2 and ρ_3 respectively. In other words, by the successive performance upon a vector of the operations $\phi-g_1$, $\phi-g_2$, $\phi-g_3$, it is deprived successively of its resolved parts in the directions of ρ_1, ρ_2, ρ_3 respectively; and is thus necessarily reduced to zero, since ρ_1, ρ_2, ρ_3 are (because we have supposed g_1, g_2, g_3 to be distinct) distinct and non-coplanar vectors.

166.] If we take ρ_1, ρ_2, ρ_3 as rectangular *unit*-vectors, we have

$$-\rho = \rho_1 S\rho_1\rho + \rho_2 S\rho_2\rho + \rho_3 S\rho_3\rho,$$

whence $$\phi\rho = -g_1\rho_1 S\rho_1\rho - g_2\rho_2 S\rho_2\rho - g_3\rho_3 S\rho_3\rho\,;$$

or, still more simply, putting i, j, k for ρ_1, ρ_2, ρ_3, we find that *any* self-conjugate function may be thus expressed

$$\phi\rho = -g_1 iSi\rho - g_2 jSj\rho - g_3 kSk\rho,$$

provided, of course, i, j, k be taken as roots of the equation

$$V\rho\phi\rho = 0.$$

167.] A very important transformation of the self-conjugate linear and vector function is easily derived from this form.

We have seen that it involves *three* scalar constants only, viz. g_1, g_2, g_3. Let us enquire, then, whether it can be reduced to the following form $$\phi\rho = f\rho + hV.(i+ek)\rho(i-ek),$$

which also involves but three scalar constants f, h, e. Here, again, i, j, k are the roots of $$V\rho\phi\rho = 0.$$

Substituting for ρ the equivalent

$$\rho = -iSi\rho - jSj\rho - kSk\rho,$$

expanding, and equating coefficients of i, j, k in the two expressions for $\phi\rho$, we find

$$-g_1 = -f + h(2-1+e^2),$$
$$-g_2 = -f - h(1-e^2),$$
$$-g_3 = -f - h(2e^2+1-e^2).$$

These give at once

$$-(g_1-g_2) = 2h,$$
$$-(g_2-g_3) = 2e^2 h.$$

Hence, as we suppose the transformation to be real, and therefore e^2 to be positive, it is evident that g_1-g_2 and g_2-g_3 have the same sign; so that we must choose as auxiliary vectors in the last term of $\phi\rho$ those two of the rectangular directions i, j, k for which the coefficients g have the greatest and least values.

We have then

$$e^2 = \frac{g_2-g_3}{g_1-g_2},$$

$$h = -\tfrac{1}{2}(g_1-g_2),$$

and $$f = \tfrac{1}{2}(g_1+g_3).$$

168.] We may, therefore, always determine definitely the vectors λ, μ, and the scalar f, in the equation

$$\phi\rho = f\rho + V.\lambda\rho\mu$$

when ϕ is self-conjugate, and the corresponding cubic has not equal roots, subject to the single restriction that

$$T.\lambda\mu$$

is known, but not the separate tensors of λ and μ. This result is important in the theory of surfaces of the second order, and will be considered in Chapter VII.

169.] Another important transformation of ϕ when self-conjugate is the following, $\phi\rho = a\alpha V\alpha\rho + b\beta S\beta\rho$, where a and b are scalars, and α and β unit-vectors. This, of course, involves six scalar constants, and belongs to the most general form $\phi\rho = -g_1\rho_1 S\rho_1\rho - g_2\rho_2 S\rho_2\rho - g_3\rho_3 S\rho_3\rho$, where ρ_1, ρ_2, ρ_3 are the rectangular unit-vectors for which ρ and $\phi\rho$ are parallel. We merely mention this form in passing, as it belongs to the *focal* transformation of the equation of surfaces of the second order, which will not be farther alluded to in this work. It will be a good exercise for the student to determine α, β, a and b, in terms of g_1, g_2, g_3, and ρ_1, ρ_2, ρ_3.

170.] We cannot afford space for a detailed account of the singular properties of these vector functions, and will therefore content ourselves with the enuntiation and proof of one or two of the most important.

In the equation $m\phi^{-1}V\lambda\mu = V\phi'\lambda\phi'\mu$ (§ 145), substitute λ for $\phi'\lambda$ and μ for $\phi'\mu$, and we have

$$mV\phi'^{-1}\lambda\phi'^{-1}\mu = \phi V\lambda\mu.$$

Change ϕ to $\phi+g$, and therefore ϕ' to $\phi'+g$, and m to m_g, we have

$$m_g V(\phi'+g)^{-1}\lambda(\phi'+g)^{-1}\mu = (\phi+g)V\lambda\mu;$$

a formula which will be found to be of considerable use.

171.] Again, by § 147,

$$\frac{m_g}{g} S.\rho(\phi+g)^{-1}\rho = \frac{m}{g} S\rho\phi^{-1}\rho + S\rho\chi\rho + g\rho^2.$$

Similarly $\quad \frac{m_h}{h} S.\rho(\phi+h)^{-1}\rho = \frac{m}{h} S\rho\phi^{-1}\rho + S\rho\chi\rho + h\rho^2.$

Hence

$$\frac{m_g}{g} S.\rho(\phi+g)^{-1}\rho - \frac{m_h}{h} S.\rho(\phi+h)^{-1}\rho = (g-h)\left\{\rho^2 - \frac{mS\rho\phi^{-1}\rho}{gh}\right\}.$$

That is, the functions

$$\frac{m_g}{g} S.\rho(\phi+g)^{-1}\rho, \quad \text{and} \quad \frac{m_h}{h} S.\rho(\phi+h)^{-1}\rho$$

are identical, i. e. *when equated to constants represent the same series of surfaces*, not merely when

$$g = h,$$

but also, whatever be g and h, if they be scalar functions of ρ which satisfy the equation $\quad mS.\rho\phi^{-1}\rho = gh\rho^2.$

This is a generalization, due to Hamilton, of a singular result obtained by the author*

172.] The equations

$$\left.\begin{array}{l} S.\rho\,(\phi+g)^{-1}\rho = 0, \\ S.\rho\,(\phi+h)^{-1}\rho = 0, \end{array}\right\} \quad \ldots\ldots\ldots\ldots\ldots\ldots \quad (1)$$

are equivalent to $\quad mS\rho\phi^{-1}\rho + gS\rho\chi\rho + g^2\rho^2 = 0,$

$$mS\rho\phi^{-1}\rho + hS\rho\chi\rho + h^2\rho^2 = 0.$$

Hence $\quad m(1-x)S\rho\phi^{-1}\rho + (g-hx)S\rho\chi\rho + (g^2-h^2x)\rho^2 = 0,$

whatever scalar be represented by x.

That is, the two equations (1) represent the *same* surface if this identity be satisfied. As particular cases let

(1) $x = 1$, in which case

$$S\rho^{-1}\chi\rho + g + h = 0.$$

(2) $g - hx = 0$, in which case

$$m\left(1-\frac{g}{h}\right)S\rho^{-1}\phi^{-1}\rho + \left(g^2 - h^2\frac{g}{h}\right) = 0,$$

or $\quad mS\rho^{-1}\phi^{-1}\rho - gh = 0.$

(3) $x = \frac{g^2}{h^2}$, giving

$$m\left(1-\frac{g^2}{h^2}\right)S\rho\phi^{-1}\rho + \left(g - h\frac{g^2}{h^2}\right)S\rho\chi\rho = 0,$$

or $\quad m(h+g)S\rho\phi^{-1}\rho + ghS\rho\chi\rho = 0.$

* Note on the Cartesian equation of the Wave-Surface. *Quarterly Math. Journal*, Oct. 1859.

173.] In various investigations we meet with the quaternion

$$q = a\phi a + \beta\phi\beta + \gamma\phi\gamma,$$

where a, β, γ are three unit-vectors at right angles to each other. It admits of being put in a very simple form, which is occasionally of considerable importance.

We have, obviously, by the properties of a rectangular unit-system

$$q = \beta\gamma\phi a + \gamma a\phi\beta + a\beta\phi\gamma.$$

As we have also $\quad S.a\beta\gamma = -1 \quad$ (§ 71 (13)),

a glance at the formulæ of § 147 shews that

$$Sq = -m_2,$$

at least if ϕ be self-conjugate. Even if it be not, still (as will be shewn in § 174) $\quad \phi\rho = \phi'\rho + V\epsilon\rho,$

and the new term disappears in Sq.

We have also, by § 90 (2),

$$\begin{aligned} Vq &= a(S\beta\phi\gamma - S\gamma\phi\beta) + \beta(S\gamma\phi a - Sa\phi\gamma) + \gamma(Sa\phi\beta - S\beta\phi a) \\ &= aS\beta(\phi - \phi')\gamma + \beta S\gamma(\phi - \phi')a + \gamma Sa(\phi - \phi')\beta \\ &= aS.\beta\epsilon\gamma + \beta S.\gamma\epsilon a + \gamma S.a\epsilon\beta \\ &= -(aSa\epsilon + \beta S\beta\epsilon + \gamma S\gamma\epsilon) = \epsilon. \end{aligned}$$

[We may note in passing that this quaternion admits of being expressed in the remarkable form

$$\nabla\phi\rho;$$

where $\quad \nabla = a\frac{d}{dx} + \beta\frac{d}{dy} + \gamma\frac{d}{dz},$

and $\quad \rho = ax + \beta y + \gamma z.$

We will recur to this towards the end of the work.]

Many similar singular properties of ϕ in connection with a rectangular system might easily be given; for instance,

$$\begin{aligned} &V(aV\phi\beta\phi\gamma + \beta V\phi\gamma\phi a + \gamma V\phi a\phi\beta) \\ &\qquad = mV(a\phi'^{-1}a + \beta\phi'^{-1}\beta + \gamma\phi'^{-1}\gamma) = mV.\nabla\phi'^{-1}\rho = \phi\epsilon; \end{aligned}$$

which the reader may easily verify by a process similar to that just given, or (more directly) by the help of § 145 (4). A few others will be found among the Examples appended to this Chapter.

174.] To conclude, we may remark that as in many of the immediately preceding investigations we have supposed ϕ to be self-conjugate, a very simple step enables us to pass from this to the non-conjugate form.

For, if ϕ' be conjugate to ϕ, we have

$$S\rho\phi'\sigma = S\sigma\phi\rho,$$

and also $\quad S\rho\phi\sigma = S\sigma\phi'\rho.$

Adding, we have

$$S\rho(\phi+\phi')\sigma = S\sigma(\phi+\phi')\rho;$$

so that the function $(\phi+\phi')$ is self-conjugate.

Again, $$S\rho\phi\rho = S\rho\phi'\rho,$$

which gives $$S\rho(\phi-\phi')\rho = 0.$$

Hence $$(\phi-\phi')\rho = V\epsilon\rho,$$

where, if ϕ be not self-conjugate, ϵ is some real vector, and therefore

$$\phi\rho = \tfrac{1}{2}(\phi+\phi')\rho + \tfrac{1}{2}(\phi-\phi')\rho$$
$$-\tfrac{1}{2}(\phi+\phi')\rho + \tfrac{1}{2}V\epsilon\rho.$$

Thus *every non-conjugate linear and vector function differs from a conjugate function solely by a term of the form*

$$V\epsilon\rho.$$

The geometric signification of this will be found in the Chapter on Kinematics.

175.] We have shewn, at some length, how a linear and vector equation containing an unknown *vector* is to be solved in the most general case; and this, by § 138, shews how to find an unknown *quaternion* from any sufficiently general linear equation containing it. That such an equation may be sufficiently general it must have both scalar and vector parts: the first gives *one*, and the second *three*, scalar equations; and these are required to determine completely the four scalar elements of the unknown quaternion.

176.] Thus $$Tq = a$$

being but one scalar equation, gives

$$q = a\,Ur,$$

where r is any quaternion whatever.

Similarly $$Sq = a$$

gives $$q = a + \theta,$$

where θ is any vector whatever. In each of these cases, only one scalar condition being given, the solution contains three scalar indeterminates. A similar remark applies to the following ·

$$TVq = a$$

gives $$q = x + a\theta$$

and $$SUq = \cos a,$$

gives $$q = x\theta^{\frac{2a}{\pi}},$$

in each of which x is any scalar, and θ any unit vector.

177.] Again, the reader may easily prove that

$$V.aVq = \beta,$$

where α is a given vector, gives, by putting $Sq = x$,

$$V\alpha q = \beta + x\alpha.$$

Hence, assuming $S\alpha q = y$,

we have $\alpha q = y + x\alpha + \beta$,

or $q = x + y\alpha^{-1} + \alpha^{-1}\beta$.

Here, the given equation being equivalent to two scalar conditions, the solution contains two scalar indeterminates.

178.] Next take the equation

$$V\alpha q = \beta.$$

Operating by $S.\alpha^{-1}$, we get

$$Sq = S\alpha^{-1}\beta,$$

so that the given equation becomes

$$V\alpha(S\alpha^{-1}\beta + Vq) = \beta,$$

or $V\alpha Vq = \beta - \alpha S\alpha^{-1}\beta = \alpha V\alpha^{-1}\beta$.

From this, by § 158, we see that

$$Vq = \alpha^{-1}(x + \alpha V\alpha^{-1}\beta),$$

whence
$$q = S\alpha^{-1}\beta + \alpha^{-1}(x + \alpha V\alpha^{-1}\beta)$$
$$= \alpha^{-1}(\beta + x),$$

and, the given equation being equivalent to three scalar conditions, but one undetermined scalar remains in the value of q.

This solution might have been obtained at once, since our equation gives merely the *vector* of the quaternion αq, and leaves its scalar undetermined.

Hence, taking x for the scalar, we have

$$\alpha q = S\alpha q + V\alpha q$$
$$= x + \beta.$$

179.] Finally, of course, from

$$\alpha q = \beta,$$

which is equivalent to four scalar equations, we obtain a definite value of the unknown quaternion in the form

$$q = \alpha^{-1}\beta.$$

180.] Before taking leave of linear equations, we may mention that Hamilton has shewn how to solve any linear equation containing an unknown *quaternion*, by a process analogous to that which he employed to determine an unknown *vector* from a linear and vector equation; and to which a large part of this Chapter has been devoted. Besides the increased complexity, the peculiar feature disclosed by this beautiful discovery is that the symbolic equation for a linear quaternion function, corresponding to the cubic in ϕ of § 162, is a *biquadratic*, so that the inverse function is given in terms of the first, second, and third powers of the direct function. In an elementary work like the present the discussion of such a question would be out of place: although it is not very difficult to derive the more general result by an application of processes already explained. But it forms a curious example of the well-known fact that a biquadratic equation depends for its solution upon a cubic. The reader is therefore referred to the *Elements of Quaternions*, p. 491.

181.] The solution of the following frequently-occurring particular form of linear quaternion equation

$$aq+qb = c,$$

where a, b, and c are any given quaternions, has been effected by Hamilton by an ingenious process, which was applied in § 133 (5) above to a simple case.

Multiply the whole *by* Ka, and *into* b, and we have

$$T^2a.q + Ka.qb = Ka.c,$$

$$\text{and} \qquad aqb+qb^2 = cb.$$

Adding, we have

$$q(T^2a+b^2+2Sa.b) = Ka.c+cb,$$

from which q is at once found.

To this form any equation such as

$$a'qb'+c'qd' = e'$$

can of course be reduced, by multiplication *by* c'^{-1} and *into* b'^{-1}.

182.] As another example, let us find the differential of the cube root of a quaternion. If

$$q^3 = r$$

we have

$$q^2dq+qdq.q+dq.q^2 = dr.$$

Multiply *by* q, and *into* q^{-1}, simultaneously, and we obtain

$$q^3dq.q^{-1}+q^2dq+qdq.q = qdr.q^{-1}.$$

Subtracting this from the preceding equation we have

$$dq.q^2-q^3dq.q^{-1} = dr-qdr.q^{-1},$$

or

$$dq.q^3-q^3dq = dr.q-qdr,$$

from which dq, or $d(r^{\frac{1}{3}})$, can be found by the process of last section.

The method here employed can be easily applied to find the differential of any root of a quaternion.

183.] To shew some of the characteristic peculiarities in the solution even of quaternion equations of the first degree when they are not sufficiently general, let us take the very simple one

$$aq = qb,$$

and give every step of the solution, as practice in transformations.

Apply Hamilton's process (§ 181), and we get

$$T^2 a.q = Ka.qb,$$

$$qb^2 = aqb.$$

These give $$q(T^2 a + b^2 - 2bSa) = 0,$$

so that the equation gives no real finite value for q unless

$$T^2 a + b^2 - 2bSa = 0,$$

or $$b = Sa + \beta TVa,$$

where β is some unit-vector.

By a similar process we may evidently shew that

$$a = Sb + \alpha TVb,$$

α being another unit-vector.

But, by the given equation,

$$Ta = Tb,$$

or $$S^2 a + T^2 Va = S^2 b + T^2 Vb;$$

from which, and the above values of a and b, we see that we may write

$$\frac{Sa}{TVa} = \frac{Sb}{TVb} = \text{a, suppose.}$$

If, then, we separate q into its scalar and vector parts, thus

$$q = r + \rho,$$

the given equation becomes

$$(\text{a} + \alpha)(r + \rho) = (r + \rho)(\text{a} + \beta). \quad \ldots\ldots\ldots\ldots\ldots\ldots (1)$$

Multiplying out we have

$$r(\alpha - \beta) = \rho\beta - \alpha\rho,$$

which gives $$S(\alpha - \beta)\rho = 0,$$

and therefore $$\rho = V\gamma(\alpha - \beta),$$

where γ is an undetermined vector.

We have now

$$\begin{aligned} r(\alpha - \beta) &= \rho\beta - \alpha\rho \\ &= V\gamma(\alpha - \beta).\beta - \alpha V\gamma(\alpha - \beta) \\ &= \gamma(S\alpha\beta + 1) - (\alpha - \beta)S\beta\gamma + \gamma(1 + S\alpha\beta) - (\alpha - \beta)S\alpha\gamma \\ &= -(\alpha - \beta)S(\alpha + \beta)\gamma. \end{aligned}$$

Having thus determined r, we have

$$\begin{aligned} q &= -S(\alpha + \beta)\gamma + V\gamma(\alpha - \beta) \\ 2q &= -(\alpha + \beta)\gamma - \gamma(\alpha + \beta) + \gamma(\alpha - \beta) - (\alpha - \beta)\gamma \\ &= -2\alpha\gamma - 2\gamma\beta. \end{aligned}$$

Here, of course, we may change the sign of γ, and write the solution of

$$aq = qb$$

in the form $$q = \alpha\gamma + \gamma\beta,$$

where γ is any vector, and

$$\alpha = UVa, \qquad \beta = UVb.$$

To verify this solution, we see by (1) that we require only to shew that
$$\alpha q = q\beta.$$
But their common value is evidently
$$-\gamma + \alpha\gamma\beta.$$

It will be excellent practice for the student to represent the terms of this equation by versor-arcs, as in § 54, and to deduce the above solution from the diagram directly. He will find that the solution may thus be obtained almost intuitively.

184.] No general method of solving quaternion equations of the second or higher degrees has yet been found; in fact, as will be shewn immediately, even those of the second degree involve (in their most general form) algebraic equations of the *sixteenth* degree. Hence, in the few remaining sections of this Chapter we shall confine ourselves to one or two of the simple forms for the treatment of which a definite process has been devised. But first, let us consider how many roots an equation of the second degree in an unknown quaternion must generally have.

If we substitute for the quaternion the expression
$$w + ix + jy + kz \quad (\S\ 80),$$
and treat the quaternion constants in the same way, we shall have (§ 80) four equations, generally of the second degree, to determine w, x, y, z. The number of roots will therefore be 2^4 or 16. And similar reasoning shews us that a quaternion equation of the mth degree has m^4 roots. It is easy to see, however, from some of the simple examples given above (§§ 175–178, &c.) that, unless the given equation is equivalent to four scalar equations, the roots will contain one or more indeterminate quantities.

185.] Hamilton has effected in a simple way the solution of the quadratic
$$q^2 = qa + b,$$
or the following, which is virtually the same (as we see by taking the conjugate of each side),
$$q^2 = aq + b.$$
He puts
$$q = \tfrac{1}{2}(a + w + \rho),$$
where w is a scalar, and ρ a vector.

Substituting this value in the first equation, we get
$$a^2 + (w+\rho)^2 + 2wa + a\rho + \rho a = 2(a^2 + wa + \rho a) + 4b,$$
$$\text{or} \qquad (w+\rho)^2 + a\rho - \rho a = a^2 + 4b.$$
If we put $Va = \alpha$, $S(a^2 + 4b) = c$, $V(a^2+4b) = 2\gamma$, this becomes
$$(w+\rho)^2 + 2Va\rho = c + 2\gamma;$$

which, by equating separately the scalar and vector parts, may be broken up into the two equations

$$w^2 + \rho^2 = c,$$
$$V(w+\alpha)\rho = \gamma.$$

The latter of these can be solved for ρ by the process of § 156, or more simply by operating at once by $S.\alpha$ which gives the value of $S(w+\alpha)\rho$. If we substitute the resulting value of ρ in the former we obtain, as the reader may easily prove, the equation

$$(w^2 - \alpha^2)(w^4 - cw^2 + \gamma^2) - (V\alpha\gamma)^2 = 0.$$

The solution of this scalar cubic gives six values of w, for each of which we find a value of ρ, and thence a value of q.

Hamilton shews (*Lectures*, p. 633) that only two of these values are real quaternions, the remaining four being biquaternions, and the other ten roots of the given equation being infinite.

Hamilton farther remarks that the above process leads, as the reader may easily see, to the solution of the two simultaneous equations

$$q + r = a,$$
$$qr = -b;$$

and he connects it also with the evaluation of certain continued fractions with quaternion constituents. (See the Miscellaneous Examples at the end of the volume.)

186.] The equation $\qquad q^2 = aq + qb,$

though apparently of the second degree, is easily reduced to the first degree by multiplying *by*, and *into*, q^{-1}, when it becomes

$$1 = q^{-1}a + bq^{-1},$$

and may be treated by the process of § 181.

187.] The equation $\qquad q^m = aqb,$

where a and b are given quaternions, gives

$$q(aqb) = (aqb)q\cdot$$

and, by § 54, it is evident that the planes of q and aqb must coincide. A little consideration will shew that the solution depends upon drawing two arcs which shall intercept given arcs upon each of two great circles; while one of them bisects the other, and is divided by it in the proportion of $m : 1$.

EXAMPLES TO CHAPTER V.

1. Solve the following equations:—

$(a.)\ \ V.\alpha\rho\beta = V.\alpha\gamma\beta.$

$(b.)\ \ \alpha\rho\beta\rho = \rho\alpha\rho\beta.$

$(c.)\ \ \alpha\rho + \rho\beta = \gamma.$

$(d.)\ \ S.\alpha\beta\rho + \beta S\alpha\rho - \alpha V\beta\rho = \gamma.$

$(e.)\ \ \rho + \alpha\rho\beta = \alpha\beta.$

$(f.)\ \ \alpha\rho\beta\rho = \rho\beta\rho\alpha.$

Do any of these impose any restriction on the generality of α and β?

2. Suppose $\qquad \rho = ix + jy + kz,$

and $\qquad \phi\rho = aiSi\rho + bjSj\rho + ckSk\rho;$

put into Cartesian cöordinates the following equations:—

$(a.)\ \ T\phi\rho = 1.$

$(b.)\ \ S\rho\phi^2\rho = -1.$

$(c.)\ \ S.\rho\,(\phi^2 - \rho^2)^{-1}\rho = -1.$

$(d.)\ \ T\rho = T.\phi U\rho.$

3. If λ, μ, ν be *any* three non-coplanar vectors, and

$$q = V\mu\nu.\phi\lambda + V\nu\lambda.\phi\mu + V\lambda\mu.\phi\nu,$$

shew that q is necessarily divisible by $S.\lambda\mu\nu$.

Also shew that the quotient is

$$m_2 - 2\,\epsilon,$$

where $V\epsilon\rho$ is the non-commutative part of $\phi\rho$.

Hamilton, *Elements*, p. 442.

4. Solve the simultaneous equations:—

$(a.)\ \ \left.\begin{array}{ll} S\alpha\rho & - 0, \\ S.\alpha\rho\phi\rho & = 0. \end{array}\right\}$

$(b.)\ \ \left.\begin{array}{ll} S\alpha\rho & - 0, \\ S\rho\phi\rho & = 0. \end{array}\right\}$

$(c.)\ \ \left.\begin{array}{ll} S\alpha\rho & - 0, \\ S.\alpha\iota\rho\kappa\rho & = 0. \end{array}\right\}$

5. If $\qquad \phi\rho = \Sigma\beta S\alpha\rho + Vr\rho,$

where r is a given quaternion, shew that

$$m = \Sigma(S.\alpha_1\alpha_2\alpha_3 S.\beta_3\beta_2\beta_1) + \Sigma S(rV\alpha_1\alpha_2.V\beta_2\beta_1) + Sr\Sigma S.\alpha\beta r - \Sigma(S\alpha rS\beta r) + SrTr^2,$$

and $\quad m\phi^{-1}\sigma = \Sigma(V\alpha_1\alpha_2 S.\beta_2\beta_1\sigma) + \Sigma V.\alpha V(V\beta\sigma.r) + V\sigma rSr - VrS\sigma r.$

Lectures, p. 561.

6. If $[pq]$ denote $pq-qp$,
(pqr) ,, $S.p[qr]$,
$[pqr]$,, $(pqr)+[rq]Sp+[pr]Sq+[qp]Sr$,
and $(pqrs)$,, $S.p[qrs]$;
shew that the following relations exist among any five quaternions

$$0=p(qrst)+q(rstp)+r(stpq)+s(tpqr)+t(pqrs),$$

and $$q(prst)=[rst]Spq-[stp]Srq+[tpr]Ssq-[prs]Stq.$$

Elements, p. 492.

7. Shew that if ϕ, ψ be any linear and vector functions, and α, β, γ rectangular unit-vectors, the vector

$$\theta=V(\phi\alpha\psi\alpha+\phi\beta\psi\beta+\phi\gamma\psi\gamma)$$

is an invariant. [This will be immediately seen if we write it in the form $$\theta=V.\phi\nabla\psi\rho,$$
which is independent of the directions of α, β, γ. But it is good practice to dispense with ∇.]

If $$\phi\rho=\Sigma\eta S\zeta\rho,$$
and $$\psi\rho=\Sigma\eta_1 S\zeta_1\rho,$$
shew that this invariant may be expressed as

$$-\Sigma V\eta\psi\zeta \quad \text{or} \quad \Sigma V\eta_1\phi\zeta_1.$$

Shew also that $$\phi\psi'\rho-\psi\phi'\rho=V\theta\rho.$$
The scalar of the same quaternion is also an invariant, and may be written as

$$-\Sigma\Sigma_1 S\eta\eta_1 S\zeta\zeta_1$$
$$=-\Sigma S\eta\psi\zeta$$
$$=-\Sigma_1 S\eta_1\phi\zeta_1.$$

8. Shew that if $$\phi\rho=\alpha S\alpha\rho+\beta S\beta\rho+\gamma S\gamma\rho,$$
where α, β, γ are any three vectors, then

$$-\phi^{-1}\rho S^2.\alpha\beta\gamma=\alpha_1 S\alpha_1\rho+\beta_1 S\beta_1\rho+\gamma_1 S\gamma_1\rho,$$

where $\alpha_1=V\beta\gamma$, &c.

9. Shew that any self-conjugate linear and vector function may in general be expressed in terms of two given ones, the expression involving terms of the second order.

Shew also that we may write

$$\phi+z=a(\varpi+x)^2+b(\varpi+x)(\omega+y)+c(\omega+y)^2,$$

where a, b, c, x, y, z are scalars, and ϖ and ω the two given functions. What character of generality is necessary in ϖ and ω? How is the solution affected by non-self-conjugation in one or both?

10. Solve the equations:—

(*a*.) $q^2=5qi+10j$.

(*b*.) $q^2=2q+i$.

(*c*.) $qaq=bq+c$.

(*d*.) $aq=qr=rb$.

11. Shew that $\phi V\nabla\phi\rho = mV\nabla\phi^{-1}\rho$.

12. If ϕ be self-conjugate, and α, β, γ a rectangular system,

$$S.V\alpha\phi\alpha V\beta\phi\beta V\gamma\phi\gamma = 0.$$

13. $\phi\psi$ and $\psi\phi$ give the same values of the invariants m, m_1, m_2.

14. If ϕ' be conjugate to ϕ, $\phi\phi'$ is self-conjugate.

15. Shew that $(V\alpha\theta)^2+(V\beta\theta)^2+(V\gamma\theta)^2 = 2\theta^2$

if α, β, γ be rectangular unit-vectors.

16. Prove that $\nabla^2(\phi-g)\rho = -\rho\nabla^2 g + 2\nabla g$.

17. Solve the equations :—

$$(a.)\quad \phi^2 = \varpi;$$

$$(b.)\quad \left.\begin{aligned}\phi+\chi &= \varpi,\\ \phi\chi &= \theta;\end{aligned}\right\}$$

where one, or two, unknown linear and vector functions are given in terms of known ones. (Tait, *Proc. R. S. E.* 1870–71.)

18. If ϕ be a self-conjugate linear and vector function, ξ and η two vectors, the two following equations are consequences one of the other, viz. :—

$$\frac{\xi}{S^{\frac{1}{3}}.\xi\phi\xi\phi^2\xi} = \frac{V.\eta\phi\eta}{S^{\frac{2}{3}}.\eta\phi\eta\phi^2\eta},$$

$$\frac{\eta}{S^{\frac{1}{3}}.\eta\phi\eta\phi^2\eta} = \frac{V\xi\phi\xi}{S^{\frac{2}{3}}.\xi\phi\xi\phi^2\xi}.$$

From either of them we obtain the equation

$$S\phi\xi\phi\eta = S^{\frac{1}{3}}.\xi\phi\xi\phi^2\xi S^{\frac{1}{3}}.\eta\phi\eta\phi^2\eta.$$

This, taken along with one of the others, gives a singular theorem when translated into ordinary algebra. What property does it give of the surface $S.\rho\phi\rho\phi^2\rho = 1$?

CHAPTER VI.

GEOMETRY OF THE STRAIGHT LINE AND PLANE.

188.] HAVING, in the five preceding Chapters, given a brief exposition of the theory and properties of quaternions, we intend to devote the rest of the work to examples of their practical application, commencing, of course, with the simplest curve and surface, the straight line and the plane. In this and the remaining Chapters of the work a few of the earlier examples will be wrought out in their fullest detail, with a reference to the first five whenever a transformation occurs; but, as each Chapter proceeds, superfluous steps will be gradually omitted, until in the later examples the full value of the quaternion processes is exhibited.

189.] Before proceeding to the proper business of the Chapter we make a digression in order to give a few instances of applications to ordinary plane geometry. These the student may multiply indefinitely with great ease.

(*a*.) *Euclid*, I. 5. Let α and β be the vector sides of an isosceles triangle; $\beta-\alpha$ is the base, and

$$T\alpha = T\beta.$$

The proposition will evidently be proved if we shew that

$$\alpha(\alpha-\beta)^{-1}=K\beta(\beta-\alpha)^{-1} \quad (\S\ 52).$$

This gives
$$\alpha(\alpha-\beta)^{-1}=(\beta-\alpha)^{-1}\beta,$$
or
$$(\beta-\alpha)\alpha=\beta(\alpha-\beta),$$
or
$$-\alpha^2=-\beta^2.$$

(*b*) *Euclid*, I. 32. Let *ABC* be the triangle, and let

$$U\frac{\overline{AC}}{\overline{AB}}=\gamma',$$

where γ is a unit-vector perpendicular to the plane of the triangle. If $l=1$, the angle CAB is a right angle (§ 74). Hence

$$A = l\frac{\pi}{2} \text{ (§ 74). Let } B = m\frac{\pi}{2}, \quad C = n\frac{\pi}{2}. \text{ We have}$$

$$U\overline{AC} = \gamma^l U\overline{AB},$$

$$U\overline{CB} = \gamma^n U\overline{CA},$$

$$U\overline{BA} = \gamma^m U\overline{BC}.$$

Hence

$$U\overline{BA} = \gamma^m . \gamma^n . \gamma^l U\overline{AB},$$

or

$$1 = \gamma^{l+m+n}$$

That is

$$l+m+n=2,$$

or

$$A+B+C=\pi.$$

This is, properly speaking, Legendre's proof; and might have been given in a far shorter form than that above. In fact we have for any three vectors whatever,

$$U.\frac{\alpha}{\beta}\frac{\beta}{\gamma}\frac{\gamma}{\alpha} = 1$$

which contains Euclid's proposition as a mere particular case.

(*c.*) *Euclid*, I. 35. Let β be the common vector-base of the parallelograms, α the conterminous vector-side of any one of them. For any other the vector-side is $\alpha + x\beta$ (§ 28), and the proposition appears as

$$TV\beta(\alpha + x\beta) = TV\beta\alpha \quad (§§ 96, 98),$$

which is obviously true.

(*d.*) In the base of a triangle find the point from which lines, drawn parallel to the sides and limited by them, are equal.

If α, β be the sides, any point in the base has the vector

$$\rho = (1 \dot{-} x)\alpha + x\beta.$$

For the required point

$$(1-x)T\alpha = xT\beta$$

which determines x.

Hence the point lies on the line

$$\rho = y(U\alpha + U\beta)$$

which bisects the vertical angle of the triangle.

This is not the only solution, for we should have written

$$T(1-x)T\alpha = TxT\beta,$$

instead of the less general form above *which tacitly assumes that* $1-x$ *and* x *are positive.* We leave this to the student.

(*e*.) If perpendiculars be erected outwards at the middle points of the sides of a triangle, each being proportional to the corresponding side, the mean point of the triangle formed by their extremities coincides with that of the original triangle. Find the ratio of each perpendicular to half the corresponding side of the old triangle that the new triangle may be equilateral.

Let 2α, 2β, and $2(\alpha+\beta)$ be the vector-sides of the triangle, i a unit-vector perpendicular to its plane, e the ratio in question. The vectors of the corners of the new triangle are (taking the corner opposite to 2β as origin)

$$\rho_1 = \alpha + ei\alpha,$$
$$\rho_2 = 2\alpha+\beta+ei\beta,$$
$$\rho_3 = \alpha+\beta-ei(\alpha+\beta).$$

From these

$$\tfrac{1}{3}(\rho_1+\rho_2+\rho_3) = \tfrac{1}{3}(4\alpha+2\beta) = \tfrac{1}{3}(2\alpha+2(\alpha+\beta)),$$

which proves the first part of the proposition.

For the second part, we must have

$$T(\rho_2-\rho_1) = T(\rho_3-\rho_2) = T(\rho_1-\rho_3).$$

Substituting, expanding, and erasing terms common to all, the student will easily find $\quad 3e^2 = 1.$

Hence, if equilateral triangles be described on the sides of any triangle, their mean points form an equilateral triangle.

190.] Such applications of quaternions as those just made are of course legitimate, but they are not always profitable. In fact, when applied to plane problems, quaternions often degenerate into mere scalars, and become (§ 33) Cartesian cöordinates of some kind, so that nothing is gained (though nothing is lost) by their use. Before leaving this class of questions we take, as an additional example, the investigation of some properties of the ellipse.

191.] We have already seen (§ 31 (*k*)) that the equation

$$\rho = \alpha\cos\theta+\beta\sin\theta$$

represents an ellipse, θ being a scalar which may have any value. Hence, for the vector-tangent at the extremity of ρ we have

$$\varpi - \frac{d\rho}{d\theta} = -\alpha\sin\theta+\beta\cos\theta,$$

which is easily seen to be the value of ρ when θ is increased by $\frac{\pi}{2}$.

Thus it appears that any two values of ρ, for which θ differs by

$\frac{\pi}{2}$, are conjugate diameters. The area of the parallelogram circumscribed to the ellipse and touching it at the extremities of these diameters is, therefore, by § 96,

$$4TV\rho\frac{d\rho}{d\theta} = 4TV(\alpha\cos\theta+\beta\sin\theta)(-\alpha\sin\theta+\beta\cos\theta)$$
$$= 4TV\alpha\beta,$$

a constant, as is well known.

192.] For equal conjugate diameters we must have

$$T(\alpha\cos\theta+\beta\sin\theta) = T(-\alpha\sin\theta+\beta\cos\theta),$$

or
$$(\alpha^2-\beta^2)(\cos^2\theta-\sin^2\theta)+4S\alpha\beta\cos\theta\sin\theta = 0,$$

or
$$\tan 2\theta = -\frac{\alpha^2-\beta^2}{2S\alpha\beta}.$$

The square of the common length of these diameters is of course

$$-\frac{\alpha^2+\beta^2}{2},$$

because we see at once from § 191 that the sum of the squares of conjugate diameters is constant.

193.] The maximum or minimum of ρ is thus found;

$$\frac{dT\rho}{d\theta} = -\frac{1}{T\rho}S\rho\frac{d\rho}{d\theta},$$
$$= -\frac{1}{T\rho}(-(\alpha^2-\beta^2)\cos\theta\sin\theta+S\alpha\beta(\cos^2\theta-\sin^2\theta)).$$

For a maximum or minimum this must vanish *, hence

$$\tan 2\theta = \frac{2S\alpha\beta}{\alpha^2-\beta^2},$$

and therefore the longest and shortest diameters are equally inclined to each of the equal conjugate diameters. Hence, also, they are at right angles to each other.

194.] Suppose for a moment α and β to be the greatest and least semidiameters. Then the equations of any two tangent-lines are

$$\rho = \alpha\cos\theta + \beta\sin\theta + x(-\alpha\sin\theta+\beta\cos\theta),$$
$$\rho = \alpha\cos\theta_1 + \beta\sin\theta_1 + x_1(-\alpha\sin\theta_1+\beta\cos\theta_1).$$

If these tangent-lines be at right angles to each other

$$S(-\alpha\sin\theta+\beta\cos\theta)(-\alpha\sin\theta_1+\beta\cos\theta_1) = 0,$$

or
$$\alpha^2\sin\theta\sin\theta_1+\beta^2\cos\theta\cos\theta_1 = 0.$$

* The student must carefully notice that here we put $\frac{dT\rho}{d\theta} = 0$, and not $\frac{d\rho}{d\theta} = 0$. A little reflection will shew him that the latter equation involves an absurdity.

Also, for their point of intersection we have, by comparing coefficients of α, β in the above values of ρ,

$$\cos\theta - x\sin\theta = \cos\theta_1 - x_1\sin\theta_1,$$
$$\sin\theta + x\cos\theta = \sin\theta_1 + x_1\cos\theta_1.$$

Determining x_1 from these equations, we easily find

$$T\rho^2 = -(\alpha^2+\beta^2),$$

the equation of a circle; if we take account of the above relation between θ and θ_1.

Also, as the equations above give $x = -x_1$, the tangents are equal multiples of the diameters parallel to them; so that the line joining the points of contact is parallel to that joining the extremities of these diameters.

195.] Finally, when the tangents

$$\rho = \alpha\cos\theta + \beta\sin\theta + x\,(-\alpha\sin\theta + \beta\cos\theta),$$
$$\rho = \alpha\cos\theta_1 + \beta\sin\theta_1 + x_1(-\alpha\sin\theta_1 + \beta\cos\theta_1),$$

meet in a given point $\rho = a\alpha + b\beta$,

we have $a = \cos\theta - x\sin\theta = \cos\theta_1 - x_1\sin\theta_1,$

$$b = \sin\theta + x\cos\theta = \sin\theta_1 + x_1\cos\theta_1.$$

Hence $x^2 = a^2 + b^2 - 1 = x_1^2$

and $a\cos\theta + b\sin\theta = 1 = a\cos\theta_1 + b\sin\theta_1$

determine the values of θ and x for the directions and lengths of the two tangents. The equation of the chord of contact is

$$\rho = y(\alpha\cos\theta + \beta\sin\theta) + (1-y)(\alpha\cos\theta_1 + \beta\sin\theta_1).$$

If this pass through the point

$$\rho = p\alpha + q\beta,$$

we have $p = y\cos\theta + (1-y)\cos\theta_1,$

$$q = y\sin\theta + (1-y)\sin\theta_1,$$

from which, by the equations which determine θ and θ_1, we get

$$ap + bq = y + 1 - y = 1.$$

Thus if either a and b, or p and q, be given, a linear relation connects the others. This, by § 30, gives all the ordinary properties of poles and polars.

196.] Although, in §§ 28–30, we have already given some of the equations of the line and plane, these were adduced merely for their applications to anharmonic cöordinates and transversals; and not for investigations of a higher order. Now that we are prepared to determine the lengths and inclinations of lines we may investigate these and other similar forms anew.

197.] *The equation of the indefinite line drawn through the origin* O, *of which the vector* $\overline{OA}$, $= a$, *forms a part*, is evidently

$$\rho = x\alpha,$$

$$\text{or} \qquad \rho \parallel \alpha,$$

$$\text{or} \qquad V\alpha\rho = 0,$$

$$\text{or} \qquad U\rho = U\alpha\,;$$

the essential characteristic of these equations being that they are linear, and involve *one* indeterminate scalar in the value of ρ.

We may put this perhaps more clearly if we take any two vectors, β, γ, which, along with α, form a non-coplanar system. Operating with $S.V\alpha\beta$ and $S.V\alpha\gamma$ upon any of the preceding equations, we get

$$\left.\begin{array}{r} S.\alpha\beta\rho = 0, \\ \text{and} \qquad S.\alpha\gamma\rho = 0. \end{array}\right\} \qquad (1)$$

Separately, these are the equations of the planes containing α, β, and α, γ; together, of course, they denote the line of intersection.

198.] Conversely, to solve equations (1), or to find ρ in terms of known quantities, we see that they may be written

$$\left.\begin{array}{r} S.\rho V\alpha\beta = 0, \\ S.\rho V\alpha\gamma = 0, \end{array}\right\}$$

so that ρ is perpendicular to $V\alpha\beta$ and $V\alpha\gamma$, and is therefore parallel to the vector of their product. That is,

$$\rho \parallel V.V\alpha\beta V\alpha\gamma,$$

$$\parallel -\alpha S.\alpha\beta\gamma,$$

$$\text{or} \qquad \rho = x\alpha.$$

199.] By putting $\rho - \beta$ for ρ we change the origin to a point B where $\overline{OB} = -\beta$, or $\overline{BO} = \beta$; so that the equation of a line parallel to α, and passing through the extremity of a vector β drawn from the origin, is

$$\rho - \beta = x\alpha,$$

$$\text{or} \qquad \rho = \beta + x\alpha.$$

Of course any two parallel lines may be represented as

$$\rho = \beta + x\alpha,$$

$$\rho = \beta_1 + x_1\alpha\,;$$

$$\text{or} \qquad V\alpha(\rho - \beta) = 0,$$

$$V\alpha(\rho - \beta_1) = 0.$$

200.] *The equation of a line, drawn through the extremity of* β, *and meeting* α *perpendicularly*, is thus found. Suppose it to be parallel to γ, its equation is

$$\rho = \beta + x\gamma.$$

To determine γ we know, *first*, that it is perpendicular to α, which gives

$$S\alpha\gamma = 0.$$

Secondly, α, β, and γ are in one plane, which gives

$$S.\alpha\beta\gamma = 0.$$

These two equations give $\gamma \parallel V.\alpha V\alpha\beta$,

whence we have $\rho = \beta + x\alpha V\alpha\beta$.

This might have been obtained in many other ways; for instance we see at once that

$$\beta = \alpha^{-1}\alpha\beta = \alpha^{-1}S\alpha\beta + \alpha^{-1}V\alpha\beta.$$

This shews that $\alpha^{-1}V\alpha\beta$ (which is evidently perpendicular to α) is coplanar with α and β, and is therefore the direction of the required line; so that its equation is

$$\rho = \beta + y\alpha^{-1}V\alpha\beta,$$

the same as before if we put $-\dfrac{y}{T\alpha^2}$ for x.

201.] By means of the last investigation we see that

$$-\alpha^{-1}V\alpha\beta$$

is the vector perpendicular drawn from the extremity of β to the line

$$\rho = x\alpha.$$

Changing the origin, we see that

$$\alpha^{-1}V\alpha(\beta - \gamma)$$

is the vector perpendicular from the extremity of β upon the line

$$\rho = \gamma + x\alpha.$$

202.] The vector joining B (where $\overline{OB} = \beta$) with any point in

$$\rho = \gamma + x\alpha$$

is

$$\gamma + x\alpha - \beta.$$

Its length is least when

$$dT(\gamma + x\alpha - \beta) = 0,$$

or

$$S\alpha(\gamma + x\alpha - \beta) = 0,$$

i. e. when it is perpendicular to α.

The last equation gives

$$x\alpha^2 + S\alpha(\gamma - \beta) = 0,$$

or

$$x\alpha = -\alpha^{-1}S\alpha(\gamma - \beta).$$

Hence the vector perpendicular is

$$\gamma - \beta - \alpha^{-1}S\alpha(\gamma - \beta),$$

or

$$\alpha^{-1}V\alpha(\gamma - \beta) = -\alpha^{-1}V\alpha(\beta - \gamma),$$

which agrees with the result of last section.

203.] *To find the shortest vector distance between two lines*

$$\rho = \beta + x\alpha,$$

and

$$\rho_1 = \beta_1 + x_1\alpha_1;$$

we must put $$dT(\rho-\rho_1) = 0,$$
or $$S(\rho-\rho_1)(d\rho-d\rho_1) = 0$$
or $$S(\rho-\rho_1)(\alpha dx-\alpha_1 dx_1) = 0.$$

Since x and x_1 are independent, this breaks up into the two conditions
$$S\alpha(\rho-\rho_1) = 0,$$
$$S\alpha_1(\rho-\rho_1) = 0;$$
proving the well-known truth that the required line is perpendicular to each of the given lines.

Hence it is parallel to $V\alpha\alpha_1$, and therefore we have
$$\rho-\rho_1 = \beta+x\alpha-\beta_1-x_1\alpha_1 = yV\alpha\alpha_1. \quad \ldots\ldots\ldots\ldots\ldots (1)$$
Operate by $S.\alpha\alpha_1$ and we get
$$S.\alpha\alpha_1(\beta-\beta_1) = y(V\alpha\alpha_1)^2.$$
This determines y, and the shortest distance required is
$$T(\rho-\rho_1) = T(yV\alpha\alpha_1) = \frac{TS.\alpha\alpha_1(\beta-\beta_1)}{TV\alpha\alpha_1} = TS.(UV\alpha\alpha_1)(\beta-\beta_1).$$

[*Note.* In the two last expressions T before S is inserted simply to ensure that the length be positive. If
$$S.\alpha\alpha_1(\beta-\beta_1) \text{ be negative,}$$
then (§ 89) $$S.\alpha_1\alpha(\beta-\beta_1) \text{ is positive.}$$
If we omit the T, we must use in the text that one of these two expressions which is positive.]

To find the extremities of this shortest distance, we must operate on (1) with $S.\alpha$ and $S.\alpha_1$. We thus obtain two equations, which determine x and x_1, as y is already known.

A somewhat different mode of treating this problem will be discussed presently.

204.] *In a given tetrahedron to find a set of rectangular cöordinate axes, such that each axis shall pass through a pair of opposite edges.*

Let α, β, γ be three (vector) edges of the tetrahedron, one corner being the origin. Let ρ be the vector of the origin of the sought rectangular system, which may be called i, j, k (unknown vectors). The condition that i, drawn from ρ, intersects α is
$$S.i\alpha\rho = 0. \quad \ldots\ldots\ldots\ldots\ldots\ldots\ldots\ldots\ldots (1)$$
That it intersects the opposite edge, whose equation is
$$\varpi = \beta+x(\beta-\gamma),$$
the condition is
$$S.i(\beta-\gamma)(\rho-\beta) = 0, \quad \text{or} \quad Si\{(\beta-\gamma)\rho-\beta\gamma\} = 0. \quad \ldots\ldots (2)$$
There are two other equations like (1), and two like (2), which can be at once written down.

Put
$$\beta-\gamma = \alpha_1, \quad \gamma-\alpha = \beta_1, \quad \alpha-\beta = \gamma_1,$$
$$V\beta\gamma = \alpha_2, \quad V\gamma\alpha = \beta_2, \quad V\alpha\beta = \gamma_2,$$
$$V\alpha_1\alpha = \alpha_3, \quad V\beta_1\beta = \beta_3, \quad V\gamma_1\gamma = \gamma_3;$$
and the six become
$$S.i\alpha\rho = 0, \qquad S.i\alpha_1\rho - Si\alpha_2 = 0,$$
$$S.j\beta\rho = 0, \qquad S.j\beta_1\rho - Sj\beta_2 = 0,$$
$$S.k\gamma\rho = 0, \qquad S.k\gamma_1\rho - Sk\gamma_2 = 0.$$
The two in i give $\quad i \parallel \alpha S\alpha_2\rho - \rho(S\alpha\alpha_2 + S\alpha_3\rho)$.
Similarly,
$$j \parallel \beta S\beta_2\rho - \rho(S\beta\beta_2 + S\beta_3\rho), \quad \text{and} \quad k \parallel \gamma S\gamma_2\rho - \rho(S\gamma\gamma_2 + S\gamma_3\rho).$$
The conditions of rectangularity, viz.,
$$Sij = 0, \quad Sjk = 0, \quad Ski = 0,$$
at once give three equations of the fourth order, the first of which is
$$0 = S\alpha\beta\, S\alpha_2\rho\, S\beta_2\rho - S\alpha\rho\, S\alpha_2\rho(S\beta\beta_2 + S\beta_3\rho) - S\beta\rho\, S\beta_2\rho(S\alpha\alpha_2 + S\alpha_3\rho)$$
$$+ \rho^2(S\alpha\alpha_2 + S\alpha_3\rho)(S\beta\beta_2 + S\beta_3\rho).$$

The required origin of the rectangular system is thus given as the intersection of three surfaces of the fourth order.

205.] The equation $\qquad S\alpha\rho = 0$

imposes on ρ the sole condition of being perpendicular to α; and therefore, being satisfied by the vector drawn from the origin to any point in a plane through the origin and perpendicular to α, is the equation of that plane.

To find this equation by a direct process similar to that usually employed in cöordinate geometry, we may remark that, by § 29, we may write
$$\rho = x\beta + y\gamma,$$
where β and γ are any two vectors perpendicular to α. In this form the equation contains two indeterminates, and is often useful; but it is more usual to eliminate them, which may be done at once by operating by $S.\alpha$, when we obtain the equation first written.

It may also be written, by eliminating one of the indeterminates only, as
$$V\beta\rho = y\alpha,$$
where the form of the equation shews that $S\alpha\beta = 0$.

Similarly we see that
$$S\alpha(\rho-\beta) = 0$$
represents a plane drawn through the extremity of β and perpendicular to α. This, of course, may, like the last, be put into various equivalent forms.

206.] The line of intersection of the two planes
$$\left.\begin{aligned} S.\alpha(\rho-\beta) &= 0, \\ \text{and} \quad S.\alpha_1(\rho-\beta_1) &= 0, \end{aligned}\right\} \qquad (1)$$

contains all points whose value of ρ satisfies both conditions. But we may write (§ 92), since α, α_1, and $V\alpha\alpha_1$ are not coplanar,

$$\rho S.\alpha\alpha_1 V\alpha\alpha_1 = V\alpha\alpha_1 S.\alpha\alpha_1\rho + V.\alpha_1 V\alpha\alpha_1 S\alpha\rho + V.V(\alpha\alpha_1)\,\alpha S\alpha_1\rho,$$

or, by the given equations,

$$-\rho T^2 V\alpha\alpha_1 = V.\alpha_1 V\alpha\alpha_1 S\alpha\beta + V.V(\alpha\alpha_1)\,\alpha S\alpha_1\beta_1 + xV\alpha\alpha_1, \qquad (2)$$

where x, a scalar indeterminate, is put for $S.\alpha\alpha_1\rho$ which may have any value. In practice, however, the two definite given scalar equations are generally more useful than the partially indeterminate vector-form which we have derived from them.

When both planes pass through the origin we have $\beta = \beta_1 = 0$, and obtain at once

$$\rho = xV\alpha\alpha_1$$

as the equation of the line of intersection.

207.] *The plane passing through the origin, and through the line of intersection of the two planes* (1), is easily seen to have the equation

$$S\alpha_1\beta_1 S\alpha\rho - S\alpha\beta S\alpha_1\rho = 0,$$

or

$$S(\alpha S\alpha_1\beta_1 - \alpha_1 S\alpha\beta)\rho = 0.$$

For this is evidently the equation of a plane passing through the origin. And, if ρ be such that

$$S\alpha\rho = S\alpha\beta,$$

we also have

$$S\alpha_1\rho = S\alpha_1\beta_1,$$

which are equations (1).

Hence we see that the vector

$$\alpha S\alpha_1\beta_1 - \alpha_1 S\alpha\beta$$

is perpendicular to the vector-line of intersection (2) of the two planes (1), and to every vector joining the origin with a point in that line.

The student may verify these statements as an exercise.

208.] *To find the vector-perpendicular from the extremity of β on the plane*

$$S\alpha\rho = 0,$$

we must note that it is necessarily parallel to α, and hence that the value of ρ for its foot is

$$\rho = \beta + x\alpha,$$

where $x\alpha$ is the vector-perpendicular in question.

Hence

$$S\alpha(\beta + x\alpha) = 0,$$

which gives

$$x\alpha^2 = -S\alpha\beta,$$

or

$$x\alpha = -\alpha^{-1}S\alpha\beta.$$

Similarly the vector-perpendicular from the extremity of β on the plane

$$S\alpha(\rho - \gamma) = 0$$

may easily be shewn to be

$$-\alpha^{-1}S\alpha(\beta - \gamma).$$

209.] *The equation of the plane which passes through the extremities of α, β, γ* may be thus found. If ρ be the vector of any point in it, $\rho-\alpha$, $\alpha-\beta$, and $\beta-\gamma$ lie in the plane, and therefore (§ 101)

$$S.(\rho-\alpha)(\alpha-\beta)(\beta-\gamma) = 0,$$

or
$$S\rho(V\alpha\beta+V\beta\gamma+V\gamma\alpha)-S.\alpha\beta\gamma = 0.$$

Hence, if
$$\delta = x(V\alpha\beta+V\beta\gamma+V\gamma\alpha)$$
be the vector-perpendicular from the origin on the plane containing the extremities of α, β, γ, we have

$$\delta = (V\alpha\beta+V\beta\gamma+V\gamma\alpha)^{-1}S.\alpha\beta\gamma.$$

From this formula, whose interpretation is easy, many curious properties of a tetrahedron may be deduced by the reader. Thus, for instance, if we take the tensor of each side, and remember the result of § 100, we see that

$$T(V\alpha\beta+V\beta\gamma+V\gamma\alpha)$$

is twice the area of the base of the tetrahedron. This may be more simply proved thus. The vector area of base is

$$\tfrac{1}{2}V(\alpha-\beta)(\gamma-\beta) = -\tfrac{1}{2}(V\alpha\beta+V\beta\gamma+V\gamma\alpha).$$

Hence the sum of the vector areas of the faces of a tetrahedron, and therefore of any solid whatever, is zero. This is the hydrostatic proposition for solids immersed in a fluid subject to no external forces.

210.] Taking any two lines whose equations are

$$\rho = \beta+x\alpha,$$
$$\rho = \beta_1+x_1\alpha_1,$$

we see that
$$S.\alpha\alpha_1(\rho-\delta) = 0$$
is the equation of a plane parallel to both. *Which* plane, of course, depends on the value of δ.

Now if $\delta = \beta$, the plane contains the first line; if $\delta = \beta_1$, the second.

Hence, if $yV\alpha\alpha_1$ be the shortest vector distance between the lines, we have

$$S.\alpha\alpha_1(\beta-\beta_1-yV\alpha\alpha_1) = 0,$$

or
$$T(yV\alpha\alpha_1) = TS.(\beta-\beta_1)UV\alpha\alpha_1,$$

the result of § 203.

211.] *Find the equation of the plane, passing through the origin, which makes equal angles with three given lines. Also find the angles in question.*

Let α, β, γ be unit-vectors in the directions of the lines, and let the equation of the plane be

$$S\delta\rho = 0.$$

Then we have evidently

$$S\alpha\delta = S\beta\delta = S\gamma\delta = x, \text{ suppose,}$$

where

$$-\frac{x}{T\delta}$$

is the sine of each of the required angles.

But (§ 92) we have

$$\delta S.\alpha\beta\gamma = x(V\alpha\beta + V\beta\gamma + V\gamma\alpha).$$

Hence

$$S.\rho(V\alpha\beta + V\beta\gamma + V\gamma\alpha) = 0$$

is the required equation; and the required sine is

$$-\frac{S.\alpha\beta\gamma}{T(V\alpha\beta + V\beta\gamma + V\gamma\alpha)}.$$

212.] *Find the locus of the middle points of a series of straight lines, each parallel to a given plane and having its extremities in two fixed lines.*

Let

$$S\gamma\rho = 0$$

be the plane, and $\rho = \beta + x\alpha, \qquad \rho = \beta_1 + x_1\alpha_1,$

the fixed lines. Also let x and x_1 correspond to the extremities of one of the variable lines, ϖ being the vector of its middle point. Then, obviously, $2\varpi = \beta + x\alpha + \beta_1 + x_1\alpha_1$.

Also

$$S\gamma(\beta - \beta_1 + x\alpha - x_1\alpha_1) = 0.$$

This gives a linear relation between x and x_1, so that, if we substitute for x_1 in the preceding equation, we obtain a result of the form

$$\varpi = \delta + x\epsilon,$$

where δ and ϵ are known vectors. The required locus is, therefore, a straight line.

213.] *Three planes meet in a point, and through the line of intersection of each pair a plane is drawn perpendicular to the third; prove that, in general, these planes pass through the same line.*

Let the point be taken as origin, and let the equations of the planes be

$$S\alpha\rho = 0, \qquad S\beta\rho = 0, \qquad S\gamma\rho = 0.$$

The line of intersection of the first two is $\parallel V\alpha\beta$, and therefore the normal to the first of the new planes is

$$V.\gamma V\alpha\beta.$$

Hence the equation of this plane is

$$S.\rho V.\gamma V\alpha\beta = 0,$$

or

$$S\beta\rho S\alpha\gamma - S\alpha\rho S\beta\gamma = 0,$$

and those of the other two planes may be easily formed from this by cyclical permutation of α, β, γ.

We see at once that any two of these equations give the third by addition or subtraction, which is the proof of the theorem.

214.] *Given any number of points A, B, C, &c., whose vectors (from the origin) are* $\alpha_1, \alpha_2, \alpha_3$, *&c., find the plane through the origin for which the sum of the squares of the perpendiculars let fall upon it from these points is a maximum or minimum.*

Let
$$S\varpi\rho = 0$$
be the required equation, with the condition (evidently allowable)
$$T\varpi = 1.$$
The perpendiculars are (§ 208) $-\varpi^{-1}S\varpi\alpha_1$, &c.

Hence
$$\Sigma S^2\varpi\alpha$$
is a maximum. This gives
$$\Sigma . S\varpi\alpha S\alpha d\varpi = 0;$$
and the condition that ϖ is a unit-vector gives
$$S\varpi d\varpi = 0.$$

Hence, as $d\varpi$ may have any of an infinite number of values, these equations cannot be consistent unless
$$\Sigma . \alpha S\alpha\varpi = x\varpi,$$
where x is a scalar.

The values of α are known, so that if we put
$$\Sigma . \alpha S\alpha\varpi = \phi\varpi,$$
ϕ is a given self-conjugate linear and vector function, and therefore x has three values (g_1, g_2, g_3, § 164) which correspond to three mutually perpendicular values of ϖ. For one of these there is a maximum, for another a minimum, for the third a maximum-minimum, in the most general case when g_1, g_2, g_3 are all different.

√ 215.] The following beautiful problem is due to Maccullagh. *Of a system of three rectangular vectors, passing through the origin, two lie on given planes, find the locus of the third.*

Let the rectangular vectors be ϖ, ρ, σ. Then by the conditions of the problem
$$S\varpi\rho = S\rho\sigma = S\sigma\varpi = 0,$$
and
$$S\alpha\varpi = 0, \quad S\beta\rho = 0.$$
The solution depends on the elimination of ρ and ϖ among these five equations. [This would, in general, be impossible, as ρ and ϖ between them involve *six* unknown scalars; but, as the tensors are (by the very form of the equations) not involved, the five given equations are necessary and sufficient to eliminate the four unknown scalars which are really involved. Formally to complete the requisite number of equations we might write
$$T\varpi = a, \quad T\rho = b,$$
but a and b may have any values whatever.]

From $\quad S\alpha\varpi = 0, \quad S\sigma\varpi = 0,$

we have $\quad \varpi = xV\alpha\sigma.$

Similarly, from $\quad S\beta\rho = 0, \quad S\sigma\rho = 0,$

we have $\quad \rho = yV\beta\sigma.$

Substitute in the remaining equation

$$S\varpi\rho = 0,$$

and we have

$$S.V\alpha\sigma V\beta\sigma = 0,$$

or

$$S\alpha\sigma S\beta\sigma - \sigma^2 S\alpha\beta = 0,$$

the required equation. As will be seen in next Chapter, this is a cone of the second order whose circular sections are perpendicular to α and β. [The disappearance of x and y in the elimination instructively illustrates the note above.]

EXAMPLES TO CHAPTER VI.

✓1. What propositions of Euclid are proved by the mere *form* of the equation $\quad \rho = (1-x)\alpha + x\beta,$
which denotes the line joining any two points in space?

·2. Shew that the chord of contact, of tangents to a parabola which meet at right angles, passes through a fixed point.

·3. Prove the chief properties of the circle (as in *Euclid*, III) from the equation $\quad \rho = \alpha\cos\theta + \beta\sin\theta\,;$
where $T\alpha = T\beta$, and $S\alpha\beta = 0$.

·4. What locus is represented by the equation

$$S^2\alpha\rho + \rho^2 = 0,$$

where $T\alpha = 1$?

·5. What is the condition that the lines

$$V\alpha\rho = \beta, \qquad V\alpha_1\rho = \beta_1,$$

intersect? If this is not satisfied, what is the shortest distance between them?

·6. Find the equation of the plane which contains the two parallel lines $\quad V\alpha(\rho-\beta) = 0, \quad V\alpha(\rho-\beta_1) = 0.$

•7. Find the equation of the plane which contains

$$V\alpha(\rho-\beta) = 0,$$

and is perpendicular to $\quad S\gamma\rho = 0.$

• 8. Find the equation of a straight line passing through a given point, and making a given angle with a given plane.

Hence form the general equation of a right cone.

9. What conditions must be satisfied with regard to a number of given lines in space that it may be possible to draw through each of them a plane in such a way that these planes may intersect in a common line?

10. Find the equation of the locus of a point the sum of the squares of whose distances from a number of given planes is constant.

11. Substitute "lines" for "planes" in (10).

12. Find the equation of the plane which bisects, at right angles, the shortest distance between two given lines.

Find the locus of a point in this plane which is equidistant from the given lines.

13. Find the conditions that the simultaneous equations

$$S\alpha\rho = a, \qquad S\beta\rho = b, \qquad S\gamma\rho = c,$$

may represent a line, and not a point.

14. What is represented by the equations

$$(S\alpha\rho)^2 = (S\beta\rho)^2 = (S\gamma\rho)^2,$$

where α, β, γ are any three vectors?

15. Find the equation of the plane which passes through two given points and makes a given angle with a given plane.

16. Find the area of the triangle whose corners have the vectors α, β, γ.

Hence form the equation of a circular cylinder whose axis and radius are given.

17. (Hamilton, *Bishop Law's Premium Ex.*, 1858).

(*a*.) Assign some of the transformations of the expression

$$\frac{V\alpha\beta}{\beta-\alpha},$$

where α and β are the vectors of two given points A and B.

(*b*.) The expression represents the vector γ, or $\overline{OC}$, of a point C in the straight line AB.

(*c*.) Assign the position of this point C.

18. (*Ibid.*)

(*a*.) If $\alpha, \beta, \gamma, \delta$ be the vectors of four points, A, B, C, D, what is the condition for those points being in one plane?

(*b*,) When these four vectors from one origin do not thus terminate upon one plane, what is the expression for the volume of the pyramid, of which the four points are the corners?

(*c*). Express the perpendicular δ let fall from the origin O on the plane ABC, in terms of α, β, γ.

19. Find the locus of a point equidistant from the three planes

$$S\alpha\rho = 0, \qquad S\beta\rho = 0, \qquad S\gamma\rho = 0.$$

20. If three mutually perpendicular vectors be drawn from a point to a plane, the sum of the reciprocals of the squares of their lengths is independent of their directions.

21. Find the general form of the equation of a plane from the condition (which is to be assumed as a definition) that any two planes intersect in a single straight line.

22. Prove that the sum of the vector areas of the faces of any polyhedron is zero.

CHAPTER VII.

THE SPHERE AND CYCLIC CONE.

216.] After that of the plane the equations next in order of simplicity are those of the sphere, and of the cone of the second order. To these we devote a short Chapter as a valuable preparation for the study of surfaces of the second order in general.

217.] The equation
$$T\rho = Ta,$$
or
$$\rho^2 = a^2,$$
denotes that the length of ρ is the same as that of a given vector a, and therefore belongs to a sphere of radius Ta whose centre is the origin. In § 107 several transformations of this equation were obtained, some of which we will repeat here with their interpretations. Thus
$$S(\rho+a)(\rho-a) = 0$$
shews that the chords drawn from any point on the sphere to the extremities of a diameter (whose vectors are a and $-a$) are at right angles to each other.
$$T(\rho+a)(\rho-a) = 2TVa\rho$$
shews that the rectangle under these chords is four times the area of the triangle two of whose sides are a and ρ.
$$\rho = (\rho+a)^{-1}a(\rho+a) \quad \text{(see § 105)}$$
shews that the angle at the centre in any circle is double that at the circumference standing on the same arc. All these are easy consequences of the processes already explained for the interpretation of quaternion expressions.

218.] If the centre of the sphere be at the extremity of a the equation may be written
$$T(\rho-a) = T\beta,$$
which is the most general form.

If
$$Ta = T\beta,$$
or
$$a^2 = \beta^2,$$

in which case the origin is a point on the surface of the sphere, this becomes

$$\rho^2 - 2S\alpha\rho = 0.$$

From this, in the form

$$S\rho(\rho - 2\alpha) = 0$$

another proof that the angle in a semicircle is a right angle is derived at once.

219.] The converse problem is—*Find the locus of the feet of perpendiculars let fall from a given point* ($\rho = \beta$) *on planes passing through the origin.*

Let $$S\alpha\rho = 0$$

be one of the planes, then (§ 208) the vector-perpendicular is

$$-\alpha^{-1}S\alpha\beta,$$

and, for the locus of its foot,

$$\rho = \beta - \alpha^{-1}S\alpha\beta,$$
$$= \alpha^{-1}V\alpha\beta.$$

[This is an example of a peculiar form in which quaternions sometimes give us the equation of a surface. The equation is a vector one, or equivalent to three scalar equations; but it involves the undetermined vector α in such a way as to be equivalent to only two indeterminates (as the tensor of α is evidently not involved). To put the equation in a more immediately interpretable form, α must be eliminated, and the remarks just made shew this to be possible.]

Now $$(\rho - \beta)^2 = \alpha^{\ 2}S^2\alpha\beta,$$

and (operating by $S.\beta$)

$$S\beta\rho - \beta^2 = -\alpha^{-2}S^2\alpha\beta.$$

Adding these equations, we get

$$\rho^2 - S\beta\rho = 0,$$

or $$T\left(\rho - \frac{\beta}{2}\right) = T\frac{\beta}{2},$$

so that, as is evident, the locus is the sphere of which β is a diameter.

220.] *To find the intersection of the two spheres*

$$T(\rho - \alpha) = T\beta,$$

and $$T(\rho - \alpha_1) = T\beta_1,$$

square the equations, and subtract, and we have

$$2S(\alpha - \alpha_1)\rho = \alpha^2 - \alpha_1^{\ 2} - (\beta^2 - \beta_1^{\ 2}),$$

which is the equation of a plane, perpendicular to $\alpha - \alpha_1$ the vector joining the centres of the spheres. This is always a real plane whether the spheres intersect or not. It is, in fact, what is called their *Radical Plane.*

221.] *Find the locus of a point the ratio of whose distances from two given points is constant.*

Let the given points be O and A, the extremities of the vector a. Also let P be the required point in any of its positions, and $\overline{OP}=\rho$.

Then, at once, if n be the ratio of the lengths of the two lines,

$$T(\rho-a) = nT\rho.$$

This gives
$$\rho^2-2Sa\rho+a^2 = n^2\rho^2,$$
or, by an easy transformation,

$$T\left(\rho-\frac{a}{1-n^2}\right) = T\left(\frac{na}{1-n^2}\right).$$

Thus the locus is a sphere whose radius is $T\left(\frac{na}{1-n^2}\right)$, and whose centre is at B, where $\overline{OB} = \frac{a}{1-n^2}$, a definite point in the line OA.

222.] *If in any line, OP, drawn from the origin to a given plane, OQ be taken such that OQ.OP is constant, find the locus of Q.*

Let
$$Sa\rho = a$$
be the equation of the plane, ϖ a vector of the required surface. Then, by the conditions,

$$T\varpi\, T\rho = \text{constant} = b^2 \text{ (suppose)},$$
and
$$U\varpi = U\rho.$$

From these
$$\rho = \frac{b^2 U\varpi}{T\varpi} = -\frac{b^2\varpi}{\varpi^2}.$$

Substituting in the equation of the plane, we have

$$a\varpi^2+b^2Sa\varpi = 0,$$

which shews that the locus is a sphere, the origin being situated on it at the point farthest from the given plane.

223.] *Find the locus of points the sum of the squares of whose distances from a set of given points is a constant quantity. Find also the least value of this constant, and the corresponding locus.*

Let the vectors from the origin to the given points be a_1, a_2, ...a_n, and to the sought point ρ, then

$$-c^2 = (\rho-a_1)^2+(\rho-a_2)^2+\ldots\ldots+(\rho-a_n)^2,$$
$$= n\rho^2-2S\rho\Sigma a+\Sigma(a^2).$$

Otherwise
$$\left(\rho-\frac{\Sigma a}{n}\right)^2 = -\frac{c^2+\Sigma(a^2)}{n}+\frac{(\Sigma a)^2}{n^2},$$

the equation of a sphere the vector of whose centre is $\frac{\Sigma a}{n}$, i.e. whose centre is the mean of the system of given points.

Suppose the origin to be placed at the mean point, the equation becomes

$$\rho^2 = -\frac{c^2+\Sigma(a^2)}{n} \quad (\text{for } \Sigma a = 0, \text{ § } 31\ (e)).$$

The right-hand side is negative, and therefore the equation denotes a real surface, if

$$c^2 > \Sigma T\alpha^2$$

as might have been expected. When these quantities are equal, the locus becomes a point, viz. the new origin, or the mean point of the system.

√ 224.] If we differentiate the equation

$$T\rho = T\alpha$$

we get

$$S\rho d\rho = 0.$$

Hence (§ 137), ρ is *normal* to the surface at its extremity, a well-known property of the sphere.

If ϖ be any point in the plane which touches the sphere at the extremity of ρ, $\varpi - \rho$ is a line in the tangent plane, and therefore perpendicular to ρ. So that

$$S\rho(\varpi - \rho) = 0,$$

or

$$S\varpi\rho = -T\rho^2 = a^2$$

is the equation of the tangent plane.

√ 225.] If this plane pass through a given point B, whose vector is β, we have

$$S\beta\rho = a^2$$

This is the equation of a plane, perpendicular to β, and cutting from it a portion whose length is

$$\frac{T\alpha^2}{T\beta}.$$

If this plane pass through a fixed point whose vector is γ we must have

$$S\beta\gamma = a^2,$$

so that the locus of β is a plane. These results contain all the ordinary properties of poles and polars with regard to a sphere.

√ 226.] A line drawn parallel to γ, from the extremity of β, has the equation

$$\rho = \beta + x\gamma.$$

This meets the sphere

$$\rho^2 = a^2$$

in points for which x has the values given by the equation

$$\beta^2 + 2xS\beta\gamma + x^2\gamma^2 = a^2.$$

The values of x are imaginary, that is, there is no intersection, if

$$a^2\gamma^2 + V^2\beta\gamma < 0.$$

The values are equal, or the line touches the sphere, if

$$a^2\gamma^2 + V^2\beta\gamma = 0,$$

or

$$S^2\beta\gamma = \gamma^2(\beta^2 - a^2).$$

This is the equation of a cone similar and similarly situated to the cone of tangent-lines drawn to the sphere, but its vertex is at the centre. That the equation represents a cone is obvious from the fact that it is *homogeneous* in $T\gamma$, i.e. that it is independent of the length of the vector γ.

[It may be remarked that from the form of the above equation we see that, if x and x' be its roots, we have

$$(xT\gamma)(x'T\gamma) = a^2 - \beta^2,$$

which is *Euclid*, III, 35, 36, extended to a sphere.]

227.] *Find the locus of the foot of the perpendicular let fall from a given point of a sphere on any tangent-plane.*

Taking the centre as origin, the equation of any tangent-plane may be written

$$S\varpi\rho = a^2.$$

The perpendicular must be parallel to ρ, so that, if we suppose it drawn from the extremity of a (which is a point on the sphere) we have as one value of ϖ

$$\varpi = a + x\rho.$$

From these equations, with the help of that of the sphere

$$\rho^2 = a^2,$$

we must eliminate ρ and x.

We have by operating on the vector equation by $S.\varpi$

$$\varpi^2 = Sa\varpi + xS\varpi\rho$$
$$- Sa\varpi + xa^2.$$

Hence $$\rho = \frac{\varpi - a}{x} = \frac{a^2(\varpi - a)}{\varpi^2 - Sa\varpi}.$$

Taking the tensors, we have

$$(\varpi^2 - Sa\varpi)^2 = a^2(\varpi - a)^2,$$

the required equation. It may be put in the form

$$S^2\varpi U(\varpi - a) = -a^2,$$

and the interpretation of this gives at once a characteristic property of the surface formed by the rotation of the *Cardioid* about its axis of symmetry.

228.] We have seen that a sphere, referred to any point whatever as origin, has the equation

$$T(\rho - a) = T\beta.$$

Hence, *to find the rectangle under the segments of a chord drawn through any point*, we may put

$$\rho = x\gamma;$$

where γ is *any* unit-vector whatever. This gives

$$x^2\gamma^2 - 2xSa\gamma + a^2 = \beta^2,$$

and the product of the two values of x is

$$-\frac{\beta^2 - a^2}{\gamma^2} = -a^2 + \beta^2$$

This is positive, or the vector-chords are drawn in the *same* direction, if

$$T\beta < Ta,$$

i.e. if the origin is outside the sphere.

229.] *A, B are fixed points; and, O being the origin and P a point in space,*

$$AP^2 + BP^2 = OP^2;$$

find the locus of **P**, *and explain the result when* $\angle AOB$ *is a right, or an obtuse, angle.*

Let $\overline{OA} = a$, $\overline{OB} = \beta$, $\overline{OP} = \rho$, then

$$(\rho - a)^2 + (\rho - \beta)^2 = \rho^2,$$

or $$\rho^2 - 2S(a+\beta)\rho = -(a^2+\beta^2),$$

or $$T\{\rho - (a+\beta)\} = \sqrt{(-2Sa\beta)}.$$

While $Sa\beta$ is negative, that is, while $\angle AOB$ is acute, the locus is a sphere whose centre has the vector $a+\beta$. If $Sa\beta = 0$, or $\angle AOB = \frac{\pi}{2}$, the locus is reduced to the point

$$\rho = a+\beta.$$

If $\angle AOB > \frac{\pi}{2}$ there is no point which satisfies the conditions.

230.] *Describe a sphere, with its centre in a given line, so as to pass through a given point and touch a given plane.*

Let xa, where x is an undetermined scalar, be the vector of the centre, r the radius of the sphere, β the vector of the given point, and

$$S\gamma\rho = a$$

the equation of the given plane.

The vector-perpendicular from the point xa on the given plane is (§ 208)

$$(a - xS\gamma a)\gamma^{-1}.$$

Hence, to determine x we have the equation

$$T.(a - xS\gamma a)\gamma^{-1} = T(xa - \beta) = r,$$

so that there are, in general, two solutions. It will be a good exercise for the student to find from this equation the condition that there may be no solution, or two coincident ones.

231.] *Describe a sphere whose centre is in a given line, and which passes through two given points.*

Let the vector of the centre be xa, as in last section, and let the vectors of the points be β and γ. Then, at once,

$$T(\gamma - xa) = T(\beta - xa) = r.$$

Here there is but *one* sphere, except in the particular case when we have

$$T\gamma = T\beta, \quad \text{and} \quad Sa\gamma = Sa\beta,$$

in which case there is an infinite number.

The student should carefully compare the results of this section and the last, so as to discover why in general two solutions are possible in the one case, and only one in the other.

232.] *A sphere touches each of two straight lines, which do not meet: find the locus of its centre.*

We may take the origin at the middle point of the shortest distance (§ 203) between the given lines, and their equations will then be

$$\rho = a + x\beta,$$
$$\rho = -a + x_1\beta_1$$

where we have, of course,

$$Sa\beta = 0, \qquad Sa\beta_1 = 0.$$

Let σ be the vector of the centre, ρ that of any point, of one of the spheres, and r its radius; its equation is

$$T(\rho - \sigma) = r.$$

Since the two given lines are tangents, the following equations in x and x_1 must have pairs of equal roots,

$$T(a + x\beta - \sigma) = r,$$
$$T(-a + x_1\beta_1 - \sigma) = r.$$

The equality of the roots in each gives us the conditions

$$S^2\beta\sigma = \beta^2((a-\sigma)^2 + r^2),$$
$$S^2\beta_1\sigma = \beta_1^2((a+\sigma)^2 + r^2).$$

Eliminating r we obtain

$$\beta^{-2}S^2\beta\sigma - \beta_1^{-2}S^2\beta_1\sigma = (a-\sigma)^2 - (a+\sigma)^2 = -4Sa\sigma,$$

which is the equation of the required locus.

[As we have not, so far, entered on the consideration of the qua ternion form of the equations of the various surfaces of the second order, we may translate this into Cartesian cöordinates to find its meaning. If we take cöordinate axes of x, y, z respectively parallel to β, β_1, a, it becomes at once

$$(x + my)^2 - (y + mx)^2 = pz,$$

where m and p are constants; and shews that the locus is a hyperbolic paraboloid. Such transformations, which are exceedingly simple in all cases, will be of frequent use to the student who is proficient in Cartesian geometry, in the early stages of his study of quaternions. As he acquires a practical knowledge of the new calculus, the need of such assistance will gradually cease to be felt.]

Simple as the above solution is, quaternions enable us to give one vastly simpler. For the problem may be thus stated—*Find the locus of the point whose distances from two given lines are equal.*

And, with the above notation, the equality of the perpendiculars is expressed (§ 201) by

$$TV.(\alpha-\sigma)U\beta = TV.(\alpha+\sigma)U\beta_1,$$

which is easily seen to be equivalent to the equation obtained above.

233.] *Two spheres being given, shew that spheres which cut them at given angles cut at right angles another fixed sphere.*

If c be the distance between the centres of two spheres whose radii are a and b, the cosine of the angle of intersection is evidently

$$\frac{a^2+b^2-c^2}{2ab}.$$

Hence, if α, α_1, and ρ be the vectors of the centres, and a, a_1, r the radii, of the two fixed, and of one of the variable, spheres; A and A_1 the angles of intersection, we have

$$(\rho-\alpha)^2+a^2+r^2 = 2ar\cos A,$$
$$(\rho-\alpha_1)^2+a_1^2+r^2 = 2a_1r\cos A_1.$$

Eliminating the first power of r, we evidently must obtain a result such as

$$(\rho-\beta)^2+b^2+r^2 = 0,$$

where (by what precedes) β is the vector of the centre, and b the radius, of a fixed sphere

$$(\rho-\beta)^2+b^2 = 0,$$

which is cut at right angles by all the varying spheres. By effecting the elimination exactly we easily find b and β in terms of given quantities.

234.] *To inscribe in a given sphere a closed polygon, plane or gauche, whose sides shall be parallel respectively to each of a series of given vectors.*

Let

$$T\rho = 1$$

be the sphere, α, β, γ, ..., η, θ the vectors, n in number, and let $\rho_1, \rho_2, \ldots\ldots \rho_n$, be the vector-radii drawn to the angles of the polygon.

Then

$$\rho_2-\rho_1 = x_1\alpha, \text{ \&c., \&c.}$$

From this, by operating by $S.(\rho_2+\rho_1)$, we get

$$\rho_2^2-\rho_1^2 = 0 = S\alpha\rho_2+S\alpha\rho_1.$$

Also

$$0 = V\alpha\rho_2-V\alpha\rho_1.$$

Adding, we get

$$0 = \alpha\rho_2+K\alpha\rho_1 = \alpha\rho_2+\rho_1\alpha.$$

Hence

$$\rho_2 = -\alpha^{-1}\rho_1\alpha.$$

[This might have been written down at once from the result of § 105.]

Similarly

$$\rho_3 = -\beta^{-1}\rho_2\beta = \beta^{-1}\alpha^{-1}\rho_1\alpha\beta, \text{ \&c.}$$

Thus, finally, since the polygon is closed,

$$\rho_{n+1} = \rho_1 = (-)^n\theta^{-1}\eta^{-1}\ldots\ldots\beta^{-1}\alpha^{-1}\rho_1\alpha\beta\ldots\ldots\eta\theta.$$

We may suppose the tensors of $\alpha, \beta \ldots\ldots \eta, \theta$ to be each unity. Hence, if

$$a = \alpha\beta \ldots\ldots \eta\theta,$$

we have

$$a^{-1} = \theta^{-1}\eta^{-1} \ldots\ldots \beta^{-1}\alpha^{-1},$$

which is a known quaternion; and thus our condition becomes

$$\rho_1 = (-)^n a^{-1}\rho_1 a.$$

This divides itself into two cases, according as n is an even or an odd number.

If n be even, we have

$$a\rho_1 = \rho_1 a.$$

Removing the common part $\rho_1 Sa$, we have

$$V\rho_1 Va = 0.$$

This gives one determinate direction, $\pm Va$, for ρ_1; and shews that there are two, and only two, solutions.

If n be odd, we have $\quad a\rho_1 = -\rho_1 a,$

which requires that we have

$$Sa = 0,$$

i.e. a must be a vector.

Hence $$Sa\rho_1 = 0,$$

and therefore ρ_1 may be drawn to any point in the great circle of the unit-sphere whose poles are on the vector a.

235.] To illustrate these results, let us take first the case of $n=3$. Here we must have $\quad S.\alpha\beta\gamma = 0,$

or the three given vectors must (as is obvious on other grounds) be parallel to one plane. Here $\alpha\beta\gamma$, which lies in this plane, is (§ 106) the vector-tangent at the first corner of each of the inscribed triangles; and is obviously perpendicular to the vector drawn from the centre to that corner.

If $n = 4$, we have $\quad \rho_1 \parallel V.\alpha\beta\gamma\delta,$

as might have been at once seen from § 106.

236.] Hamilton has given (*Lectures*, p. 674) an ingenious and simple process by which the above investigation is rendered applicable to the more difficult problem in which each side of the inscribed polygon is to pass through a given point instead of being parallel to a given line. His process depends upon the integration of a linear equation in finite differences. By an immediate application of the linear and vector function of Chapter V, the above solutions may be at once extended to any central surface of the second order.

237.] *To find the equation of a cone of revolution, whose vertex is the origin.*

Suppose α, where $T\alpha = 1$, to be its axis, and e the cosine of its semi-vertical angle; then, if ρ be the vector of any point in the cone,

$$S\alpha U\rho = +e,$$

or

$$S^2\alpha\rho = -e^2\rho^2.$$

238.] Change the origin to the point in the axis whose vector is $x\alpha$, and the equation becomes

$$(-x+S\alpha\varpi)^2 = -e^2(x\alpha+\varpi)^2.$$

Let the radius of the section of the cone made by

$$S\alpha\varpi = 0$$

retain a constant value b, while x changes; this necessitates

$$\frac{x}{\sqrt{b^2+x^2}} = e,$$

so that when x is infinite, e is unity. In this case the equation becomes

$$S^2\alpha\varpi+\varpi^2+b^2 = 0,$$

which must therefore be the equation of a circular cylinder of radius b, whose axis is the vector α. To verify this we have only to notice that if ϖ be the vector of a point of such a cylinder we must (§ 201) have

$$TV\alpha\varpi = b,$$

which is the same equation as that above.

239.] *To find, generally, the equation of a cone which has a circular section:—*

Take the origin as vertex, and let the circular section be the intersection of the plane

$$S\alpha\rho = 1$$

with the sphere (passing through the origin)

$$\rho^2 = S\beta\rho.$$

These equations may be written thus,

$$S\alpha U\rho = \frac{1}{T\rho},$$

$$-T\rho = S\beta U\rho.$$

Hence, eliminating $T\rho$, we find the following equation which $U\rho$ must satisfy—

$$S\alpha U\rho S\beta U\rho = -1,$$

or

$$\rho^2 - S\alpha\rho S\beta\rho = 0,$$

which is therefore the required equation of the cone.

As α and β are similarly involved, the mere *form* of this equation proves the existence of the subcontrary section discovered by Apollonius.

240.] The equation just obtained may be written

$$S.U\alpha U\rho S.U\beta U\rho = -\frac{1}{T.\alpha\beta},$$

or, since α and β are perpendicular to the cyclic arcs (§ 59*),

$$\sin p \sin p' = \text{constant},$$

where p and p' are arcs drawn from any point of a spherical conic perpendicular to the cyclic arcs. This is a well-known property of such curves.

241.] If we cut the cyclic cone by any plane passing through the origin, as
$$S\gamma\rho = 0,$$
then $V\alpha\gamma$ and $V\beta\gamma$ are the traces on the cyclic planes, so that

$$\rho = xUV\alpha\gamma + yUV\beta\gamma \quad (\S\ 29).$$

Substitute in the equation of the cone, and we get

$$-x^2 - y^2 + Pxy = 0,$$

where **P** is a known scalar. Hence the values of x and y are the same pair of numbers. This is a very elementary proof of the proposition in § 59*, that $PL = MQ$ (in the last figure of that section).

242.] When x and y are *equal*, the transversal arc becomes a tangent to the spherical conic, and is evidently bisected at the point of contact. Here we have

$$\mathbf{P} = 2 = 2S.UV\alpha\gamma UV\beta\gamma + \frac{(S.\alpha\beta\gamma)^2}{T.V\alpha\gamma V\beta\gamma}.$$

This is the equation of the cone whose sides are perpendiculars (through the origin) to the planes which touch the cyclic cone, and from this property the same equation may readily be deduced.

243.] It may be well to observe that the property of the Stereographic projection of the sphere, viz. that the projection of a circle is a circle, is an immediate consequence of the above form of the equation of a cyclic cone.

244.] That § 239 gives the most general form of the equation of a cone of the second order, when the vertex is taken as origin, follows from the early results of next Chapter. For it is shewn in § 249 that the equation of a cone of the second order can always be put in the form
$$2\Sigma.S\alpha\rho S\beta\rho + A\rho^2 = 0.$$
This may be written
$$S\rho\phi\rho = 0,$$
where ϕ is the self-conjugate linear and vector function

$$\phi\rho = \Sigma V.\alpha\rho\beta + (A + \Sigma S\alpha\beta)\rho.$$

By § 168 this may be transformed to

$$\phi\rho = p\rho + V.\lambda\rho\mu,$$

and the general equation of the cone becomes

$$(p - S\lambda\mu)\rho^2 + 2S\lambda\rho S\mu\rho = 0,$$

which is the form obtained in § 239.

245.] Taking the form $S\rho\phi\rho = 0$
as the simplest, we find by differentiation

$$Sd\rho\phi\rho + S\rho d\phi\rho = 0,$$

or

$$2Sd\rho\phi\rho = 0.$$

Hence $\phi\rho$ is perpendicular to the tangent-plane at the extremity of ρ. The equation of this plane is therefore (ϖ being the vector of any point in it)

$$S\phi\rho(\varpi - \rho) = 0,$$

or, by the equation of the cone,

$$S\varpi\phi\rho = 0.$$

246.] *The equation of the cone of normals to the tangent-planes of a given cone can be easily formed from that of the cone itself.* For we may write it in the form

$$S(\phi^{-1}\phi\rho)\phi\rho = 0,$$

and if we put $\phi\rho = \sigma$, a vector of the new cone, the equation becomes

$$S\sigma\phi^{-1}\sigma = 0.$$

Numerous curious properties of these connected cones, and of the corresponding spherical conics, follow at once from these equations. But we must leave them to the reader.

247.] As a final example, let us *find the equation of a cyclic cone when five of its vector-sides are given*—i. e. *find the cone of the second order whose vertex is the origin, and on whose surface lie the vectors* $\alpha, \beta, \gamma, \delta, \epsilon$.

If we write

$$0 = S.V(V\alpha\beta V\delta\epsilon)V(V\beta\gamma V\epsilon\rho)V(V\gamma\delta V\rho\alpha), \text{ .... } (1)$$

we have the equation of a *cone* whose vertex is the origin—for the equation is not altered by putting $x\rho$ for ρ. Also it is the equation of a cone of the second degree, since ρ occurs only twice. Moreover the vectors $\alpha, \beta, \gamma, \delta, \epsilon$ are sides of the cone, because if any one of them be put for ρ the equation is satisfied. Thus if we put β for ρ the equation becomes

$$0 = S.V(V\alpha\beta V\delta\epsilon)V(V\beta\gamma V\epsilon\beta)V(V\gamma\delta V\beta\alpha)$$
$$= S.V(V\alpha\beta V\delta\epsilon)\{V\beta\alpha S.V\gamma\delta V\beta\gamma V\epsilon\beta - V\gamma\delta S.V\beta\alpha V\beta\gamma V\epsilon\beta\}.$$

The first term vanishes because

$$S.V(V\alpha\beta V\delta\epsilon)V\beta\alpha = 0,$$

and the second because

$$S.V\beta\alpha V\beta\gamma V\epsilon\beta = 0,$$

since the three vectors $V\beta\alpha$, $V\beta\gamma$, $V\epsilon\beta$, being each at right angles to β, must be in one plane.

As is remarked by Hamilton, this is a very simple proof of Pascal's

Theorem—for (1) is the condition that the intersections of the planes of α, β and δ, ϵ; β, γ and ϵ, ρ; γ, δ and ρ, α; shall lie in one plane; or, making the statement for any plane section of the cone, that the points of intersection of the three pairs of opposite sides, of a hexagon inscribed in a curve, may always lie in one straight line, the curve must be a conic section.

EXAMPLES TO CHAPTER VII.

1. On the vector of a point P in the plane

$$S\alpha\rho = 1$$

a point Q is taken, such that $QO.OP$ is constant; find the equation of the locus of Q.

2. What spheres cut the loci of P and Q in (1) so that both lines of intersection lie on a cone whose vertex is O?

3. A sphere touches a fixed plane, and cuts a fixed sphere. If the point of contact with the plane be given, the plane of the intersection of the spheres contains a fixed line.

Find the locus of the centre of the variable sphere, if the plane of its intersection with the fixed sphere passes through a given point.

4. Find the radii of the spheres which touch, simultaneously, the four given planes

$$S\alpha\rho = 0, \qquad S\beta\rho = 0, \qquad S\gamma\rho = 0, \qquad S\delta\rho = 1.$$

[What is the volume of the tetrahedron enclosed by these planes?]

5. If a moveable line, passing through the origin, make with any number of fixed lines angles θ, θ_1, θ_2, &c., such that

$$a \cos.\theta + a_1 \cos.\theta_1 + \ldots\ldots - \text{constant},$$

where a, a_1,...... are constant scalars, the line describes a right cone.

6. Determine the conditions that

$$S\rho\phi\rho = 0$$

may represent a *right* cone.

7. What property of a cone (or of a spherical conic) is given directly by the following form of its equation,

$$S.\iota\rho\kappa\rho = 0?$$

8. What are the conditions that the surfaces represented by

$$S\rho\phi\rho = 0, \quad \text{and} \quad S.\iota\rho\kappa\rho = 0,$$

may degenerate into pairs of planes?

9. Find the locus of the vertices of all right cones which have a common ellipse as base.

10. Two right circular cones have their axes parallel, shew that the orthogonal projection of their curve of intersection on the plane containing their axes is a parabola.

11. Two spheres being given in magnitude and position, every sphere which intersects them in given angles will touch two other fixed spheres and cut a third at right angles.

12. If a sphere be placed on a table, the breadth of the elliptic shadow formed by rays diverging from a fixed point is independent of the position of the sphere.

13. Form the equation of the cylinder which has a given circular section, and a given axis. Find the direction of the normal to the subcontrary section.

14. Given the base of a spherical triangle, and the product of the cosines of the sides, the locus of the vertex is a spherical conic, the poles of whose cyclic arcs are the extremities of the given base.

15. (Hamilton, *Bishop Law's Premium Ex.*, 1858.)

(*a.*) What property of a sphero-conic is most immediately indicated by the equation

$$S\frac{\rho}{\alpha}S\frac{\beta}{\rho} = 1\,?$$

(*b.*) The equation $(V\lambda\rho)^2 + (S\mu\rho)^2 = 0$
also represents a cone of the second order; λ is a focal line, and μ is perpendicular to the director-plane corresponding.

(*c.*) What property of a sphero-conic does the equation most immediately indicate?

16. Shew that the areas of all triangles, bounded by a tangent to a spherical conic and the cyclic arcs, are equal.

17. Shew that the locus of a point, the sum of whose arcual dis tances from two given points on a sphere is constant, is a spherical conic.

18. If two tangent planes be drawn to a cyclic cone, the four lines in which they intersect the cyclic planes are sides of a right cone.

19. Find the equation of the cone whose sides are the intersections of pairs of mutually perpendicular tangent planes to a given cyclic cone.

20. Find the condition that five given points may lie on a sphere.

21. What is the surface denoted by the equation

$$\rho^2 = x\alpha^2 + y\beta^2 + z\gamma^2,$$

where

$$\rho = x\alpha + y\beta + z\gamma,$$

α, β, γ being given vectors, and x, y, z variable scalars?

Express the equation of the surface in terms of $\rho, \alpha, \beta, \gamma$ alone.

22. Find the equation of the cone whose sides bisect the angles between a fixed line and any line, in a given plane, which meets the fixed line.

What property of a spherical conic is most directly given by this result?

248.] THE general scalar equation of the second order in a vector ρ must evidently contain a term independent of ρ, terms of the form $S.a\rho b$ involving ρ to the first degree, and others of the form $S.a\rho b\rho c$ involving ρ to the second degree, a, b, c, &c. being constant quaternions. Now the term $S.a\rho b$ may be written as

$$S\rho V(ba),$$

or as $S.(Sa+Va)\rho(Sb+Vb) = Sa\,S\rho Vb + Sb\,S\rho Va + S.\rho Vb Va,$
each of which may evidently be put in the form $S\gamma\rho$, where γ is a known vector.

Similarly * the term $S.a\rho b\rho c$ may be reduced to a set of terms, each of which has one of the forms

$$A\rho^2, \quad (Sa\rho)^2, \quad Sa\rho S\beta\rho,$$

the second being merely a particular case of the third. Thus (the numerical factors 2 being introduced for convenience) we may write the general scalar equation of the second degree as follows :—

$$2\Sigma.Sa\rho S\beta\rho + A\rho^2 + 2S\gamma\rho = C. \quad \ldots\ldots\ldots\ldots\ldots\ldots \quad (1)$$

249.] Change the origin to D where $\overline{OD} = \delta$, then ρ becomes $\rho+\delta$, and the equation takes the form

$$2\Sigma.Sa\rho S\beta\rho + A\rho^2 + 2\Sigma(Sa\rho S\beta\delta + S\beta\rho Sa\delta) + 2AS\delta\rho + 2S\gamma\rho$$
$$+ 2\Sigma.Sa\delta S\beta\delta + A\delta^2 + 2S\gamma\delta - C = 0\,;$$

from which the first power of ρ disappears, that is *the surface is referred to its centre,* if

$$\Sigma(aS\beta\delta + \beta Sa\delta) + A\delta + \gamma = 0, \quad \ldots\ldots\ldots\ldots\ldots\ldots\ldots \quad (2)$$

* For $S.a\rho b\rho c = S.ca\rho b\rho = S.a'\rho b\rho = (2Sa'Sb - Sa'b)\rho^2 + 2Sa'\rho\, Sb\rho$; and in particular cases we may have $Va' = Vb$.

a vector equation of the first degree, which in general gives a single definite value for δ, by the processes of Chapter V. [It would lead us beyond the limits of an elementary treatise to consider the special cases in which (2) represents a line, or a plane, any point of which is a centre of the surface. The processes to be employed in such special cases have been amply illustrated in the Chapter referred to.]

With this value of δ, and putting

$$D = C - 2S\gamma\delta - A\delta^2 - 2\Sigma.S\alpha\delta S\beta\delta,$$

the equation becomes

$$2\Sigma.S\alpha\rho S\beta\rho + A\rho^2 = D.$$

If $D = 0$, the surface is conical (a case treated in last Chapter); if not, it is an ellipsoid or hyperboloid. Unless expressly stated not to be, the surface will, when D is not zero, be considered an ellipsoid. By this we avoid for the time some rather delicate considerations.

By dividing by D, and thus altering only the tensors of the constants, we see that the equation of central surfaces of the second order, referred to the centre, is (excluding cones)

$$2\Sigma(S\alpha\rho S\beta\rho) + g\rho^2 = 1. \quad \ldots\ldots\ldots\ldots\ldots\ldots (3)$$

250.] Differentiating, we obtain

$$2\Sigma\{S\alpha d\rho S\beta\rho + S\alpha\rho S\beta d\rho\} + 2gS\rho d\rho = 0,$$

or $$S.d\rho\{\Sigma(\alpha S\beta\rho + \beta S\alpha\rho) + g\rho\} = 0,$$

and therefore, by § 137, the tangent plane is

$$S(\varpi - \rho)\,\{\Sigma(\alpha S\beta\rho + \beta S\alpha\rho) + g\rho\} = 0,$$

i. e. $$S.\varpi\{\Sigma(\alpha S\beta\rho + \beta S\alpha\rho) + g\rho\} = 1, \text{ by (3)}.$$

Hence, if $$\nu = \Sigma(\alpha S\beta\rho + \beta S\alpha\rho) + g\rho, \quad \ldots\ldots\ldots\ldots\ldots\ldots (4)$$

the tangent plane is $S\nu\varpi = 1$,

and the surface itself is $S\nu\rho = 1$.

And, as ν^{-1} (being perpendicular to the tangent plane, and satisfying its equation) is evidently the vector-perpendicular from the origin on the tangent plane, ν is called the *vector of proximity*.

251.] Hamilton uses for ν, which is obviously a linear and vector function of ρ, the notation $\phi\rho$, ϕ expressing a functional operation, as in Chapter V. But, for the sake of clearness, we will go over part of the ground again, especially for the benefit of students who have mastered only the more elementary parts of that Chapter.

We have, then, $\phi\rho = \Sigma(\alpha S\beta\rho + \beta S\alpha\rho) + g\rho.$

With this definition of ϕ, it is easy to see that

(*a.*) $\phi(\rho+\sigma) = \phi\rho+\phi\sigma$, &c., for *any* two or *more* vectors.

(*b.*) $\phi(x\rho) = x\phi\rho$, a particular case of (*a*), x being a scalar.

(*c.*) $d\phi\rho = \phi(d\rho)$.

(*d.*) $S\sigma\phi\rho = \Sigma(S\alpha\sigma S\beta\rho + S\beta\sigma S\alpha\rho) + gS\rho\sigma = S\rho\phi\sigma$,

or ϕ is, in this case, self-conjugate.

This last property is of great importance.

252.] Thus the general equation of central surfaces of the second degree (excluding cones) may now be written

$$S\rho\phi\rho = 1. \qquad \text{..............................} \quad (1)$$

Differentiating, $\quad Sd\rho\phi\rho + S\rho d\phi\rho = 0,$

which, by applying (*c.*) and then (*d.*) to the last term on the left, gives

$$2S\phi\rho d\rho = 0,$$

and therefore, as in § 250, though now much more simply, the tangent plane at the extremity of ρ is

$$S(\varpi-\rho)\phi\rho = 0,$$

or $\quad S\varpi\phi\rho = S\rho\phi\rho = 1.$

If this pass through $A(\overline{OA} = \alpha)$, we have

$$S\alpha\phi\rho = 1,$$

or, by (*d.*), $\quad S\rho\phi\alpha = 1,$

for all possible points of contact.

This is therefore the equation of the plane of contact of tangent planes drawn from A.

253.] *To find the enveloping cone whose vertex is A*, notice that

$$(S\rho\phi\rho-1) + p(S\rho\phi\alpha-1)^2 = 0,$$

where p is any scalar, is the equation of a surface of the second order *touching* the ellipsoid along its intersection with the plane. If this pass through A we have

$$(S\alpha\phi\alpha-1) + p(S\alpha\phi\alpha+1)^2 = 0,$$

and p is found. Then our equation becomes

$$(S\rho\phi\rho-1)(S\alpha\phi\alpha-1) - (S\rho\phi\alpha-1)^2 = 0, \qquad \text{...............} \quad (1)$$

which is the cone required. To assure ourselves of this, transfer the origin to A, by putting $\rho+\alpha$ for ρ. The result is, using (*a.*) and (*d.*),

$$(S\rho\phi\rho + 2S\rho\phi\alpha + S\alpha\phi\alpha - 1)(S\alpha\phi\alpha-1) - (S\rho\phi\alpha + S\alpha\phi\alpha - 1)^2 = 0,$$

or $\quad S\rho\phi\rho(S\alpha\phi\alpha-1) - (S\rho\phi\alpha)^2 = 0,$

which is homogeneous in $T\rho^2$, and is therefore the equation of a cone.

Suppose A infinitely distant, then we may put in (1) $x\alpha$ for α, where x is infinitely great, and, omitting all but the higher terms, the equation of the cylinder formed by tangent lines parallel to α is

$$(S\rho\phi\rho-1)S\alpha\phi\alpha-(S\rho\phi\alpha)^2 = 0.$$

254.] To study the nature of the surface more closely, let us *find the locus of the middle points of a system of parallel chords.*

Let them be parallel to α, then, if ϖ be the vector of the middle point of one of them, $\varpi+x\alpha$ and $\varpi-x\alpha$ are simultaneous values of ρ which ought to satisfy (1) of § 252.

That is $$S.(\varpi \pm x\alpha)\phi(\varpi \pm x\alpha) = 1.$$

Hence, by (*a.*) and (*d.*), as before,

$$S\varpi\phi\varpi + x^2 S\alpha\phi\alpha = 1,$$

$$S\varpi\phi\alpha = 0. \qquad \text{(1)}$$

The latter equation shews that the locus of the extremity of ϖ, the middle point of a chord parallel to α, is a plane through the centre, whose normal is $\phi\alpha$; that is, a plane parallel to the tangent plane at the point where OA cuts the surface. And (*d.*) shews that this relation is reciprocal—so that if β be *any* value of ϖ, i. e. be any vector in the plane (1), α will be a vector in a diametral plane which bisects all chords parallel to β. The equations of these planes are

$$S\varpi\phi\alpha = 0,$$

$$S\varpi\phi\beta = 0,$$

so that if $V.\phi\alpha\phi\beta = \gamma$ (suppose) is their line of intersection, we have

$$\left.\begin{aligned} S\gamma\phi\alpha &= 0 = S\alpha\phi\gamma, \\ S\gamma\phi\beta &= 0 = S\beta\phi\gamma, \\ \text{and (1) gives}\quad S\beta\phi\alpha &= 0 = S\alpha\phi\beta. \end{aligned}\right\} \qquad \text{(2)}$$

Hence there is *an infinite number of sets of three vectors* α, β, γ, *such that all chords parallel to any one are bisected by the diametral plane containing the other two.*

255.] It is evident from § 23 that any vector may be expressed as a linear function of any three others not in the same plane, let then

$$\rho = x\alpha + y\beta + z\gamma,$$

where, by last section,

$$S\alpha\phi\beta = S\beta\phi\alpha = 0,$$

$$S\alpha\phi\gamma = S\gamma\phi\alpha = 0,$$

$$S\beta\phi\gamma = S\gamma\phi\beta = 0.$$

And let

$$\left.\begin{aligned} S\alpha\phi\alpha &= 1, \\ S\beta\phi\beta &= 1, \\ S\gamma\phi\gamma &= 1, \end{aligned}\right\}$$

so that α, β, and γ are vector conjugate semi-diameters of the surface we are engaged on.

Substituting the above value of ρ in the equation of the surface, and attending to the equations in a, β, γ and to (*a.*), (*b.*), and (*d.*), we have

$$S\rho\phi\rho = S(xa + y\beta + z\gamma)\phi(xa + y\beta + z\gamma),$$
$$= x^2 + y^2 + z^2 = 1.$$

To transform this equation to Cartesian cöordinates, we notice that x is the ratio which the projection of ρ on a bears to a itself, &c. If therefore we take the conjugate diameters as axes of ξ, η, ζ, and their lengths as a, b, c, the above equation becomes at once

$$\frac{\xi^2}{a^2} + \frac{\eta^2}{b^2} + \frac{\zeta^2}{c^2} = 1,$$

the ordinary equation of the ellipsoid referred to conjugate diameters.

√256.] If we write $-\psi^2$ instead of ϕ, these equations assume an interesting form. We take for granted, what we shall afterwards prove, that this halving or extracting the root of the vector function is lawful, and that the new linear and vector function has the same properties (*a.*), (*b.*), (*c.*), (*d.*) (§ 251) as the old. The equation of the surface now becomes

$$S\rho\psi^2\rho = -1,$$

or

$$S\psi\rho\psi\rho = -1,$$

or, finally,

$$T\psi\rho = 1.$$

If we compare this with the equation of the unit-sphere

$$T\rho = 1$$

we see at once the analogy between the two surfaces. *The sphere can be changed into the ellipsoid, or vice versâ, by a linear deformation of each vector, the operator being the function ψ or its inverse.* See the Chapter on Kinematics.

√257.] Equations (2) § 254 now become

$$Sa\psi^2\beta = 0 = S\psi a\psi\beta, \text{ \&c.}, \quad \dots\dots\dots\dots\dots\dots \quad (1)$$

so that ψa, $\psi\beta$, $\psi\gamma$, *the vectors of the unit-sphere which correspond to semi-conjugate diameters of the ellipsoid, form a rectangular system.*

We may remark here, that, as the equation of the ellipsoid referred to its principal axes is a case of § 255, we may now suppose i, j, and k to have these directions, and the equation is $\frac{x^2}{a^2} + \frac{y^2}{b^2} + \frac{z^2}{c^2} = 1$, which, in quaternions, is

$$S\rho\phi\rho = \frac{(Si\rho)^2}{a^2} + \frac{(Sj\rho)^2}{b^2} + \frac{(Sk\rho)^2}{c^2} = 1.$$

We here tacitly assume the existence of such axes, but in all cases, by the help of Hamilton's method, developed in Chapter V, we at once arrive at the cubic equation which gives them.

It is evident from the last-written equation that

$$\phi\rho = +\frac{iSi\rho}{a^2} + \frac{jSj\rho}{b^2} + \frac{kSk\rho}{c^2},$$

and
$$\psi\rho = -\left(\frac{iSi\rho}{a} + \frac{jSj\rho}{b} + \frac{kSk\rho}{c}\right),$$

which latter may be easily proved by shewing that

$$\psi^2\rho = -\phi\rho.$$

And this expression enables us to verify the assertion of last section about the properties of ψ.

As $Si\rho = -x$, &c., x, y, z being the Cartesian cöordinates referred to the principal axes, we have now the means of at once transforming any quaternion result connected with the ellipsoid into the ordinary one.

258.] Before proceeding to other forms of the equation of the ellipsoid, we may use those already given in solving a few problems.

Find the locus of a point when the perpendicular from the centre on its polar plane is of constant length.

If ϖ be the vector of the point, the polar plane is

$$S\rho\phi\varpi = 1,$$

and the length of the perpendicular from O is $\dfrac{1}{T\phi\varpi}$ (§ 208).

Hence the required locus is

$$T\phi\varpi = C,$$

or
$$S\varpi\phi^2\varpi = -C^2,$$

a concentric ellipsoid, with its axes in the same direction as those of the first. By § 257 its Cartesian equation is

$$\frac{x^2}{a^4} + \frac{y^2}{b^4} + \frac{z^2}{c^4} = C^2.$$

259.] *Find the locus of a point whose distance from a given point is always in a given ratio to its distance from a given line.*

Let $\rho = x\beta$ be the given line, and $A\,(OA = \alpha)$ the given point, and let $S\alpha\beta = 0$. Then for any one of the required points

$$T(\rho - \alpha) = eTV\beta\rho,$$

a surface of the second order, which may be written

$$\rho^2 - 2S\alpha\rho + \alpha^2 = e^2(S^2\beta\rho - \beta^2\rho^2).$$

Let the centre be at δ, and make it the origin, then

$$\rho^2 + 2S\rho(\delta-\alpha) + (\delta-\alpha)^2 = e^2\{S^2.\beta(\rho+\delta) - \beta^2(\rho+\delta)^2\},$$

and, that the first power of ρ may disappear,

$$(\delta - \alpha) = e^2(\beta S\beta\delta - \beta^2\delta),$$

a linear equation for δ. To solve it, note that $S\alpha\beta = 0$, operate by $S.\beta$ and we get $(1 - e^2\beta^2 + e^2\beta^2)S\beta\delta = S\beta\delta = 0.$

Hence $$\delta - a = -e^2\beta^2\delta,$$

or $$\delta = \frac{a}{1+e^2\beta^2}.$$

Referred to this point as origin the equation becomes

$$(1+e^2\beta^2)\rho^2 - e^2 S^2\beta\rho + \frac{e^2\beta^2 a^2}{1+e^2\beta^2} = 0,$$

which shews that it belongs to a surface of revolution (of the second order) whose axis is parallel to β, as its intersection with a plane $S\beta\rho = a$, perpendicular to that axis, lies also on the sphere

$$\rho^2 = \frac{e^2a^2}{1+e^2\beta^2} - \frac{e^2\beta^2 a^2}{(1+e^2\beta^2)^2}.$$

In fact, if the point be the focus of any meridian section of an oblate spheroid, the line is the directrix of the same.

√260.] *A sphere, passing through the centre of an ellipsoid, is cut by a series of spheres whose centres are on the ellipsoid and which pass through the centre thereof; find the envelop of the planes of intersection.*

Let $(\rho - a)^2 = a^2$ be the first sphere, i.e.

$$\rho^2 - 2Sa\rho = 0.$$

One of the others is $$\rho^2 - 2S\varpi\rho = 0,$$

where $$S\varpi\phi\varpi = 1.$$

The plane of intersection is

$$S(\varpi - a)\rho = 0.$$

Hence, for the envelop, (see next Chapter,)

$$\left.\begin{aligned} S\varpi'\phi\varpi &= 0, \\ S\varpi'\rho &= 0, \end{aligned}\right\} \text{ where } \varpi' = d\varpi,$$

or $$\phi\varpi = x\rho, \quad \{Vx = 0\},$$

i.e. $$\varpi = x\phi^{-1}\rho.$$

Hence $$x^2 S\rho\phi^{-1}\rho = 1,$$

and $$xS\rho\phi^{-1}\rho = Sa\rho,$$

and, eliminating x,

$$S\rho\phi^{-1}\rho = (Sa\rho)^2,$$

a cone of the second order.

√261.] *From a point in the outer of two concentric ellipsoids a tangent cone is drawn to the inner, find the envelop of the plane of contact.*

If $S\varpi\phi\varpi = 1$ be the outer, and $S\rho\psi\rho = 1$ be the inner, ϕ and ψ being any two self-conjugate linear and vector functions, the plane of contact is $$S\varpi\psi\rho = 1.$$

Hence, for the envelop, $$\left.\begin{aligned} S\varpi'\psi\rho &= 0, \\ S\varpi'\phi\varpi &= 0, \end{aligned}\right\}$$

therefore $\phi\varpi = x\psi\rho$,

or $\varpi = x\phi^{-1}\psi\rho$.

This gives

$$\left.\begin{aligned} xS.\psi\rho\phi^{-1}\psi\rho &= 1, \\ \text{and} \quad x^2 S.\psi\rho\phi^{-1}\psi\rho &= 1, \end{aligned}\right\}$$

and therefore, eliminating x,

$$S.\psi\rho\phi^{-1}\psi\rho = 1,$$

or

$$S.\rho\psi\phi^{-1}\psi\rho = 1,$$

another concentric ellipsoid, as $\psi\phi^{-1}\psi$ is a linear and vector function $=\chi$ suppose; so that the equation may be written

$$S\rho\chi\rho = 1.$$

262.] *Find the locus of intersection of tangent planes at the extremities of conjugate diameters.*

If a, β, γ be the vector semi-diameters, the planes are

$$\left.\begin{aligned} S\varpi\psi^2 a &= -1, \\ S\varpi\psi^2 \beta &= -1, \\ S\varpi\psi^2 \gamma &= -1, \end{aligned}\right\}$$

with the conditions § 257.

Hence $-\psi\varpi S.\psi a\psi\beta\psi\gamma = \psi\varpi = \psi a + \psi\beta + \psi\gamma$, by § 92,

therefore $T\psi\varpi = \sqrt{3}$,

since ψa, $\psi\beta$, $\psi\gamma$ form a rectangular system of unit-vectors.

This may also evidently be written

$$S\varpi\psi^2\varpi = -3,$$

shewing that the locus is similar and similarly situated to the given ellipsoid, but larger in the ratio $\sqrt{3} : 1$.

263.] *Find the locus of the intersection of three spheres whose diameters are semi-conjugate diameters of an ellipsoid.*

If a be one of the semi-conjugate diameters

$$Sa\psi^2 a = -1.$$

And the corresponding sphere is

$$\rho^2 - Sa\rho = 0,$$

or

$$\rho^2 - S\psi a\psi^{-1}\rho = 0,$$

with similar equations in β and γ. Hence, by § 92,

$$\psi^{-1}\rho S.\psi a\psi\beta\psi\gamma = -\psi^{-1}\rho = \rho^2(\psi a + \psi\beta + \psi\gamma),$$

and, taking tensors, $T\psi^{-1}\rho = \sqrt{3}\,T\rho^2$,

or $T\psi^{-1}\rho^{-1} = \sqrt{3}$,

or, finally, $S\rho\psi^{-2}\rho = -3\rho^4$.

This is Fresnel's *Surface of Elasticity* in the Undulatory Theory.

264.] Before going farther we may prove some useful properties of the function ϕ in the form we are at present using—viz.

$$\phi\rho = \frac{iSi\rho}{a^2} + \frac{jSj\rho}{b^2} + \frac{kSk\rho}{c^2}.$$

We have $$\rho = -iSi\rho - jSj\rho - kSk\rho,$$
and it is evident that

$$\phi i = -\frac{i}{a^2}, \quad \phi j = -\frac{j}{b^2}, \quad \phi k = -\frac{k}{c^2}.$$

Hence $$\phi^2\rho = -\frac{iSi\rho}{a^4} - \frac{jSj\rho}{b^4} - \frac{kSk\rho}{c^4}.$$

Also $$\phi^{-1}\rho = a^2 iSi\rho + b^2 jSj\rho + c^2 kSk\rho,$$
and so on.

Again, if α, β, γ be *any* rectangular unit-vectors

$$S\alpha\phi\alpha = \frac{(Si\alpha)^2}{a^2} + \frac{(Sj\alpha)^2}{b^2} + \frac{(Sk\alpha)^2}{c^2},$$

$$\&c. = \&c.$$

But as $$(Si\rho)^2 + (Sj\rho)^2 + (Sk\rho)^2 = -\rho^2,$$

we have $$S\alpha\phi\alpha + S\beta\phi\beta + S\gamma\phi\gamma = \frac{1}{a^2} + \frac{1}{b^2} + \frac{1}{c^2}.$$

Again,

$$S.\phi\alpha\phi\beta\phi\gamma = S.\left(\frac{iSi\alpha}{a^2} + \ldots\right)\left(\frac{iSi\beta}{a^2} + \ldots\right)\left(\frac{iSi\gamma}{a^2} + \ldots\right)$$

$$= -\begin{vmatrix} \frac{Si\alpha}{a^2}, & \frac{Sj\alpha}{b^2}, & \frac{Sk\alpha}{c^2} \\ \frac{Si\beta}{a^2}, & \frac{Sj\beta}{b^2}, & \frac{Sk\beta}{c^2} \\ \frac{Si\gamma}{a^2}, & \frac{Sj\gamma}{b^2}, & \frac{Sk\gamma}{c^2} \end{vmatrix} = \frac{-1}{a^2b^2c^2}\begin{vmatrix} Si\alpha, & Sj\alpha, & Sk\alpha \\ Si\beta, & Sj\beta, & Sk\beta \\ Si\gamma, & Sj\gamma, & Sk\gamma \end{vmatrix} = \pm\frac{1}{a^2b^2c^2}.$$

And so on. These elementary investigations are given here for the benefit of those who have not read Chapter V. The student may easily obtain all such results in a far more simple manner by means of the formulae of that Chapter.

265.] *Find the locus of intersection of a rectangular system of three tangents to an ellipsoid.*

If ϖ be the vector of the point of intersection, α, β, γ the tangents, then, since $\varpi + x\alpha$ should give equal values of x when substituted in the equation of the surface, giving

$$S(\varpi + x\alpha)\phi(\varpi + x\alpha) = 1,$$

or $$x^2 S\alpha\phi\alpha + 2xS\varpi\phi\alpha + (S\varpi\phi\varpi - 1) = 0,$$

we have $$(S\varpi\phi\alpha)^2 = S\alpha\phi\alpha(S\varpi\phi\varpi - 1).$$

Adding this to the two similar equations in β and γ

$$(S\alpha\phi\varpi)^2 + (S\beta\phi\varpi)^2 + (S\gamma\phi\varpi)^2 = (S\alpha\phi\alpha + S\beta\phi\beta + S\gamma\phi\gamma)(S\varpi\phi\varpi - 1),$$

or $$-(\phi\varpi)^2 = \left(\frac{1}{a^2} + \frac{1}{b^2} + \frac{1}{c^2}\right)(S\varpi\phi\varpi - 1),$$

or $$S.\varpi\left\{\left(\frac{1}{a^2} + \frac{1}{b^2} + \frac{1}{c^2}\right)\phi + \phi^2\right\}\varpi = \frac{1}{a^2} + \frac{1}{b^2} + \frac{1}{c^2},$$

an ellipsoid concentric with the first.

266.] *If a rectangular system of chords be drawn through any point within an ellipsoid, the sum of the reciprocals of the rectangles under the segments into which they are divided is constant.*

With the notation of the solution of the preceding problem, ϖ giving the intersection of the vectors, it is evident that the product of the values of x is one of the rectangles in question taken negatively.

Hence the required sum is

$$-\frac{\Sigma S\alpha\phi\alpha}{S\varpi\phi\varpi - 1} = -\frac{\frac{1}{a^2} + \frac{1}{b^2} + \frac{1}{c^2}}{S\varpi\phi\varpi - 1}.$$

This evidently depends on $S\varpi\phi\varpi$ only and not on the particular directions of α, β, γ: and is therefore unaltered if ϖ be the vector of any point of an ellipsoid similar, and similarly situated, to the given one. [The expression is interpretable even if the point be exterior to the ellipsoid.]

267.] *Shew that if any rectangular system of three vectors be drawn from a point of an ellipsoid, the plane containing their other extremities passes through a fixed point. Find the locus of the latter point as the former varies.*

With the same notation as before, we have

$$S\varpi\phi\varpi = 1,$$

and $$S(\varpi + x\alpha)\,\phi\,(\varpi + x\alpha) = 1;$$

therefore $$x = -\frac{2S\alpha\phi\varpi}{S\alpha\phi\alpha}.$$

Hence the required plane passes through the extremity of

$$\varpi - 2\alpha\frac{S\alpha\phi\varpi}{S\alpha\phi\alpha},$$

and those of two other vectors similarly determined. It therefore passes through the point whose vector is

$$\theta = \varpi - 2\frac{\alpha S\alpha\phi\varpi + \beta S\beta\phi\varpi + \gamma S\gamma\phi\varpi}{S\alpha\phi\alpha + S\beta\phi\beta + S\gamma\phi\gamma},$$

or $$\theta = \varpi + \frac{2\phi\varpi}{m_2} \quad (\S\, 173).$$

Thus the first part of the proposition is proved.

But we have also $\varpi = \frac{m_2}{2}\left(\phi + \frac{m_2}{2}\right)^{-1}\theta$,

whence by the equation of the ellipsoid we obtain

$$\frac{m_2^2}{4} S.\theta \left(\phi + \frac{m_2}{2}\right)^{-1} \phi \left(\phi + \frac{m_2}{2}\right)^{-1} \theta = 1,$$

the equation of a concentric ellipsoid.

268.] *Find the directions of the three vectors which are parallel to a set of conjugate diameters in each of two central surfaces of the second degree.*

Transferring the centres of both to the origin, let their equations be

$$\left.\begin{aligned} &S\rho\phi\rho = 1 \text{ or } 0, \\ \text{and} \quad &S\rho\psi\rho = 1 \text{ or } 0. \end{aligned}\right\} \qquad (1)$$

If α, β, γ be vectors in the required directions, we must have (§ 254)

$$\left.\begin{aligned} S\alpha\phi\beta &= 0, & S\alpha\psi\beta &= 0, \\ S\beta\phi\gamma &= 0, & S\beta\psi\gamma &= 0, \\ S\gamma\phi\alpha &= 0, & S\gamma\psi\alpha &= 0. \end{aligned}\right\} \qquad (2)$$

From these equations $\phi\alpha \parallel V\beta\gamma \parallel \psi\alpha$, &c.

Hence the three required directions are the roots of

$$V.\phi\rho\psi\rho = 0. \qquad (3)$$

This is evident on other grounds, for it means that *if one of the surfaces expand or contract uniformly till it meets the other, it will touch it successively at points on the three sought vectors.*

We may put (3) in either of the following forms—

$$\left.\begin{aligned} &V.\rho\phi^{-1}\psi\rho = 0, \\ \text{or} \quad &V.\rho\psi^{-1}\phi\rho = 0, \end{aligned}\right\} \qquad (4)$$

and, as ϕ and ψ are given functions, we find the solutions by the processes of Chapter V.

[*Note.* As $\phi^{-1}\psi$ and $\psi^{-1}\phi$ are not, in general, self-conjugate functions, equations (4) do not signify that α, β, γ are vectors parallel to the principal axes of the surfaces

$$S.\rho\phi^{-1}\psi\rho = 1, \qquad S.\rho\psi^{-1}\phi\rho = 1.$$

In *these* equations it does not matter whether $\phi^{-1}\psi$ is self-conjugate or not; but it does most particularly matter when they are differentiated, so as to find axes, &c.]

Given two surfaces of the second degree, there exists in general a set of Cartesian axes, whose directions are those of conjugate diameters in every one of the surfaces of the second degree passing through the intersection of the two surfaces given.

For any surface through the intersection of

$$S\rho\phi\rho = 1 \quad \text{and} \quad S(\rho-\alpha)\psi(\rho-\alpha) = e,$$

is $$fS\rho\phi\rho - S(\rho-\alpha)\psi(\rho-\alpha) = f-e,$$

where f and e are scalars.

The axes of this depend only on the term

$$S\rho(f\phi-\psi)\rho.$$

Hence the set of conjugate diameters which are the same in all are the roots of

$$V(f\phi-\psi)\rho(f_1\phi-\psi)\rho = 0, \quad \text{or} \quad V\phi\rho\psi\rho = 0,$$

as we might have seen without analysis.

The locus of the centres is given by the equation

$$(\psi-f\phi)\rho-\psi\alpha = 0,$$

where f is a scalar variable.

269.] *Find the equation of the ellipsoid of which three conjugate semi-diameters are given.*

Let the vector semi-diameters be α, β, γ, and let

$$S\rho\phi\rho = 1$$

be the equation of the ellipsoid. Then (§ 255) we have

$$S\alpha\phi\alpha = 1, \qquad S\alpha\phi\beta = 0,$$
$$S\beta\phi\beta = 1, \qquad S\beta\phi\gamma = 0,$$
$$S\gamma\phi\gamma = 1, \qquad S\gamma\phi\alpha = 0;$$

the six scalar conditions requisite (§ 139) for the determination of the linear and vector function ϕ.

They give $$\alpha \parallel V\phi\beta\phi\gamma,$$

or $$x\alpha = \phi^{-1}V\beta\gamma.$$

Hence $$x = xS\alpha\phi\alpha = S.\alpha\beta\gamma,$$

and similarly for the other combinations. Thus, as we have

$$\rho S.\alpha\beta\gamma = \alpha S.\beta\gamma\rho + \beta S.\gamma\alpha\rho + \gamma S.\alpha\beta\rho,$$

we find at once

$$\phi\rho S^2.\alpha\beta\gamma = V\beta\gamma S.\beta\gamma\rho + V\gamma\alpha S.\gamma\alpha\rho + V\alpha\beta S.\alpha\beta\rho;$$

and the required equation may be put in the form

$$S^2.\alpha\beta\gamma = S^2.\alpha\beta\rho + S^2.\beta\gamma\rho + S^2.\gamma\alpha\rho.$$

The immediate interpretation is that *if four tetrahedra be formed by grouping, three and three, a set of semi-conjugate vector axes of an ellipsoid and any other vector of the surface, the sum of the squares of the volumes of three of these tetrahedra is equal to the square of the volume of the fourth.*

270.] When the equation of a surface of the second order can be put in the form $$S\rho\phi^{-1}\rho = 1, \quad (1)$$
where $$(\phi-g)(\phi-g_1)(\phi-g_2) = 0,$$
we know that g, g_1, g_2 are the squares of the principal semi-diameters. Hence, if we put $\phi+h$ for ϕ we have a second surface, the differences of the squares of whose principal semiaxes are the same as for the first. That is, $$S\rho(\phi+h)^{-1}\rho = 1 \quad (2)$$
is a surface *confocal* with (1). From this simple modification of the equation all the properties of a series of confocal surfaces may easily be deduced. We give one as an example.

271.] *Any two confocal surfaces of the second order, which meet, intersect at right angles.*

For the normal to (2) is, evidently,
$$(\phi+h)^{-1}\rho;$$
and that to another of the series, if it passes through the common point whose vector is ρ, is there
$$(\phi+h_1)^{-1}\rho.$$
But $$S.(\phi+h)^{-1}\rho(\phi+h_1)^{-1}\rho = S.\rho\frac{1}{(\phi+h)(\phi+h_1)}\rho$$
$$= \frac{1}{h-h_1}S\rho\left((\phi+h_1)^{-1}-(\phi+h)^{-1}\right)\rho,$$
and this evidently vanishes if h and h_1 are different, as they must be unless the surfaces are identical.

272.] *To find the conditions of similarity of two central surfaces of the second order.*

Referring them to their centres, let their equations be
$$\left.\begin{aligned} S\rho\phi\rho &= 1, \\ S\rho\phi'\rho &= 1. \end{aligned}\right\} \quad (1)$$
Now the obvious conditions are that the axes of the one are proportional to those of the other. Hence, if
$$\left.\begin{aligned} g^3+m_2g^2+m_1g+m &= 0, \\ g'^3+m'_2g'^2+m'_1g'+m' &= 0, \end{aligned}\right\} \quad (2)$$
be the equations for determining the squares of the reciprocals of the semiaxes, we must have
$$\frac{m'_2}{m_2} = \mu, \quad \frac{m'_1}{m_1} = \mu^2, \quad \frac{m'}{m} = \mu^3, \quad (3)$$
where μ is an undetermined scalar. Thus it appears that there are but two scalar conditions necessary. Eliminating μ we have
$$\frac{m'^2_2}{m^2_2} = \frac{m'_1}{m_1}, \qquad \frac{m'm'_2}{mm_2} = \frac{m'^2_1}{m^2_1}, \quad (4)$$
which are equivalent to the ordinary conditions.

273.] *Find the greatest and least semi-diameters of a central plane section of an ellipsoid.*

Here
$$\left.\begin{array}{r} S\rho\phi\rho = 1 \\ S\alpha\rho = 0 \end{array}\right\} \quad \text{.............................} \quad (1)$$
together represent the elliptic section; and our additional condition is that $T\rho$ is a maximum or minimum.

Differentiating the equations of the ellipse, we have
$$S\phi\rho d\rho = 0,$$
$$S\alpha d\rho = 0,$$
and the maximum condition gives
$$dT\rho = 0$$
or
$$S\rho d\rho = 0.$$
Eliminating the indeterminate vector $d\rho$ we have
$$S.\alpha\rho\phi\rho = 0. \quad \text{.............................} \quad (2)$$
This shews that *the maximum or minimum vector, the normal at its extremity, and the perpendicular to the plane of section, lie in one plane.* It also shews that there are but two vector-directions which satisfy the conditions, and that they are perpendicular to each other, for (2) is satisfied if $\alpha\rho$ be substituted for ρ.

We have now to solve the three equations (1) and (2), to find the vectors of the two (four) points in which the ellipse (1) intersects the cone (2). We obtain at once
$$\phi\rho = xV.\phi^{-1}\alpha V\alpha\rho.$$
Operating by $S.\rho$ we have
$$1 = x\rho^2 S\alpha\phi^{-1}\alpha.$$
Hence
$$\rho^2\phi\rho = \rho - \alpha\frac{S\rho\phi^{-1}\alpha}{S\alpha\phi^{-1}\alpha}$$
or
$$\rho = \frac{S\rho\phi^{-1}\alpha}{S\alpha\phi^{-1}\alpha}(1-\rho^2\phi)^{-1}\alpha; \quad \text{..................} \quad (3)$$
from which
$$S.\alpha(1-\rho^2\phi)^{-1}\alpha = 0; \quad \text{.......................} \quad (4)$$
a quadratic equation in ρ^2, from which the lengths of the maximum and minimum vectors are to be determined. By § 147 it may be written
$$m\rho^4 S\alpha\phi^{-1}\alpha - \rho^2 S.\alpha(m_2-\phi)\alpha + \alpha^2 = 0. \quad \text{..............} \quad (5)$$

[If we had operated by $S.\phi^{-1}\alpha$ or by $S.\phi^{-1}\rho$, instead of by $S.\rho$, we should have obtained an equation apparently different from this, but easily reducible to it. To prove their identity is a good exercise for the student.]

Substituting the values of ρ^2 given by (5) in (3) we obtain the vectors of the required diameters. [The student may easily prove directly that $\quad (1-\rho_1^2\phi)^{-1}\alpha \quad$ and $\quad (1-\rho_2^2\phi)^{-1}\alpha$

are necessarily perpendicular to each other, if both be perpendicular to α, and if ρ_1^2 and ρ_2^2 be different. See § 271.]

274.] By (5) of last section we see that

$$\rho_1^2 \rho_2^2 = \frac{\alpha^2}{mS\alpha\phi^{-1}\alpha}.$$

Hence the area of the ellipse (1) is

$$\frac{\pi T\alpha}{\sqrt{-mS\alpha\phi^{-1}\alpha}}.$$

Also the locus of normals to all diametral sections of an ellipsoid, whose areas are equal, is the cone

$$S\alpha\phi^{-1}\alpha = C\alpha^2.$$

When the roots of (5) are equal, i.e. when

$$(m_2\alpha^2 - S\alpha\phi\alpha)^2 = 4m\alpha^2 S\alpha\phi^{-1}\alpha, \qquad (6)$$

the section is a circle. It is not difficult to prove that this equation is satisfied by only two values of $U\alpha$, but another quaternion form of the equation gives the solution of this and similar problems by inspection. (See § 275 below.)

275.] By § 168 we may write the equation

$$S\rho\phi\rho = 1$$

in the new form $$S.\lambda\rho\mu\rho + p\rho^2 = 1,$$

where p is a known scalar, and λ and μ are definitely known (with the exception of their tensors, whose product alone is given) in terms of the constants involved in ϕ. [The reader is referred again also to §§ 121, 122.] This may be written

$$2S\lambda\rho S\mu\rho + (p - S\lambda\mu)\rho^2 = 1. \qquad (1)$$

From this form it is obvious that the surface is cut by any plane perpendicular to λ or μ in a circle. For, if we put

$$S\lambda\rho = a,$$

we have $$2aS\mu\rho + (p - S\lambda\mu)\rho^2 = 1,$$

the equation of a sphere which passes through the plane curve of intersection.

Hence λ and μ of § 168 are the values of α in equation (6) of the preceding section.

276.] *Any two circular sections of a central surface of the second order, whose planes are not parallel, lie on a sphere.*

For the equation $$(S\lambda\rho - a)(S\mu\rho - b) = 0,$$

where a and b are any scalar constants whatever, is that of a system of two non-parallel planes, cutting the surface in circles. Eliminating the product $S\lambda\rho S\mu\rho$ between this and equation (1) of last section, there remains the equation of a sphere.

277.] *To find the generating lines of a central surface of the second order.*

Let the equation be $\quad S\rho\phi\rho = 1$;

then, if α be the vector of any point on the surface, and ϖ a vector parallel to a generating line, we must have

$$\rho = \alpha + x\varpi$$

for all values of the scalar x.

Hence $\quad S(\alpha + x\varpi)\phi(\alpha + x\varpi) = 1,$

which gives the two equations

$$\left.\begin{array}{l} S\alpha\phi\varpi = 0, \\ S\varpi\phi\varpi = 0. \end{array}\right\}$$

The first is the equation of a plane through the origin parallel to the tangent plane at the extremity of α, the second is the equation of the asymptotic cone. The generating lines are therefore parallel to the intersections of these two surfaces, as is well known.

From these equations we have

$$y\phi\varpi = V\alpha\varpi$$

where y is a scalar to be determined. Operating on this by $S.\beta$ and $S.\gamma$, where β and γ are any two vectors not coplanar with α, we have

$$S\varpi(y\phi\beta + V\alpha\beta) = 0, \qquad S\varpi(y\phi\gamma - V\gamma\alpha) = 0. \;\ldots\ldots\ldots (1)$$

Hence $\quad S.\phi\alpha(y\phi\beta + V\alpha\beta)(y\phi\gamma - V\gamma\alpha) = 0,$

or $\quad my^2 S.\alpha\beta\gamma - S\alpha\phi\alpha S.\alpha\beta\gamma = 0.$

Thus we have the two values

$$y = \pm\sqrt{\frac{S\alpha\phi\alpha}{m}} = \pm\sqrt{\frac{1}{m}},$$

belonging to the two generating lines.

278.] But by equation (1) we have

$$\begin{aligned} z\varpi &= V.(y\phi\beta + V\alpha\beta)(y\phi\gamma - V\gamma\alpha) \\ &= my^2\phi^{-1}V\beta\gamma + yV.\phi\alpha V\beta\gamma - \alpha S.\alpha V\beta\gamma; \end{aligned}$$

which, according to the sign of y, gives one or other generating line.

Here $V\beta\gamma$ may be any vector whatever, provided it is not perpendicular to α (a condition assumed in last section), and we may write for it θ.

Substituting the value of y before found, we have

$$\begin{aligned} z\varpi &= \phi^{-1}\theta - \alpha S\alpha\theta \pm \sqrt{\frac{1}{m}}V\phi\alpha\theta, \\ &= V.\phi\alpha V.\alpha\phi^{-1}\theta \pm \sqrt{\frac{1}{m}}V\phi\alpha\theta, \end{aligned}$$

or, as we may evidently write it,

$$= \phi^{-1}(V.aV\phi a\theta) \pm \sqrt{\frac{1}{m}}\, V\phi a\theta. \quad \ldots\ldots\ldots\ldots\ldots\ldots (2)$$

Put $\tau = V\phi a\theta,$

and we have $$z\varpi = \phi^{-1}Va\tau + \sqrt{\frac{1}{m}}\,\tau,$$

with the condition $S\tau\phi a = 0.$

[Any one of these sets of values forms the complete solution of the problem; but more than one have been given, on account of their singular nature and the many properties of surfaces of the second order which immediately follow from them. It will be excellent practice for the student to shew that

$$\psi\theta = U\left(V.\phi aVa\phi^{-1}\theta + \sqrt{\frac{1}{m}}\, V\phi a\theta\right)$$

is an invariant. This may most easily be done by proving that

$$V.\psi\theta\psi\theta_1 = 0 \text{ identically.}]$$

Perhaps, however, it is simpler to write a for $V\beta\gamma$, and we thus obtain

$$z\varpi = -\phi^{-1}VaVa\phi a \mp \sqrt{\frac{1}{m}}\, Va\phi a.$$

[The reader need hardly be reminded that we are dealing with the *general* equation of the central surfaces of the second order—the centre being origin.]

EXAMPLES TO CHAPTER VIII.

1. Find the locus of points on the surface

$$S\rho\phi\rho = 1$$

where the generating lines are at right angles to one another.

2. Find the equation of the surface described by a straight line which revolves about an axis, which it does not meet, but with which it is rigidly connected.

3. Find the conditions that

$$S\rho\phi\rho = 1$$

may be a surface of revolution, with axis parallel to a given vector.

4. Find the equations of the right cylinders which circumscribe a given ellipsoid.

5. Find the equation of the locus of the extremities of perpendiculars to central plane sections of an ellipsoid, erected at the

centre, their lengths being the principal semi-axes of the sections. [Fresnel's *Wave-Surface*. See Chap. XI.]

6. The cone touching central plane sections of an ellipsoid, which are of equal area, is asymptotic to a confocal hyperboloid.

7. Find the envelop of all non-central plane sections of an ellipsoid whose area is constant.

8. Find the locus of the intersection of three planes, perpendicular to each other, and touching, respectively, each of three confocal surfaces of the second order.

9. Find the locus of the foot of the perpendicular from the centre of an ellipsoid upon the plane passing through the extremities of a set of conjugate diameters.

10. Find the points in an ellipsoid where the inclination of the normal to the radius-vector is greatest.

11. If four similar and similarly situated surfaces of the second order intersect, the planes of intersection of each pair pass through a common point.

12. If a parallelepiped be inscribed in a central surface of the second degree its edges are parallel to a system of conjugate diameters.

13. Shew that there is an infinite number of sets of axes for which the Cartesian equation of an ellipsoid becomes

$$x^2 + y^2 + z^2 = e^2.$$

14. Find the equation of the surface of the second order which circumscribes a given tetrahedron so that the tangent plane at each angular point is parallel to the opposite face; and shew that its centre is the mean point of the tetrahedron.

15. Two similar and similarly situated surfaces of the second order intersect in a plane curve, whose plane is conjugate to the vector joining their centres.

16. Find the locus of all points on

$$S\rho\phi\rho = 1,$$

where the normals meet the normal at a given point.

Also the locus of points on the surface, the normals at which meet a given line in space.

17. Normals drawn at points situated on a generating line are parallel to a fixed plane.

18. Find the envelop of the planes of contact of tangent planes drawn to an ellipsoid from points of a concentric sphere. Find the locus of the point from which the tangent planes are drawn if the envelop of the planes of contact is a sphere.

19. The sum of the reciprocals of the squares of the perpendiculars from the centre upon three conjugate tangent planes is constant.

20. Cones are drawn, touching an ellipsoid, from any two points of a similar, similarly situated, and concentric ellipsoid. Shew that they intersect in two plane curves.

Find the locus of the vertices of the cones when these plane sections are at right angles to one another.

21. Find the locus of the points of contact of tangent planes which are equidistant from the centre of a surface of the second order.

22. From a fixed point A, on the surface of a given sphere, draw any chord AD; let D' be the second point of intersection of the sphere with the secant BD drawn from any point B; and take a radius vector AE, equal in length to BD', and in direction either coincident with, or opposite to, the chord AD: the locus of E is an ellipsoid, whose centre is A, and which passes through B. (Hamilton, *Elements*, p. 227.)

23. Shew that the equation

$$l^2(e^2-1)(e+Saa') = (Sa\rho)^2 - 2e\,Sa\rho Sa'\rho + (Sa'\rho)^2 + (1-e^2)\rho^2,$$

where e is a variable (scalar) parameter, and a, a' unit-vectors, represents a system of confocal surfaces. (*Ibid.* p. 644.)

24. Shew that the locus of the diameters of

$$S\rho\phi\rho = 1$$

which are parallel to the chords bisected by the tangent planes to the cone

$$S\rho\psi\rho = 0$$

is the cone

$$S.\rho\phi\psi^{-1}\phi\rho = 0.$$

25. Find the equation of a cone, whose vertex is one summit of a given tetrahedron, and which passes through the circle circumscribing the opposite side.

26. Shew that the locus of points on the surface

$$S\rho\phi\rho = 1,$$

the normals at which meet that drawn at the point $\rho = \varpi$, is on the cone

$$S.(\rho-\varpi)\,\phi\varpi\phi\rho = 0.$$

27. Find the equation of the locus of a point the square of whose distance from a given line is proportional to its distance from a given plane.

28. Shew that the locus of the pole of the plane

$$Sa\rho = 1,$$

with respect to the surface

$$S\rho\phi\rho = 1,$$

is a sphere, if a be subject to the condition

$$Sa\phi^{-2}a = C.$$

29. Shew that the equation of the surface generated by lines drawn through the origin parallel to the normals to

$$S\rho\phi^{-1}\rho = 1$$

along its lines of intersection with

$$S\rho(\phi+k)^{-1}\rho = 1,$$

is $$\varpi^2 - kS\varpi(\phi+k)^{-1}\varpi = 0.$$

30. Common tangent planes are drawn to

$$2S\lambda\rho S\mu\rho + (p - S\lambda\mu)\rho^2 = 1, \quad \text{and} \quad T\rho = h,$$

find the value of h that the lines of contact with the former surface may be plane curves. What are they, in this case, on the sphere? Discuss the case of $p^2 - S^2\lambda\mu = 0$.

31. If tangent cones be drawn to

$$S\rho\phi^2\rho = 1,$$

from every point of $$S\rho\phi\rho = 1,$$

the envelop of their planes of contact is

$$S\rho\phi^3\rho = 1.$$

32. Tangent cones are drawn from every point of

$$S(\rho - a)\phi(\rho - a) = n^2,$$

to the similar and similarly situated surface

$$S\rho\phi\rho = 1,$$

shew that their planes of contact envelop the surface

$$(Sa\phi\rho - 1)^2 = n^2 S\rho\phi\rho.$$

33. Find the envelop of planes which touch the parabolas

$$\rho = at^2 + \beta t, \qquad \rho = a\tau^2 + \gamma\tau,$$

where a, β, γ form a rectangular system, and t and τ are scalars.

34. Find the equation of the surface on which lie the lines of contact of tangent cones drawn from a fixed point to a series of similar, similarly situated, and concentric ellipsoids.

35. Discuss the surfaces whose equations are

$$Sa\rho S\beta\rho = S\gamma\rho,$$

and $$S^2 a\rho + S.a\beta\rho = 1.$$

36. Shew that the locus of the vertices of the right cones which touch an ellipsoid is a hyperbola.

37. If a_1, a_2, a_3 be vector conjugate diameters of

$$S\rho\phi\rho = 1,$$

where $$\phi^3 - m_2\phi^2 + m_1\phi - m = 0,$$

shew that $\Sigma(a^2) - \frac{m_1}{m}, \quad \Sigma(Va_1a_2)^2 = -\frac{m_2}{m}, \quad S^2.a_1a_2a_3 = -\frac{1}{m},$

and $$\Sigma(\phi a)^2 = m_2.$$

CHAPTER IX.

GEOMETRY OF CURVES AND SURFACES.

279.] We have already seen (§ 31 (l)) that the equations

$$\rho = \phi t = \Sigma . af(t),$$

and $$\rho = \phi(t, u) = \Sigma . af(t, u),$$

where a represents one of a set of given vectors, and f a scalar function of scalars t and u, represent respectively a curve and a surface. We commence the present too brief Chapter with a few of the immediate deductions from these forms of expression. We shall then give a number of examples, with little attempt at systematic development or even arrangement.

280.] What may be denoted by t and u in these equations is, of course, quite immaterial: but in the case of curves, considered geometrically, t is most conveniently taken as the length, s, of the curve, measured from some fixed point. In the Kinematical investigations of the next Chapter t may, with great convenience, be employed to denote *time*.

281.] Thus we may write the equation of any curve in space as

$$\rho = \phi s,$$

where ϕ is a vector function of the length, s, of the curve. Of course it is only a *linear* function when the equation (as in § 31 (l)) represents a straight line.

282.] We have also seen (§§ 38, 39) that

$$\frac{d\rho}{ds} = \frac{d}{ds}\phi s = \phi' s$$

is a vector of *unit* length in the direction of the tangent at the extremity of ρ.

At the proximate point, denoted by $s + \delta s$, this unit tangent vector becomes $$\phi' s + \phi'' s\, \delta s + \&c.$$

But, because $$T\phi' s = 1,$$
we have $$S.\phi' s\,\phi'' s = 0.$$
Hence $\phi'' s$ is a vector in the osculating plane of the curve, and perpendicular to the tangent.

Also, if $\delta\theta$ be the angle between the successive tangents $\phi' s$ and $\phi' s + \phi'' s\,\delta s + \ldots\ldots$, we have
$$\mathfrak{L}\frac{\delta\theta}{\delta s} = T\phi'' s;$$
so that *the tensor of* $\phi'' s$ *is the reciprocal of the radius of absolute curvature at the point s.*

283.] Thus, if $\overline{OP} = \phi s$ be the vector of any point P of the curve, and if C be the centre of curvature at P, we have
$$\overline{PC} = -\frac{1}{\phi'' s};$$
and thus $$\overline{OC} = \phi s - \frac{1}{\phi'' s}$$
is the equation of the locus of the centre of curvature.

Hence also $$V.\phi' s\,\phi'' s \quad \text{or} \quad \phi' s\,\phi'' s$$
is the vector perpendicular to the osculating plane; and
$$T\frac{d}{ds}(\phi' s\,U\phi'' s)$$
is the *tortuosity* of the given curve, or the rate of rotation of its osculating plane per unit of length.

284.] As an example of the use of these expressions let us *find the curve whose curvature and tortuosity are both constant.*

We have $$\text{curvature} = T\phi'' s = T\rho'' = c.$$
Hence $$\phi' s\,\phi'' s = \rho'\rho'' = ca,$$
where a is a unit vector perpendicular to the osculating plane. This gives
$$\rho'\rho''' + \rho''^2 = c\,\mathfrak{L}\frac{\delta a}{\delta s} - cc_1 U\rho'' = c_1\rho'',$$
if c_1 represent the tortuosity.

Integrating we get $$\rho'\rho'' = c_1\rho' + \beta, \quad \ldots\ldots\ldots\ldots\ldots\ldots\ldots\ldots\ldots\ (1)$$
where β is a constant vector. Squaring both sides of this equation, we get
$$\begin{aligned} c^2 &= c_1^2 - \beta^2 - 2c_1 S\beta\rho' \\ &= -c_1^2 - \beta^2 \end{aligned}$$
(for by operating with $S.\rho'$ upon (1) we get $+c_1 = S\beta\rho'$),
or $$T\beta = \sqrt{c^2 + c_1^2}.$$

Multiply (1) by ρ', remembering that

$$T\rho' = 1,$$

and we obtain $$-\rho'' = -c_1 + \rho'\beta,$$

or, by integration, $$\rho' = c_1 s - \rho\beta + a, \quad \text{.......................} \quad (2)$$

where a is a constant quaternion. Eliminating ρ', we have

$$\rho'' = -c_1 + c_1 s\beta - \rho\beta^2 + a\beta,$$

of which the vector part is

$$\rho'' - \rho\beta^2 = -c_1 s\beta - Va\beta.$$

The complete integral of this equation is evidently

$$\rho = \xi\cos.sT\beta + \eta\sin.sT\beta - \frac{1}{T\beta^2}(c_1 s\beta + Va\beta), \quad \text{...........} \quad (3)$$

ξ and η being any two constant vectors. We have also by (2),

$$S\beta\rho = c_1 s + Sa,$$

which requires that $S\beta\xi = 0, \quad S\beta\eta = 0.$

The farther test, that $T\rho' = 1$, gives us

$$-1 = T\beta^2(\xi^2\sin^2.sT\beta + \eta^2\cos^2.sT\beta - 2S\xi\eta\sin.sT\beta\cos.sT\beta) - \frac{c_1^2}{c^2 + c_1^2}.$$

This requires, of course,

$$S\xi\eta = 0, \qquad T\xi = T\eta = \frac{c}{c^2 + c_1^2},$$

so that (3) becomes the general equation of a helix traced on a right cylinder. (Compare § 31 (m).)

285.] The vector perpendicular from the origin on the tangent to the curve $$\rho = \phi s$$

is, of course, $$\frac{1}{\rho'}V\rho'\rho, \quad \text{or} \quad \rho' V\rho\rho'$$

(since ρ' is a unit vector).

To find a common property of curves whose tangents are all equidistant from the origin.

Here $$TV\rho\rho' = c,$$

which may be written $$-\rho^2 - S^2\rho\rho' = c^2. \quad \text{.......................} \quad (1)$$

This equation shews that, as is otherwise evident, *every curve on a sphere whose centre is the origin* satisfies the condition. For obviously $-\rho^2 = c^2$ gives $S\rho\rho' = 0$,

and these satisfy (1).

If $S\rho\rho'$ does not vanish, the integral of (1) is

$$\sqrt{T\rho^2 - c^2} = s, \quad \text{.........................} \quad (2)$$

an arbitrary constant not being necessary, as we may measure s from any point of the curve. The equation of an involute which commences at this assumed point is

$$\varpi = \rho - s\rho'.$$

This gives
$$
\begin{aligned}
T\varpi^2 &= T\rho^2 + s^2 + 2sS\rho\rho' \\
&= T\rho^2 + s^2 - 2s\sqrt{T\rho^2 - c^2}, \quad \text{by (1)}, \\
&= c^2, \quad \text{by (2)}.
\end{aligned}
$$
This includes *all curves whose involutes lie on a sphere about the origin.*

286.] *Find the locus of the foot of the perpendicular drawn to a tangent to a right helix from a point in the axis.*

The equation of the helix is
$$\rho = \alpha \cos\frac{s}{a} + \beta \sin\frac{s}{a} + \gamma s,$$
where the vectors α, β, γ are at right angles to each other, and
$$T\alpha = T\beta = b, \quad \text{while} \quad aT\gamma = \sqrt{a^2 - b^2}.$$

The equation of the required locus is, by last section,
$$
\begin{aligned}
\varpi &= \rho' V\rho\rho' \\
&= \alpha\left(\cos\frac{s}{a} + \frac{a^2-b^2}{a^3} s \sin\frac{s}{a}\right) + \beta\left(\sin\frac{s}{a} - \frac{a^2-b^2}{a^3} s\cos\frac{s}{a}\right) + \gamma\frac{b^2}{a^2}s.
\end{aligned}
$$
This curve lies on the hyperboloid whose equation is
$$S^2\alpha\varpi + S^2\beta\varpi - a^2 S^2\gamma\varpi = b^4,$$
as the reader may easily prove for himself.

287.] *To find the least distance between consecutive tangents to a tortuous curve.*

Let one tangent be $\quad \varpi = \rho + x\rho'$,

then a consecutive one, at a distance δs along the curve, is
$$\varpi = \rho + \rho'\delta s + \rho''\frac{\delta s^2}{1.2} + \&\text{c.} + y\left(\rho' + \rho''\delta s + \rho'''\frac{\delta s^2}{1.2} + \ldots\right).$$
The magnitude of the least distance between these lines is, by §§ 203, 210,
$$
\begin{aligned}
S.&\left(\rho'\delta s + \rho''\frac{\delta s^2}{1.2} + \rho'''\frac{\delta s^3}{1.2.3} + \ldots\right)UV.\rho'\left(\rho' + \rho''\delta s + \rho'''\frac{\delta s^2}{1.2} + \ldots\right) \\
&= \frac{-\dfrac{\delta s^4}{12}S.\rho'\rho''\rho'''}{TV\rho'\rho''\delta s},
\end{aligned}
$$
if we neglect terms of higher orders.

It may be written, since $\rho'\rho''$ is a vector, and $T\rho' = 1$,
$$\frac{\delta s^3}{12}S.U\rho''V\rho'\rho'''.$$

But (§ 133 (2)) $\quad \dfrac{\delta UV\rho'\rho''}{UV\rho'\rho''} = V.\dfrac{V\rho'\rho'''}{V\rho'\rho''}\delta s - \dfrac{\delta s}{\rho''^2}\rho' S.\rho'\rho''\rho'''.$

Hence $\quad \dfrac{\delta s}{T\rho''}S.U\rho''V\rho'\rho'''$

is the small angle, $\delta\phi$, between the two successive positions of the osculating plane. [See also § 283.]

Thus the shortest distance between two consecutive tangents is expressed by the formula
$$\frac{\delta\phi\,\delta s^2}{12r},$$
where $r, = \frac{1}{T\rho''}$, is the radius of absolute curvature of the tortuous curve.

288.] Let us recur for a moment to the equation of the parabola (§ 31 (*f*.))
$$\rho = \alpha t + \frac{\beta t^2}{2}.$$

Here
$$\rho' = (\alpha + \beta t)\frac{dt}{ds},$$
whence, if we assume $S\alpha\beta = 0$,
$$\frac{ds}{dt} = \sqrt{-\alpha^2 - \beta^2 t^2},$$
from which the length of the arc of the curve can be derived in terms of t by integration.

Again,
$$\rho'' = (\alpha + \beta t)\frac{d^2 t}{ds^2} + \beta\left(\frac{dt}{ds}\right).$$

But
$$\frac{d^2 t}{ds^2} = \frac{d}{ds}\cdot\frac{1}{T(\alpha + \beta t)} = +\frac{dt}{ds}\frac{S.\beta(\alpha + \beta t)}{T(\alpha + \beta t)^3}.$$

Hence
$$\rho'' = -\frac{(\alpha + \beta t)V\alpha\beta}{T(\alpha + \beta t)^4},$$
and therefore, for the vector of the centre of curvature we have (§ 283),
$$\varpi = \alpha t + \frac{\beta t^2}{2} - (\alpha^2 + \beta^2 t^2)^2(-\beta\alpha^2 + \alpha\beta^2 t)^{-1},$$
$$= \beta\left(\frac{3t^2}{2} + \frac{\alpha^2}{\beta^2}\right) - \alpha\frac{t^3\beta^2}{\alpha^2};$$
which is the quaternion equation of the evolute.

289.] One of the simplest forms of the equation of a tortuous curve is
$$\rho = \alpha t + \frac{\beta t^2}{2} + \frac{\gamma t^3}{6},$$
where α, β, γ are any three non-coplanar vectors, and the numerical factors are introduced for convenience. This curve lies on a parabolic cylinder whose generating lines are parallel to γ; and also on cylinders whose bases are a cubical and a semi-cubical parabola, their generating lines being parallel to β and α respectively. We have by the equation of the curve
$$\rho' = \left(\alpha + \beta t + \frac{\gamma t^2}{2}\right)\frac{dt}{ds},$$

from which, by $T\rho'=1$, the length of the curve can be found in terms of t; and

$$\rho''=\left(a+\beta t+\frac{\gamma t^2}{2}\right)\frac{d^2t}{ds^2}+(\beta+\gamma t)\left(\frac{dt}{ds}\right)^2,$$

from which ρ'' can be expressed in terms of s. The investigation of various properties of this curve is very easy, and will be of great use to the student.

[*Note.*—It is to be observed that in this equation t cannot stand for s, the length of the curve. It is a good exercise for the student to shew that such an equation as

$$\rho=as+\beta s^2+\gamma s^3,$$

or even the simpler form

$$\rho=as+\beta s^2,$$

involves an absurdity.]

290.] The equation $\rho=\phi^t\epsilon,$

where ϕ is a given self-conjugate linear and vector function, t a scalar variable, and ϵ an arbitrary vector constant, belongs to a curious class of curves.

We have at once $\frac{d\rho}{dt}=\phi^t\log\phi\epsilon,$

where $\log\phi$ is another self-conjugate linear and vector function, which we may denote by χ. These functions are obviously commutative, as they have the same principal set of rectangular vectors, hence we may write

$$\frac{d\rho}{dt}=\chi\rho,$$

which of course gives $\frac{d^2\rho}{dt^2}=\chi^2\rho$, &c.,

since χ does not involve t.

As a verification, we should have

$$\begin{aligned}\phi^{t+\delta t}\epsilon &= \rho+\frac{d\rho}{dt}\delta t+\frac{d^2\rho}{dt^2}\,\frac{\delta t^2}{1.2}+\&c.\\ &-\left(1+\delta t\chi+\frac{\delta t^2}{1.2}\chi^2+\ldots\ldots\right)\rho\\ &-e^{\delta t\chi}\rho,\end{aligned}$$

where e is the base of Napier's Logarithms.

This is obviously true if $\phi^{\delta t}=e^{\delta t\chi},$

or $\phi=e^\chi,$

or $\log\phi=\chi,$

which is our assumption.

[The above process is, at first sight, rather startling, but the student may easily verify it by writing, in accordance with the results of Chapter V,

$$\phi\epsilon = -g_1 \alpha S\alpha\epsilon - g_2 \beta S\beta\epsilon - g_3 \gamma S\gamma\epsilon,$$

whence
$$\phi^t\epsilon = -g_1^t \alpha S\alpha\epsilon - g_2^t \beta S\beta\epsilon - g_3^t \gamma S\gamma\epsilon.$$

He will find at once

$$\chi\epsilon = -\log g_1 \alpha S\alpha\epsilon - \log g_2 \beta S\beta\epsilon - \log g_3 \gamma S\gamma\epsilon,$$

and the results just given follow immediately.]

291.] That the equation

$$\rho = \phi(t, u) = \Sigma . \alpha f(t, u)$$

represents a surface is obvious from the fact that it becomes the equation of a definite curve whenever *either* t or u has a particular value assigned to it. Hence the equation at once furnishes us with two systems of curves, lying wholly on the surface, and such that one of each system can, in general, be drawn through any assigned point on the surface. Tangents drawn to these curves at their point of intersection must, of course, lie in the tangent plane, whose equation we have thus the means of forming.

292.] By the equation we have

$$d\rho = \left(\frac{d\phi}{dt}\right) dt + \left(\frac{d\phi}{du}\right) du,$$

where the brackets are inserted to indicate partial differential coefficients. If we write this as

$$d\rho = \phi'_t \, dt + \phi'_u \, du,$$

the normal to the tangent plane is evidently

$$V\phi'_t \phi'_u,$$

and the equation of that plane

$$S.(\varpi - \phi)\phi'_t \phi'_u = 0.$$

293.] As a simple example, suppose a straight line to move along a fixed straight line, remaining always perpendicular to it, while rotating about it through an angle proportional to the space it has advanced; the equation of the ruled surface described will evidently be

$$\rho = \alpha t + u(\beta \cos t + \gamma \sin t), \quad \ldots\ldots\ldots\ldots\ldots\ldots\ldots \quad (1)$$

where α, β, γ are rectangular vectors, and

$$T\beta = T\gamma.$$

This surface evidently intersects the right cylinder

$$\rho = a(\beta \cos t + \gamma \sin t) + v\alpha,$$

in a helix (§§ 31 (m), 284) whose equation is

$$\rho = \alpha t + a(\beta \cos t + \gamma \sin t).$$

These equations illustrate very well the remarks made in §§ 31 (l), 291 as to the curves or surfaces represented by a vector equation according as it contains one or two scalar variables.

From (1) we have

$$d\rho = [\alpha - u(\beta \sin t - \gamma \cos t)]\, dt + (\beta \cos t + \gamma \sin t)\, du,$$

so that the normal at the extremity of ρ is

$$T\alpha(\gamma \cos t - \beta \sin t) - uT\beta^2 U\alpha.$$

Hence, as we proceed along a generating line of the surface, for which t is constant, we see that the direction of the normal changes. This, of course, proves that the surface is not developable.

294.] Hence the criterion for a developable surface is that if it be expressed by an equation of the form

$$\rho = \phi t + u\psi t,$$

where ϕt and ψt are vector functions, we must have the *direction* of the normal $\quad V\{\phi' t + u\psi' t\}\,\psi t$

independent of u.

This requires either $\quad V\psi t \psi' t = 0,$

which would reduce the surface to a cylinder, all the generating lines being parallel to each other; or

$$V\phi' t \psi t = 0.$$

This is the criterion we seek, and it shews that we may write, for a developable surface in general, the equation

$$\rho = \phi t + u\phi' t. \quad \ldots\ldots\ldots\ldots\ldots\ldots \quad (1)$$

Evidently $\quad \rho = \phi t$

is a curve (generally tortuous) and $\phi' t$ is a tangent vector. Hence *a developable surface is the locus of all tangent lines to a tortuous curve.*

Of course the tangent plane to the surface is the osculating plane at the corresponding point of the curve; and this is indicated by the fact that the normal to (1) is parallel to

$$V\phi' t \phi'' t. \quad \text{(See § 283.)}$$

To find the form of the section of the surface made by a normal plane through a point in the curve.

The equation of the surface is

$$\varpi = \rho + s\rho' + \frac{s^2}{2}\rho'' + \&c. + x(\rho' + s\rho'' + \&c.).$$

The part of $\varpi - \rho$ which is parallel to ρ' is

$$-\rho' S(\varpi - \rho)\rho' = -\rho'\left(-(s+x) - \rho''^2\left(\frac{s^3}{6} + \frac{xs^2}{2}\right) + \ldots\right);$$

therefore $\quad \varpi - \rho = A\rho' + \left(\frac{s^2}{2} + xs\right)\rho'' - \left(\frac{s^3}{6} + \frac{xs^2}{2}\right)\rho' V\rho'\rho''' + \ldots.$

And, when $A=0$, i.e. in the normal section, we have approximately

$$x = -s,$$

so that
$$\varpi - \rho = -\frac{s^2}{2}\rho'' - \frac{s^3}{3}\rho' V\rho'\rho'''.$$

Hence the curve has an equation of the form

$$\sigma = s^2\alpha + s^3\beta,$$

a semicubical parabola.

295.] A *Geodetic* line is a curve drawn on a surface so that its osculating plane at any point contains the normal to the surface. Hence, if ν be the normal at the extremity of ρ, ρ' and ρ'' the first and second differentials of the vector of the geodetic,

$$S.\nu\rho'\rho'' = 0,$$

which may be easily transformed into

$$V.\nu dU\rho' = 0.$$

296.] In the sphere $T\rho = a$ we have

$$\nu \parallel \rho,$$

hence
$$S.\rho\rho'\rho'' = 0,$$

which shews of course that ρ is confined to a plane passing through the origin, the centre of the sphere.

For a formal proof, we may proceed as follows—

The above equation is equivalent to the three

$$S\theta\rho = 0, \qquad S\theta\rho' = 0, \qquad S\theta\rho'' = 0,$$

from which we see at once that θ is a constant vector, and therefore the first expression, which includes the others, is the complete integral.

Or we may proceed thus—

$$0 = -\rho S.\rho\rho'\rho'' + \rho'' S.\rho^2\rho' = V.V\rho\rho' V\rho\rho'' = V.V\rho\rho' dV\rho\rho',$$

whence by § 133 (2) we have at once

$$UV\rho\rho' = \text{const.} = \theta \text{ suppose},$$

which gives the same results as before.

297.] In any cone we have, of course,

$$S\nu\rho = 0,$$

since ρ lies in the tangent plane. But we have also

$$S\nu\rho' = 0.$$

Hence, by the general equation of § 295, eliminating ν we get

$$0 = S.\rho\rho' V\rho'\rho'' = S\rho dU\rho' \quad \text{by § 133 (2)}.$$

Integrating
$$C = S\rho U\rho' - \int S d\rho U\rho' = S\rho U\rho' + \int T d\rho.$$

The interpretation of this is, that the length of any arc of the geodetic is equal to the projection of the side of the cone (drawn to its extremity) upon the tangent to the geodetic. In other words, ***when the cone is developed the geodetic becomes a straight line.*** A similar result may easily be obtained for the geodetic lines on any developable surface whatever.

298.] *To find the shortest line connecting two points on a given surface.*

Here $\int Td\rho$ is to be a minimum, subject to the condition that $d\rho$ lies in the given surface.

Now
$$\delta\int Td\rho = \int \delta Td\rho = -\int \frac{Sd\rho\, d\delta\rho}{Td\rho} = -\int S.Ud\rho d\delta\rho$$
$$= -[S.Ud\rho\delta\rho] + \int S.\delta\rho\, dUd\rho,$$
where the term in brackets vanishes at the limits, as the extreme points are fixed, and therefore $\delta\rho = 0$.

Hence our only conditions are

$$\int S.\delta\rho dUd\rho = 0, \quad \text{and} \quad S\nu\delta\rho = 0, \quad \text{giving}$$
$$V.\nu dUd\rho = 0, \quad \text{as in § 295.}$$

If the extremities of the curve are not given, but are to lie on given curves, we must refer to the integrated portion of the expression for the variation of the length of the arc. And its form
$$S.Ud\rho\,\delta\rho$$
shews that the shortest line cuts each of the given curves at right angles.

299.] The osculating plane of the curve
$$\rho = \phi t$$
is
$$S.\phi' t\phi'' t(\varpi - \rho) = 0, \quad \text{.......................... (1)}$$
and is, of course, the tangent plane to the surface
$$\rho = \phi t + u\phi' t. \quad \text{.......................... (2)}$$
Let us attempt the converse of the process we have, so far, pursued, and endeavour to find (2) as the envelop of the variable plane (1).

Differentiating (1) with respect to t only, we have
$$S.\phi'\phi'''(\varpi - \rho) = 0.$$
By this equation, combined with (1), we have
$$\varpi - \rho \parallel V.V\phi'\phi'' V\phi'\phi''' \parallel \phi',$$
or
$$\varpi = \rho + u\phi' = \phi + u\phi',$$
which is equation (2).

300.] This leads us to the consideration of envelops generally, and the process just employed may easily be extended to the problem of *finding the envelop of a series of surfaces whose equation contains one scalar parameter.*

When the given equation is a scalar one, the process of finding the envelop is precisely the same as that employed in ordinary Cartesian geometry, though the work is often shorter and simpler.

If the equation be given in the form

$$\rho = \psi(t, u, v),$$

where ψ is a vector function, t and u the scalar variables for any one surface, v the scalar parameter, we have for a proximate surface

$$\rho_1 = \psi(t_1, u_1, v_1) = \rho + \psi'_t \delta t + \psi'_u \delta u + \psi'_v \delta v.$$

Hence at all points on the intersection of two successive surfaces of the series we have

$$\psi'_t \delta t + \psi'_u \delta u + \psi'_v \delta v = 0,$$

which is equivalent to the following scalar equation connecting the quantities t, u, and v;

$$S.\psi'_t \psi'_u \psi'_v = 0.$$

This equation, along with

$$\rho = \psi(t, u, v),$$

enables us to eliminate t, u, v, and the resulting scalar equation is that of the required envelop.

301.] As an example, let us find the envelop of the osculating plane of a tortuous curve. Here the equation of the plane is (§ 299),

$$S.(\varpi - \rho)\phi' t \phi'' t = 0,$$

or $\varpi = \phi t + x\phi' t + y\phi'' t = \psi(x, y, t),$

if $\rho = \phi t$

be the equation of the curve.

Our condition is, by last section,

$$S.\psi'_x \psi'_y \psi'_t = 0,$$

or $S.\phi' t \phi'' t[\phi' t + x\phi'' t + y\phi''' t] = 0,$

or $yS.\phi' t \phi'' t \phi''' t = 0.$

Now the second factor cannot vanish, unless the given curve be plane, so that we must have

$$y = 0,$$

and the envelop is $\varpi = \phi t + x\phi' t$

the developable surface, of which the given curve is the edge of regression, as in § 299.

302.] When the equation contains two scalar parameters its differential coefficients with respect to them must vanish, and we have thus three equations from which to eliminate two numerical quantities.

A very common form in which these two parameters appear in quaternions is that of an unknown unit-vector. In this case the problem may be thus stated—*Find the envelop of the surface whose scalar equation is*

$$F(\rho, \alpha) = 0,$$

where α is subject to the one condition

$$T\alpha = 1.$$

Differentiating with respect to α alone, we have

$$S\nu d\alpha = 0, \qquad S\alpha d\alpha = 0,$$

where ν is a known vector function of ρ and α. Since $d\alpha$ may have any of an infinite number of values, these equations shew that

$$V\alpha\nu = 0.$$

This is equivalent to two scalar conditions only, and these, in addition to the two given scalar equations, enable us to eliminate α.

With the brief explanation we have given, and the examples which follow, the student will easily see how to deal with any other set of data he may meet with in a question of envelops.

303.] *Find the envelop of a plane whose distance from the origin is constant.*

Here $$S\alpha\rho = -c,$$
with the condition $$T\alpha = 1.$$
Hence, by last section, $$V\rho\alpha = 0,$$
and therefore $$\rho = c\alpha,$$
or $$T\rho = c,$$
the sphere of radius c, as was to be expected.

If we seek the *envelop of those only of the planes which are parallel to a given vector* β, we have the additional relation

$$S\alpha\beta = 0.$$

In this case the three differentiated equations are

$$S\rho d\alpha = 0, \qquad S\alpha d\alpha = 0, \qquad S\beta d\alpha = 0,$$

and they give $$S.\alpha\beta\rho = 0.$$
Hence $$\alpha = U.\beta V\beta\rho,$$
and the envelop is $$TV\beta\rho = cT\beta,$$
the circular cylinder of radius c and axis coinciding with β.

By putting $S\alpha\beta = e$, where e is a constant different from zero, we pick out all the planes of the series which have a definite inclination to β, and of course get as their envelop a right cone.

304.] The equation $$S^2\alpha\rho + 2S.\alpha\beta\rho = b$$
represents a parabolic cylinder, whose generating lines are parallel to the vector $\alpha V\alpha\beta$. For the equation is of the second degree, and

is not altered by increasing ρ by the vector $xaVa\beta$; also the surface cuts planes perpendicular to a in one line, and planes perpendicular to $Va\beta$ in two parallel lines. Its form and position of course depend upon the values of a, β, and b. *It is required to find its envelop* if β and b be constant, and a be subject to the one scalar condition

$$Ta = 1.$$

The process of § 302 gives, by inspection,

$$\rho Sa\rho + V\beta\rho = xa.$$

Operating by $S.a$, we get

$$S^2 a\rho + S.a\beta\rho = -x,$$

which gives $$S.a\beta\rho = x + b.$$

But, by operating successively by $S.V\beta\rho$ and by $S.\rho$, we have

$$(V\beta\rho)^2 = xS.a\beta\rho,$$

and $$(\rho^2 - x) Sa\rho = 0.$$

Omitting, for the present, the factor $Sa\rho$, these three equations give, by elimination of x and a,

$$(V\beta\rho)^2 = \rho^2 (\rho^2 + b),$$

which is the equation of the envelop required.

This is evidently a surface of revolution of the fourth order whose axis is β; but, to get a clearer idea of its nature, put

$$c^2 \rho^{-1} = \varpi,$$

and the equation becomes $$(V\beta\varpi)^2 = c^4 + b\varpi^2,$$

which is obviously a surface of revolution of the second degree, referred to its centre. Hence the required envelop is the *reciprocal* of such a surface, in the sense that *the rectangle under the lengths of condirectional radii of the two is constant.*

We have a curious particular case if the constants are so related that $$b + \beta^2 = 0,$$

for then the envelop breaks up into the two equal spheres, touching each other at the origin, $$\rho^2 = + S\beta\rho,$$

while the corresponding surface of the second order becomes the two parallel planes $$S\beta\varpi = + c^2.$$

305.] The particular solution above met with, viz.

$$Sa\rho = 0,$$

limits the original problem, which now becomes one of finding the envelop of a line instead of a surface. In fact this equation, taken in conjunction with that of the parabolic cylinder, belongs to that generating line of the cylinder which is the locus of the vertices of the principal parabolic sections.

Our equations become $2S.\alpha\beta\rho = b,$

$$S\alpha\rho = 0,$$

$$T\alpha = 1;$$

whence $V\beta\rho = x\alpha,$ giving

$$x = -S.\alpha\beta\rho = \frac{b}{2},$$

and thence $TV\beta\rho = \frac{b}{2};$

so that the envelop is a circular cylinder whose axis is β. [It is to be remarked that the equations above require that

$$S\alpha\beta = 0$$

so that the problem now solved is merely that of *the envelop of a parabolic cylinder which rotates about its focal line.* This discussion has been entered into merely for the sake of explaining a peculiarity in a former result, because of course the present results can be obtained immediately by an exceedingly simple process.]

306.] The equation $S\alpha\rho\, S.\alpha\beta\rho = a^2,$
with the condition $T\alpha = 1,$
represents a series of hyperbolic cylinders. *It is required to find their envelop.*

As before, we have $\rho S.\alpha\beta\rho + V\beta\rho\, S\alpha\rho = x\alpha,$
which by operating by $S.\alpha$, $S.\rho$, and $S.V\beta\rho$, gives

$$2a^2 = -x,$$

$$\rho^2 S.\alpha\beta\rho = x\, S\alpha\rho,$$

$$(V\beta\rho)^2 S\alpha\rho = x S.\alpha\beta\rho.$$

Eliminating α and x we have, as the equation of the envelop,

$$\rho^2 (V\beta\rho)^2 = 4a^4.$$

Comparing this with the equations

$$\rho^2 = -2a^2,$$

and $(V\beta\rho)^2 = -2a^2,$

which represent a sphere and one of its circumscribing cylinders, we see that, if condirectional radii of the three surfaces be drawn from the origin, that of the new surface is a geometric mean between those of the two others.

307.] *Find the envelop of all spheres which touch one given line and have their centres in another.*

Let $\rho = \beta + y\gamma$

be the line touched by all the spheres, and let $x\alpha$ be the vector of the centre of any one of them, the equation is (by § 200, or § 201)

$$\gamma^2 (\rho - x\alpha)^2 = -(V.\gamma(\beta - x\alpha))^2$$

or, putting for simplicity, but without loss of generality,

$$T\gamma = 1, \qquad Sa\beta = 0, \qquad S\beta\gamma = 0,$$

so that β is the least vector distance between the given lines,

$$(\rho - xa)^2 = (\beta - xa)^2 + x^2 S^2 a\gamma,$$

and, finally, $$\rho^2 - \beta^2 - 2xSa\rho = x^2 S^2 a\gamma.$$

Hence, by § 300, $$-2Sa\rho = 2xS^2 a\gamma.$$

[This gives no definite envelop if

$$Sa\gamma = 0,$$

i.e. if the line of centres is perpendicular to the line touched by all the spheres.]

Eliminating x, we have for the equation of the envelop

$$S^2 a\rho + S^2 a\gamma(\rho^2 - \beta^2) = 0,$$

which denotes a surface of revolution of the second degree, whose axis is a.

Since, from the form of the equation, $T\rho$ may have any magnitude not less than $T\beta$, and since the section by the plane

$$Sa\rho = 0$$

is a real circle, on the sphere

$$\rho^2 - \beta^2 = 0,$$

the surface is a hyperboloid of one sheet.

[It will be instructive to the student to find the signs of the values of g_1, g_2, g_3 as in § 165, and thence to prove the above conclusion.]

308.] As a final example let us find the envelop of the hyperbolic cylinder $$Sa\rho S\beta\rho - c = 0,$$

where the vectors a and β are subject to the conditions

$$Ta = T\beta = 1,$$

$$Sa\gamma = 0, \qquad S\beta\delta = 0,$$

γ and δ being given vectors.

[It will be easily seen that two of the six scalars involved in a, β still remain as variable parameters.]

We have $$Sada = 0, \qquad S\gamma da = 0,$$

so that $$da = xVa\gamma.$$

Similarly $$d\beta = yV\beta\delta.$$

But, by the equation of the cylinders,

$$Sa\rho S\rho d\beta + S\rho da S\beta\rho = 0,$$

or $$ySa\rho S.\beta\delta\rho + xS.a\gamma\rho S\beta\rho = 0.$$

Now by the nature of the given equation, neither $Sa\rho$ nor $S\beta\rho$ can vanish, so that the independence of da and $d\beta$ requires

$$S.a\gamma\rho = 0, \qquad S.\beta\delta\rho = 0.$$

Hence $\alpha = U.\gamma V\gamma\rho, \quad \beta = U.\delta V\delta\rho,$
and the envelop is $T.V\gamma\rho V\delta\rho - cT\gamma\delta = 0,$
a surface of the fourth order, which may be constructed by laying off mean proportionals between the lengths of condirectional radii of two equal right cylinders whose axes meet in the origin.

309.] We may now easily see the truth of the following general statement.

Suppose the given equation of the series of surfaces, whose envelop is required, to contain m vector, and n scalar, parameters; and that the latter are subject to p vector, and q scalar, conditions.

In all there are $3m+n$ scalar parameters, subject to $3p+q$ scalar conditions.

That there may be an envelop we must therefore in general have

$$(3m+n)-(3p+q) = 1, \quad \text{or} \quad = 2.$$

In the former case the enveloping surface is given as the locus of a series of *curves*, in the latter of a series of *points*.

Differentiation of the equations gives us $3p+q+1$ equations, linear and homogeneous in the $3m+n$ differentials of the scalar parameters, so that by the elimination of these we have *one* final scalar equation in the first case, *two* in the second; and thus in each case we have just equations enough to eliminate all the arbitrary parameters.

310.] *To find the locus of the foot of the perpendicular drawn from the origin to a tangent plane to any surface.*

If

$$S\nu d\rho = 0$$

be the differentiated equation of the surface, the equation of the tangent plane is

$$S(\varpi-\rho)\,\nu = 0.$$

We may introduce the condition

$$S\nu\rho = 1,$$

which in general alters the tensor of ν, so that ν^{-1} becomes the required vector perpendicular, as it satisfies the equation

$$S\varpi\nu = 1.$$

It remains that we eliminate ρ between the equation of the given surface, and the vector equation

$$\varpi = \nu^{-1}.$$

The result is the scalar equation (in ϖ) required.

For example, if the given surface be the ellipsoid

$$S\rho\phi\rho = 1,$$

we have

$$\varpi^{-1} = \nu = \phi\rho,$$

so that the required equation is

$$S\varpi^{-1}\phi^{-1}\varpi^{-1} - 1,$$

or
$$S\varpi\phi^{-1}\varpi = \varpi^4,$$

which is Fresnel's *Surface of Elasticity*. (§ 263)

It is well to remark that this equation is derived from that of the reciprocal ellipsoid
$$S\rho\phi^{-1}\rho = 1$$
by putting ϖ^{-1} for ρ.

311.] *To find the reciprocal of a given surface with respect to the unit sphere whose centre is the origin.*

With the condition
$$S\rho\nu = 1,$$
of last section, we see that $-\nu$ is the vector of the pole of the tangent plane
$$S(\varpi-\rho)\nu = 0.$$
Hence we must put
$$\varpi = -\nu,$$
and eliminate ρ by the help of the equation of the given surface.

Take the ellipsoid of last section, and we have

$$\varpi = -\phi\rho,$$

so that the reciprocal surface is represented by

$$S\varpi\phi^{-1}\varpi = 1.$$

It is obvious that the former ellipsoid can be reproduced from this by a second application of the process.

And the property is general, for

$$S\rho\nu = 1$$

gives, by differentiation, and attention to the condition

$$S\nu d\rho = 0,$$

the new relation
$$S\rho d\nu = 0,$$
so that ρ and ν are corresponding vectors of the two surfaces: either being that of the pole of a tangent plane drawn at the extremity of the other.

312.] If the given surface be a cone with its vertex at the origin, we have a peculiar case. For here every tangent plane passes through the origin, and therefore the required locus is wholly at an infinite distance. The difficulty consists in $S\rho\nu$ becoming in this case a numerical multiple of the quantity which is equated to zero in the equation of the cone, so that of course we cannot put as above

$$S\rho\nu = 1.$$

313.] The properties of the normal vector ν enable us to write the partial differential equations of families of surfaces in a very simple form.

Thus the distinguishing property of *Cylinders* is that all their generating lines are parallel. Hence all positions of ν must be parallel to a given plane—or

$$S\alpha\nu = 0,$$

which is the quaternion form of the well-known equation

$$l\frac{dF}{dx} + m\frac{dF}{dy} + n\frac{dF}{dz} = 0.$$

To integrate it, remember that we have always

$$S\nu d\rho = 0,$$

and that as ν is perpendicular to α it may be expressed in terms of any two vectors, β and γ, each perpendicular to α.

Hence $$\nu = x\beta + y\gamma,$$

and $$xS\beta d\rho + yS\gamma d\rho = 0.$$

This shews that $S\beta\rho$ and $S\gamma\rho$ are together constant or together variable, so that $$S\beta\rho = f(S\gamma\rho),$$

where f is any scalar function whatever.

314.] In *Surfaces of Revolution* the normal intersects the axis. Hence, taking the origin in the axis α, we have

$$S.\alpha\rho\nu = 0,$$

or $$\nu = x\alpha + y\rho.$$

Hence $$xS\alpha d\rho + yS\rho d\rho = 0,$$

whence the integral $$T\rho = f(S\alpha\rho).$$

The more common form, which is easily derived from that just written, is $$TV\alpha\rho = F(S\alpha\rho).$$

In *Cones* we have $$S\nu\rho = 0,$$

and therefore

$$S\nu d\rho = S.\nu(T\rho dU\rho + U\rho dT\rho) = T\rho S\nu dU\rho.$$

Hence $$S\nu dU\rho = 0,$$

so that ν must be a function of $U\rho$, and therefore the integral is

$$f(U\rho) = 0,$$

which simply expresses the fact that the equation does not involve the tensor of ρ, i. e. that in Cartesian cöordinates it is homogeneous.

315.] *If equal lengths be laid off on the normals drawn to any surface, the new surface formed by their extremities is normal to the same lines.*

For we have $$\varpi = \rho + aU\nu,$$

and $$S\nu d\varpi = S\nu d\rho + aS\nu dU\nu = 0,$$

which proves the proposition.

Take, for example, the surface

$$S\rho\phi\rho = 1;$$

the above equation becomes

$$\varpi = \rho + \frac{a\phi\rho}{T\phi\rho};$$

so that

$$\rho = \left(\frac{a\phi}{T\phi\rho} + 1\right)^{1} \varpi,$$

and the equation of the new surface is to be found by eliminating $\frac{a}{T\phi\rho}$ (written x) between the equations

$$1 = S.(x\phi + 1)^{-1}\varpi\phi(x\phi + 1)^{-1}\varpi,$$

and

$$-\frac{a^2}{x^2} = S.\phi(x\phi + 1)^{-1}\varpi\phi(x\phi + 1)^{-1}\varpi.$$

316.] It appears from last section that if one orthogonal surface can be drawn cutting a given system of straight lines, an indefinitely great number may be drawn: and that the portions of these lines intercepted between any two selected surfaces of the series are all equal.

Let

$$\rho = \sigma + x\tau,$$

where σ and τ are vector functions of ρ, and x is any scalar, be the general equation of a system of lines: we have

$$S\tau d\rho = 0 = S(\rho - \sigma)\, d\rho$$

as the differentiated equation of the series of orthogonal surfaces, if it exist. Hence the following problem.

317.] It is required *to find the criterion of integrability of the equation*

$$S\nu\, d\rho = 0 \quad \ldots\ldots\ldots\ldots\ldots\ldots\ldots\ldots\ldots\ldots \quad (1)$$

as the complete differential of the equation of a series of surfaces.

Hamilton has given (*Elements*, p. 702) an extremely elegant solution of this problem, by means of the properties of linear and vector functions. We adopt a different and somewhat less rapid process, on account of some results it offers which will be useful to us in the next Chapter; and also because it will shew the student the connection of our methods with those of ordinary differential equations.

If we assume

$$F\rho = C$$

to be the integral, and apply to it the very singular operator de vised by Hamilton,

$$\nabla = i\frac{d}{dx} + j\frac{d}{dy} + k\frac{d}{dz},$$

we have

$$\nabla F = i\frac{dF}{dx} + j\frac{dF}{dy} + k\frac{dF}{dz}.$$

But $\rho = ix + jy + kz,$
whence $d\rho = i\,dx + j\,dy + k\,dz,$
and $0 = dF = \frac{dF}{dx}dx + \frac{dF}{dy}dy + \frac{dF}{dz}dz = -Sd\rho\nabla F.$

Comparing with the given equation, we see that the latter represents a series of surfaces if ν, or a scalar multiple of it, can be ex pressed as ∇F.

If $\nu = \nabla F,$
we have
$$\nabla\nu = \nabla^2 F = -\left(\frac{d^2F}{dx^2} + \frac{d^2F}{dy^2} + \frac{d^2F}{dz^2}\right),$$
a well-known and most important expression, to which we shall return in next Chapter. Meanwhile we need only remark that the last-written quantities are necessarily scalars, so that the only requisite condition of the integrability of (1) is
$$V\nabla\nu = 0. \quad\quad (2)$$
If ν do not satisfy this criterion, it may when multiplied by a scalar. Hence the farther condition
$$V\nabla(w\nu) = 0,$$
which may be written
$$V\nu\nabla w - wV\nabla\nu = 0. \quad\quad (3)$$
This requires that
$$S\nu\nabla\nu = 0. \quad\quad (4)$$
If then (2) be not satisfied, we must try (4). If (4) be satisfied w will be found from (3); and in either case (1) is at once integrable.

[If we put $d\nu = \phi d\rho$
where ϕ is a linear and vector function, not necessarily self-conjugate, we have
$$V\nabla\nu = V\left(i\frac{d\nu}{dx} + \ldots\right) = V(i\phi i + \ldots) = -\epsilon,$$
by § 173. Thus, if ϕ be self-conjugate, $\epsilon = 0$, and the criterion (2) is satisfied. If ϕ be not self-conjugate we have by (4) for the criterion
$$S\epsilon\nu = 0.$$
These results accord with Hamilton's, lately referred to, but the mode of obtaining them is quite different from his.]

318.] As a simple example let us first take *lines diverging from a point*. Here $\nu \parallel \rho$, and we see that if $\nu = \rho$
$$\nabla\nu = -3,$$
so that (2) is satisfied. And the equation is
$$S\rho\,d\rho = 0,$$
whose integral
$$T\rho = C$$
gives a series of concentric spheres.

Lines perpendicular to, and intersecting, a fixed line.

If α be the fixed line, β any of the others, we have

$$S.\alpha\beta\rho = 0, \qquad S\alpha\beta = 0, \qquad S\beta d\rho = 0.$$

Here $$\nu \parallel \alpha V\alpha\rho,$$

and therefore equal to it, because (2) is satisfied.

Hence $$S.d\rho\alpha V\alpha\rho = 0,$$

or $$S.V\alpha\rho V\alpha d\rho = 0,$$

whose integral is the equation of a series of right cylinders

$$T^2 V\alpha\rho = C.$$

319.] *To find the orthogonal trajectories of a series of circles whose centres are in, and their planes perpendicular to, a given line.*

Let α be a unit-vector in the direction of the line, then one of the circles has the equations

$$\left.\begin{aligned} T\rho &= C, \\ S\alpha\rho &= C', \end{aligned}\right\}$$

where C and C' are any constant scalars whatever.

Hence, for the required surfaces

$$\nu \parallel d_1\rho \parallel V\alpha\rho,$$

where $d_1\rho$ is an element of one of the circles, ν the normal to the orthogonal surface. Now let $d\rho$ be an element of a tangent to the orthogonal surface, and we have

$$S\nu d\rho = S.\alpha\rho d\rho = 0.$$

This shews that $d\rho$ is in the same plane as α and ρ, i.e. that the orthogonal surfaces are planes passing through the common axis.

[To integrate the equation $S.\alpha\rho d\rho = 0$

evidently requires, by § 317, the introduction of a factor. For

$$\begin{aligned} V\nabla V\alpha\rho &= V(iV\alpha i + jV\alpha j + kV\alpha k) \\ &= 2\alpha, \end{aligned}$$

so that the first criterion is not satisfied. But

$$S.V\alpha\rho V\nabla V\alpha\rho = 2S.\alpha V\alpha\rho = 0,$$

so that the second criterion holds. It gives, by (3) of § 317,

$$V.\nabla w V\alpha\rho + 2w\alpha = 0,$$

or $$\rho S\alpha\nabla w - \alpha S\rho\nabla w + 2w\alpha = 0.$$

That is $$\left.\begin{aligned} S\alpha\nabla w &= 0, \\ S\rho\nabla w &= 2w. \end{aligned}\right\}$$

These equations are satisfied by

$$w = \frac{1}{V^2\alpha\rho}.$$

But a simpler mode of integration is easily seen. Our equation may be written

$$0 = S.\alpha V\frac{d\rho}{\rho} = S\alpha\frac{dU\rho}{U\rho} = d.S\alpha \log U\frac{\rho}{\beta},$$

which is immediately integrable, β being an arbitrary but constant vector.

As we have not introduced into this work the *logarithms* of versors, nor the corresponding *angles* of quaternions, we must refer to Hamilton's *Elements* for a farther development of this point.]

320.] *To find the orthogonal trajectories of a given series of surfaces.*

If the equation $$F\rho = C,$$
give $$S\nu d\rho = 0,$$
the equation of the orthogonal curves is
$$V\nu d\rho = 0.$$
This is equivalent to two scalar differential equations (§ 197), which, when the problem is possible, belong to surfaces on each of which the required lines lie. The finding of the requisite criterion we leave to the student.

Let the surfaces be concentric spheres.

Here $$\rho^2 = C,$$
and therefore $$V\rho d\rho = 0.$$
Hence $$T\rho^2 dU\rho = -U\rho V\rho d\rho = 0,$$
and the integral is $$U\rho = \text{constant},$$
denoting straight lines through the origin.

Let the surfaces be spheres touching each other at a common point.
The equation is (§ 218)
$$S\alpha\rho^{-1} = C,$$
whence $$V.\rho\alpha\rho d\rho = 0.$$
The integrals may be written
$$S.\alpha\beta\rho = 0, \qquad \rho^2 + hTV\alpha\rho = 0,$$
the first (β being any vector) is a plane through the common diameter; the second represents a series of rings or *tores* (§ 323) formed by the revolution, about α, of circles *touching* that line at the point common to the spheres.

Let the surfaces be similar, similarly situated, and concentric, surfaces of the second order.

Here $$S\rho\chi\rho = C,$$
therefore $$V\chi\rho d\rho = 0.$$
But, by § 290, the integral of this equation is
$$\rho = e^{t\chi}\epsilon$$
$$= \phi^t\epsilon,$$
where ϕ and χ are related to each other, as in § 290; and ϵ is any constant vector.

321.] *To integrate the linear partial differential equation of a family of surfaces.*

The equation (see § 313)

$$P\frac{du}{dx}+Q\frac{du}{dy}+R\frac{du}{dz}=0$$

may be put in the very simple form

$$S(\sigma\nabla)u=0,$$

if we write $\sigma=iP+jQ+kR,$

and $$\nabla=i\frac{d}{dx}+j\frac{d}{dy}+k\frac{d}{dz}.$$

This gives, at once, $\nabla u=mV\theta\sigma,$

where m is a scalar and θ a vector (in whose tensor m might have been included, but is kept separate for a special purpose). Hence

$$\begin{aligned} du &= -S(d\rho\nabla)u \\ &= -mS.\theta\sigma d\rho \\ &= -S.\theta d\tau, \end{aligned}$$

if we put $d\tau=mV.\sigma d\rho$

so that m is an integrating factor of $V.\sigma d\rho$. If a value of m can be found, it is obvious, from the form of the above equation, that θ must be a function of τ alone; and the integral is therefore

$$u=F(\tau)=\text{const.}$$

where F is an arbitrary scalar function.

Thus the differential equation of *Cylinders* is

$$S(\alpha\nabla)u=0,$$

where α is a constant vector. Here $m=1$, and

$$u=F(V\alpha\rho)=\text{const.}$$

That of *Cones* referred to the vertex is

$$S(\rho\nabla)u=0.$$

Here the expression to be made integrable is

$$V.\rho d\rho.$$

But Hamilton long ago shewed that (§ 133 (2))

$$\frac{dU\rho}{U\rho}=V\frac{d\rho}{\rho}=\frac{V.\rho d\rho}{(T\rho)^2},$$

which indicates the value of m, and gives

$$u=F(U\rho)=\text{const.}$$

It is obvious that the above is only one of a great number of different processes which may be applied to integrate the differential equation. It is quite easy, for instance, to pass from it to the assumption of a vector integrating factor instead of the scalar m, and to derive the usual criterion of integrability. There is no difficulty in modifying the process to suit the case when the right-hand member is a multiple of u. In fact it seems to throw a very clear light upon the whole subject of the integration of partial differential equations. If, instead of $S(\sigma\nabla)$, we employ other operators as $S(\sigma\nabla)S(\tau\nabla)$, $S.\sigma\nabla\tau\nabla$, &c. (where ∇ may or may not operate on u alone), we can pass to linear partial differential equations of the second and higher orders. Similar theorems can be obtained from vector operations, as $V(\sigma\nabla)$*

322.] *Find the general equation of surfaces described by a line which always meets, at right angles, a fixed line.*

If α be the fixed line, β and γ forming with it a rectangular unit system, then
$$\rho = x\alpha + y(\beta + z\gamma),$$
where y may have all values, but x and z are mutually dependent, is one form of the equation.

Another, expressing the arbitrary relation between x and z is
$$\frac{S\gamma\rho}{S\beta\rho} = f(S\alpha\rho).$$
But we may also write
$$\rho = \alpha F(x) + y\alpha^x\beta,$$
as it obviously expresses the same conditions.

The simplest case is when $F(x) = hx$. The surface is one which cuts, in a right helix, every cylinder which has α for its axis.

323.] *The centre of a sphere moves in a given circle, find the equation of the ring described.*

Let α be the unit-vector axis of the circle, its centre the origin, r its radius, a that of the sphere.

Then
$$(\rho - \beta)^2 = -a^2$$
is the equation of the sphere in any position, where
$$S\alpha\beta = 0, \qquad T\beta = r.$$
These give $S.\alpha\beta\rho = 0$, and β must now be eliminated. The result is that
$$\beta = r\alpha UV\alpha\rho,$$
giving
$$(\rho^2 - r^2 + a^2)^2 = 4r^2T^2V\alpha\rho,$$
$$= 4r^2(-\rho^2 - S^2\alpha\rho),$$
which is the required equation. It may easily be changed to
$$(\rho^2 - a^2 + r^2)^2 = -4a^2\rho^2 - 4r^2S^2\alpha\rho, \quad \ldots\ldots\ldots\ldots\ldots \quad (1)$$
and in this form it enables us to give an immediate proof of the very singular property of the ring (or *tore*) discovered by Villarceau.

* Tait, *Proc. R. S. E.*, 1869–70.

For the planes $S.\rho\left(a \pm \frac{a\beta}{r\sqrt{r^2-a^2}}\right) = 0,$

which together are represented by

$$r^2(r^2-a^2)S^2a\rho - a^2S^2\beta\rho = 0,$$

evidently pass through the origin and touch (and cut) the ring.

The latter equation may be written

$$r^2S^2a\rho - a^2(S^2a\rho + S^2\rho U\beta) = 0,$$

or $$r^2S^2a\rho + a^2(\rho^2 + S^2.a\rho U\beta) = 0. \quad\ldots\ldots\ldots\ldots\ldots\ldots (2)$$

The plane intersections of (1) and (2) lie obviously on the new surface

$$(\rho^2 - a^2 + r^2)^2 = 4a^2S^2.a\rho U\beta,$$

which consists of two spheres of radius r, as we see by writing its separate factors in the form

$$(\rho \pm a\,a U\beta)^2 + r^2 = 0.$$

324.] It may be instructive to work out this problem from a different point of view, especially as it affords excellent practice in transformations.

A circle revolves about an axis passing within it, the perpendicular from the centre on the axis lying in the plane of the circle: shew that, for a certain position of the axis, the same solid may be traced out by a circle revolving about an external axis in its own plane.

Let $a = \sqrt{b^2+c^2}$ be the radius of the circle, i the vector axis of rotation, $-ca$ (where $Ta = 1$) the vector perpendicular from the centre on the axis i, and let the vector

$$bi + cia$$

be perpendicular to the plane of the circle.

The equations of the circle are

$$\left.\begin{aligned}(\rho - ca)^2 + b^2 + c^2 &= 0,\\ S\left(i + \frac{c}{b}ia\right)\rho &= 0.\end{aligned}\right\}$$

Also $$\rho^2 = S^2i\rho + S^2a\rho + S^2.ia\rho,$$

$$- S^2i\rho + S^2a\rho + \frac{b^2}{c^2}S^2i\rho$$

by the second of the equations of the circle. But, by the first,

$$(\rho^2 + b^2)^2 = 4c^2S^2a\rho = -4(c^2\rho^2 + a^2S^2i\rho),$$

which is easily transformed into

$$(\rho^2 - b^2)^2 = -4a^2(\rho^2 + S^2i\rho),$$

or $$\rho^2 - b^2 = -2aTVi\rho.$$

If we put this in the forms

$$\rho^2 - b^2 = 2aS\beta\rho,$$

and $$(\rho - a\beta)^2 + c^2 = 0,$$

where β is a unit-vector perpendicular to i and in the plane of i and ρ, we see at once that the surface will be traced out by a circle of radius c, revolving about i, an axis in its own plane, distant a from its centre.

This problem is not well adapted to shew the gain in brevity and distinctness which generally follows the use of quaternions; as, from its very nature, it hints at the adoption of rectangular axes and scalar equations for its treatment, so that the solution we have given is but little different from an ordinary Cartesian one.

325.] *A surface is generated by a straight line which intersects two fixed lines: find the general equation.*

If the given lines intersect, there is no surface but the plane containing them.

Let then their equations be,

$$\rho = \alpha + x\beta, \qquad \rho = \alpha_1 + x_1\beta_1.$$

Hence every point of the surface satisfies the condition, § 30,

$$\rho = y(\alpha + x\beta) + (1-y)(\alpha_1 + x_1\beta_1). \quad \ldots\ldots\ldots\ldots\ldots\ldots \quad (1)$$

Obviously y may have any value whatever: so that to specify a particular surface we must have a relation between x and x_1. By the help of this, x_1 may be eliminated from (1), which then takes the usual form of the equation of a surface

$$\rho = \phi(x, y).$$

Or we may operate on (1) by $V.(\alpha + x\beta - \alpha_1 - x_1\beta_1)$, so that we get a vector equation equivalent to *two* scalar equations (§§ 98, 116), and not containing y. From this x and x_1 may easily be found in terms of ρ, and the general equation of the possible surfaces may be written

$$f(x, x_1) = 0,$$

where f is an arbitrary scalar function, and the values of x and x_1 are expressed in terms of ρ.

This process is obviously applicable if we have, instead of two straight lines, any two given curves through which the line must pass; and even when the tracing line is itself a given curve, situated in a given manner. But an example or two will make the whole process clear.

326.] *Suppose the moveable line to be restricted by the condition that it is always parallel to a fixed plane.*

Then, in addition to (1), we have the condition

$$S\gamma(\alpha_1 + x_1\beta_1 - \alpha - x\beta) = 0,$$

γ being a vector perpendicular to the fixed plane.

We lose no generality by assuming α and α_1, which are any vectors drawn from the origin to the fixed lines, to be each perpendicular to γ; for, if for instance we could not assume $S\gamma\alpha = 0$, it would follow that $S\gamma\beta = 0$, and the required surface would either be impossible, or would be a plane, cases which we need not consider. Hence

$$x_1 S\gamma\beta_1 - x S\gamma\beta = 0.$$

Eliminating x_1, by the help of this equation, from (1) of last section, we have

$$\rho = y(\alpha + x\beta) + (1-y)\left(\alpha_1 + x\beta_1 \frac{S\gamma\beta}{S\gamma\beta_1}\right).$$

Operating by any three non-coplanar vectors and with the characteristic S, we obtain three equations from which to eliminate x and y. Operating by $S.\gamma$ we find

$$S\gamma\rho = xS\beta\gamma.$$

Eliminating x by means of this, we have finally

$$S.\rho\left(\alpha + \frac{\beta S\gamma\rho}{S\beta\gamma}\right)\left(\alpha_1 + \frac{\beta_1 S\gamma\rho}{S\beta_1\gamma}\right) = 0,$$

which appears to be of the third order. It is really, however, only of the second order, since, in consequence of our assumptions, we have

$$V\alpha\alpha_1 \parallel \gamma,$$

and therefore $S\gamma\rho$ is a spurious factor of the left-hand side.

327.] *Let the fixed lines be perpendicular to each other, and let the moveable line pass through the circumference of a circle, whose centre is in the common perpendicular, and whose plane bisects that line at right angles.*

Here the equations of the fixed lines may be written

$$\rho = \alpha + x\beta, \qquad \rho = -\alpha + x_1\gamma,$$

where α, β, γ, form a rectangular system, and we may assume the two latter to be unit-vectors.

The circle has the equations

$$\rho^2 = -a^2, \qquad S\alpha\rho = 0.$$

Equation (1) of § 325 becomes

$$\rho = y(\alpha + x\beta) + (1-y)(-\alpha + x_1\gamma).$$

Hence $\qquad S\alpha^{-1}\rho = y - (1-y) = 0, \quad \text{or} \quad y = \frac{1}{2}.$

Also $\qquad \rho^2 = -a^2 = (2y-1)^2\alpha^2 - x^2y^2 - x_1^2(1-y)^2,$

$$\text{or} \qquad 4a^2 = (x^2 + x_1^2),$$

so that if we now suppose the tensors of β and γ to be each $2a$, we may put $x = \cos\theta$, $x_1 = \sin\theta$, from which

$$\rho = (2y-1)\alpha + y\beta\cos\theta + (1-y)\gamma\sin\theta;$$

and finally

$$\frac{S^2\beta\rho}{(1+S\alpha^{-1}\rho)^2} + \frac{S^2\gamma\rho}{(1-S\alpha^{-1}\rho)^2} = 4a^4.$$

For this very simple case the solution is not better than the ordinary Cartesian one; but the student will easily see that we may by very slight changes adapt the above to data far less symmetrical than those from which we started. Suppose, for instance, β and γ not to be at right angles to one another; and suppose the plane of the circle not to be parallel to their plane, &c., &c. But farther, operate on every line in space by the linear and vector function ϕ, and we distort the circle into an ellipse, the straight lines remaining straight. If we choose a form of ϕ whose principal axes are parallel to α, β, γ, the data will remain symmetrical, but not unless. This subject will be considered again in the next Chapter.

328.] *To find the curvature of a normal section of a central surface of the second order.*

In this, and the few similar investigations which follow, it will be simpler to employ infinitesimals than differentials; though for a thorough treatment of the subject the latter method, as may be seen in Hamilton's *Elements*, is preferable.

We have, of course, $S\rho\phi\rho = 1$,

and, if $\rho + \delta\rho$ be also a vector of the surface, we have rigorously, *whatever be the tensor of* $\delta\rho$,

$$S(\rho + \delta\rho)\phi(\rho + \delta\rho) = 1.$$

Hence $$2S\delta\rho\phi\rho + S\delta\rho\phi\delta\rho = 0. \qquad (1)$$

Now $\phi\rho$ is normal to the tangent plane at the extremity of ρ, so that if t denote the distance of the point $\rho + \delta\rho$ from that plane

$$t = -S\delta\rho U\phi\rho,$$

and (1) may therefore be written

$$2tT\phi\rho - T^2\delta\rho S.U\delta\rho\phi U\delta\rho = 0.$$

But the curvature of the section is evidently

$$\mathfrak{L}\frac{2t}{T^2\delta\rho},$$

or, by the last equation,

$$\frac{1}{T\phi\rho}\mathfrak{L}S.U\delta\rho\phi U\delta\rho.$$

In the limit, $\delta\rho$ is a vector in the tangent plane; let ϖ be the vector semidiameter of the surface which is parallel to it, and the equation of the surface gives $T^2\varpi S.U\varpi\phi U\varpi = 1$,

so that the curvature of the normal section, at the point ρ, in the direction of ϖ, is

$$\frac{1}{T\phi\rho T^2\varpi},$$

directly as the perpendicular from the centre on the tangent plane, and inversely as the square of the semidiameter parallel to the tangent line, a well-known theorem.

329.] By the help of the known properties of the central section parallel to the tangent plane, this theorem gives us all the ordinary properties of the directions of maximum and minimum curvature, their being at right angles to each other, the curvature in any normal section in terms of the chief curvatures and the inclination to their planes, &c., &c., without farther analysis. And when, in a future section, we shew how to find an *osculating* surface of the second order at any point of a given surface, the same properties will be at once established for surfaces in general. Meanwhile we may prove another curious property of the surfaces of the second order, which similar reasoning extends to all surfaces.

The equation of the normal at the point $\rho + \delta\rho$ in the surface treated in last section is

$$\varpi = \rho + \delta\rho + x\phi(\rho + \delta\rho). \quad \ldots\ldots\ldots\ldots\ldots\ldots \quad (1)$$

This intersects the normal at ρ if (§§ 203, 210)

$$S.\delta\rho\phi\rho\phi\delta\rho = 0,$$

that is, by the result of § 273, if $\delta\rho$ be parallel to the maximum or minimum diameter of the central section parallel to the tangent plane.

Let σ_1 and σ_2 be those diameters, then we may write in general

$$\delta\rho = p\sigma_1 + q\sigma_2,$$

where p and q are scalars, infinitely small.

If we draw through a point P in the normal at ρ a line parallel to σ_1, we may write its equation

$$\varpi = \rho + a\phi\rho + y\sigma_1.$$

The proximate normal (1) passes this line at a distance (see § 203)

$$S.(a\phi\rho - \delta\rho)UV\sigma_1\phi(\rho + \delta\rho),$$

or, neglecting terms of the second order,

$$\frac{1}{TV\sigma_1\phi\rho}(apS.\phi\rho\sigma_1\phi\sigma_1 + aqS.\phi\rho\sigma_1\phi\sigma_2 + qS.\sigma_1\sigma_2\phi\rho).$$

The first term in the bracket vanishes because σ_1 is a principal vector of the section parallel to the tangent plane, and thus the expression becomes

$$q\left(\frac{a}{T\sigma_2} - T\sigma_2\right).$$

Hence, if we take $a = T\sigma_2^2$, the distance of the normal from the new line is of the second order only. This makes the distance of P from the point of contact $T\phi\rho T\sigma_2^2$, i.e. the principal radius of curvature along the tangent line parallel to σ_2. That is, *the group of normals drawn near a point of a central surface of the second order pass ultimately through two lines each parallel to the tangent to one principal section, and passing through the centre of curvature of the other.* The student may form a notion of the nature of this proposition by considering a small square plate, with normals drawn at every point, to be slightly bent, but by different amounts, in planes perpendicular to its edges. The first bending will make all the normals pass through the axis of the cylinder of which the plate now forms part; the second bending will not sensibly disturb this arrangement, except by lengthening or shortening the line in which the normals meet, but it will make them meet also in the axis of the new cylinder, at right angles to the first. A small pencil of light, with its focal lines, presents this appearance, due to the fact that a series of rays originally normal to a surface remain normals to a surface after any number of reflections and refractions. (See § 315).

330.] To extend these theorems to surfaces in general, it is only necessary, as Hamilton has shewn, to prove that if we write

$$d\nu = \phi d\rho,$$

ϕ is a *self-conjugate* function; and then the properties of ϕ, as explained in preceding Chapters, are applicable to the question.

As the reader will easily see, this is merely another form of the investigation contained in § 317. But it is given here to shew what a number of very simple demonstrations may be given of almost all quaternion theorems.

The vector ν is defined by an equation of the form

$$df\rho = S\nu d\rho,$$

where f is a scalar function. Operating on this by another independent symbol of differentiation, δ, we have

$$\delta df\rho = S\delta\nu d\rho + S\nu\delta d\rho.$$

In the same way we have

$$d\delta f\rho = Sd\nu\delta\rho + S\nu d\delta\rho.$$

But, as d and δ are independent, the left-hand members of these equations, as well as the second terms on the right (if these exist at all), are equal, so that we have

$$Sd\nu\delta\rho = S\delta\nu d\rho.$$

This becomes, putting $\qquad d\nu = \phi d\rho,$

and therefore $\qquad \delta\nu = \phi\delta\rho,$

$$S\delta\rho\phi d\rho = Sd\rho\phi\delta\rho,$$

which proves the proposition.

331.] If we write the differential of the equation of a surface in the form
$$df\rho = 2S\nu d\rho,$$
then it is easy to see that
$$f(\rho+d\rho) = f\rho + 2S\nu d\rho + Sd\nu d\rho + \&c.,$$
the remaining terms containing as factors the third and higher powers of $Td\rho$. To the second order, then, we may write, except for certain singular points,
$$0 = 2S\nu d\rho + Sd\nu d\rho,$$
and, as before, (§ 328), the curvature of the normal section whose tangent line is $d\rho$ is
$$\frac{1}{T\nu} S \frac{d\nu}{d\rho}.$$

332.] The step taken in last section, although a very simple one, virtually implies that the first three terms of the expansion of $f(\rho+d\rho)$ are to be formed in accordance with Taylor's Theorem, whose applicability to the expansion of scalar functions of quaternions has not been proved in this work, (see § 135); we therefore give another investigation of the curvature of a normal section, employing for that purpose the formulae of § (282).

We have, treating $d\rho$ as an element of a curve,
$$S\nu d\rho = 0,$$
or, making s the independent variable,
$$S\nu\rho' = 0.$$
From this, by a second differentiation,
$$S\frac{d\nu}{ds}\rho' + S\nu\rho'' = 0.$$
The curvature is, therefore, since $\nu \parallel \rho''$ and $T\rho' = 1$,
$$T\rho'' = -\frac{1}{T\nu} S\frac{d\nu}{d\rho}\rho'^2 = \frac{1}{T\nu} S\frac{d\nu}{d\rho}, \text{ as before.}$$

333.] Since we have shewn that
$$d\nu = \phi d\rho$$
where ϕ is a self-conjugate linear and vector function, whose constants depend only upon the nature of the surface, and the position of the point of contact of the tangent plane; so long as we do not alter these we must consider ϕ as possessing the properties explained in Chapter V.

Hence, as the expression for $T\rho''$ does not involve the tensor of $d\rho$, we may put for $d\rho$ any unit-vector τ, subject of course to the condition
$$S\nu\tau = 0. \quad \ldots\ldots\ldots\ldots\ldots\ldots \quad (1)$$
And the curvature of the normal section whose tangent is τ is
$$\frac{1}{T\nu} S\frac{\phi\tau}{\tau} = -\frac{1}{T\nu} S\tau\phi\tau.$$

If we consider the central section of the surface of the second order

$$S\varpi\phi\varpi + Tv = 0,$$

made by the plane $$Sv\varpi = 0,$$

we see at once that the *curvature of the given surface along the normal section touched by* τ *is inversely as the square of the parallel radius in the auxiliary surface.* This, of course, includes Euler's and other well-known Theorems.

334.] *To find the directions of maximum and minimum curvature,* we have $$S\tau\phi\tau = \text{max. or min.}$$

with the conditions, $$Sv\tau = 0,$$

$$T\tau = 1.$$

By differentiation, as in § 273, we obtain the farther equation

$$S.v\tau\phi\tau = 0. \quad\ldots\ldots\ldots\ldots\ldots\ldots \quad (1)$$

If τ be one of the two required directions, $\tau' = \tau Uv$ is the other, for the last-written equation may be put in the form

$$S.\tau Uv\phi(v\tau Uv) = 0,$$

$$\text{i.e.} \qquad S.\tau'\phi(v\tau') = 0,$$

$$\text{or} \qquad S.v\tau'\phi\tau' = 0.$$

Hence the sections of greatest and least curvature are perpendicular to one another.

We easily obtain, as in § 273, the following equation

$$S.v(\phi + S\tau\phi\tau)^{-1}v = 0,$$

whose roots divided by Tv are the required curvatures.

335.] Before leaving this very brief introduction to a subject, an exhaustive treatment of which will be found in Hamilton's *Elements,* we may make a remark on equation (1) of last section

$$S.v\tau\phi\tau = 0,$$

or, as it may be written, by returning to the notation of § 333,

$$S.vd\rho dv = 0.$$

This is the general equation of lines of curvature. For, if we define a line of curvature on any surface as a line such that normals drawn at contiguous points in it intersect, then, $\delta\rho$ being an element of such a line, the normals

$$\varpi = \rho + xv \quad\text{and}\quad \varpi = \rho + \delta\rho + y(v + \delta v)$$

must intersect. This gives, by § 203, the condition

$$S.\delta\rho v\delta v = 0,$$

as above.

EXAMPLES TO CHAPTER IX.

1. Find the length of any arc of a curve drawn on a sphere so as to make a constant angle with a fixed diameter.

2. Shew that, if the normal plane of a curve always contains a fixed line, the curve is a circle.

3. Find the radius of spherical curvature of the curve

$$\rho = \phi t.$$

Also find the equation of the locus of the centre of spherical curvature.

4. (Hamilton, *Bishop Law's Premium Examination*, 1854.)

(*a*.) If ρ be the variable vector of a curve in space, and if the differential $d\kappa$ be treated as $= 0$, then the equation

$$dT(\rho-\kappa) = 0$$

obliges κ to be the vector of some point in the normal plane to the curve.

(*b*.) In like manner the system of two equations, where $d\kappa$ and $d^2\kappa$ are each $= 0$,

$$dT(\rho-\kappa) = 0, \qquad d^2T(\rho-\kappa) = 0,$$

represents the axis of the element, or the right line drawn through the centre of the osculating circle, perpendicular to the osculating plane.

(*c*.) The system of the three equations, in which κ is treated as constant,

$$dT(\rho-\kappa) = 0, \qquad d^2T(\rho-\kappa) = 0, \qquad d^3T(\rho-\kappa) = 0,$$

determines the vector κ of the centre of the osculating sphere.

(*d*) For the three last equations we may substitute the following·

$$S.(\rho-\kappa)\,d\rho = 0,$$

$$S.(\rho-\kappa)\,d^2\rho + d\rho^2 = 0,$$

$$S.(\rho-\kappa)\,d^3\rho + 3S.d\rho d^2\rho = 0.$$

(*e*.) Hence, generally, whatever the independent and scalar variable may be, on which the variable vector ρ of the curve depends, the vector κ of the centre of the oseulating sphere admits of being thus expressed :

$$\kappa = \rho + \frac{3V.d\rho d^2\rho S.d\rho d^2\rho - d\rho^2 V.d\rho d^3\rho}{S.d\rho d^2\rho d^3\rho}.$$

(*f.*) In general,

$$d(d^{-2}V_{\iota}d\rho U\rho) = d(T\rho^{-3}V.\rho d\rho)$$
$$= T\rho^{-5}(3V.\rho d\rho S.\rho d\rho - \rho^2 V.\rho d^2\rho);$$

whence,

$$3V.\rho d\rho S.\rho d\rho - \rho^2 V.\rho d^2\rho = \rho^4 T\rho d(\rho^{-2}V.d\rho U\rho);$$

and, therefore, the recent expression for κ admits of being thus transformed,

$$\kappa = \rho + \frac{d\rho^4 d(d\rho^{-2}V.d^2\rho Ud\rho)}{S.d^2\rho d^3\rho Ud\rho}.$$

(*g.*) If the length of the element of the curve be constant, $dTd\rho = 0$, this last expression for the vector of the centre of the osculating sphere to a curve of double curvature becomes, more simply,

$$\kappa = \rho + \frac{d.d^2\rho d\rho^3}{S.d\rho d^2\rho d^3\rho};$$

or

$$\kappa = \rho + \frac{V.d^3\rho d\rho^3}{S.d\rho d^2\rho d^3\rho}.$$

(*h.*) Verify that this expression gives $\kappa = 0$, for a curve described on a sphere which has its centre at the origin of vectors; or shew that whenever $dT\rho = 0$, $d^2T\rho = 0$, $d^3T\rho = 0$, as well as $dTd\rho = 0$, then

$$\rho S.d\rho^{-1}d^2\rho d^3\rho = V.d\rho d^3\rho.$$

5. Find the curve from every point of which three given spheres appear of equal magnitude.

6. Shew that the locus of a point, the difference of whose distances from each two of three given points is constant, is a plane curve.

7. Find the equation of the curve which cuts at a given angle all the sides of a cone of the second order.

Find the length of any arc of this curve in terms of the distances of its extremities from the vertex.

8. Why is the centre of spherical curvature, of a curve described on a sphere, not necessarily the centre of the sphere?

9. Find the equation of the developable surface whose generating lines are the intersections of successive normal planes to a given tortuous curve.

10. Find the length of an arc of a tortuous curve whose normal planes are equidistant from the origin.

11. The reciprocals of the perpendiculars from the origin on the tangent planes to a developable surface are vectors of a tortuous

curve; from whose osculating planes the cusp-edge of the original surface may be reproduced by the same process.

12. The equation $\rho = V\alpha^t\beta,$

where α is a unit-vector not perpendicular to β, represents an ellipse. If we put $\gamma = V\alpha\beta$, shew that the equations of the locus of the centre of curvature are

$$S.\beta\gamma\rho = 0,$$

$$S^{\frac{2}{3}}\beta\rho + S^{\frac{2}{3}}\gamma\rho = (\beta SU\alpha\beta)^{\frac{4}{3}}.$$

13. Find the radius of absolute curvature of a spherical conic.

14. If a cone be cut in a circle by a plane perpendicular to a side, the axis of the right cone which osculates it, along that side, passes through the centre of the section.

15. Shew how to find the vector of an umbilicus. Apply your method to the surfaces whose equations are

$$S\rho\phi\rho = 1,$$

and $$S\alpha\rho S\beta\rho S\gamma\rho = 1.$$

16. Find the locus of the umbilici of the surfaces represented by the equation $$S\rho(\phi + h)^{-1}\rho = 1,$$

where h is an arbitrary parameter.

17. Shew how to find the equation of a tangent plane which touches a surface along a line, straight or curved. Find such planes for the following surfaces

$$S\rho\phi\rho = 1,$$

$$S\rho(\phi - \rho^2)^{-1}\rho = 1,$$

and $$(\rho^2 - a^2 + b^2)^2 + 4(a^2\rho^2 + b^2S^2\alpha\rho) = 0.$$

18. Find the condition that the equation

$$S(\rho + \alpha)\phi\rho = 1,$$

where ϕ is a self-conjugate linear and vector function, may represent a cone.

19. Shew from the general equation that cones and cylinders are the only developable surfaces of the second order.

20. Find the equation of the envelop of planes drawn at each point of an ellipsoid perpendicular to the radius vector from the centre.

21. Find the equation of the envelop of spheres whose centres lie on a given sphere, and which pass through a given point.

22. Find the locus of the foot of the perpendicular from the centre to the tangent plane of a hyperboloid of one, or of two, sheets.

23. Hamilton, *Bishop Law's Premium Examination*, 1852.

(*a*.) If ρ be the vector of a curve in space, the length of the element of that curve is $Td\rho$; and the variation of the length of a finite arc of the curve is

$$\delta\int Td\rho = -\int SUd\rho\,\delta d\rho = -\Delta SUd\rho\,\delta\rho + \int SdUd\rho\,\delta\rho.$$

(*b*.) Hence, if the curve be a shortest line on a given surface, for which the normal vector is ν, so that $S\nu\delta\rho = 0$, this shortest or geodetic curve must satisfy the differential equation, $\quad V\nu dUd\rho = 0.$

Also, for the extremities of the arc, we have the limiting equations,

$$SUd\rho_0\,\delta\rho_0 = 0; \qquad SUd\rho_1\,\delta\rho_1 = 0.$$

Interpret these results.

(*c*.) For a spheric surface, $V\nu\rho = 0$, $\rho dUd\rho = 0$; the integrated equation of the geodetics is $\rho Ud\rho = \varpi$, giving $S\varpi\rho = 0$ (great circle).

For an arbitrary cylindric surface,

$$S\alpha\nu = 0, \qquad \alpha dUd\rho = 0;$$

the integral shews that the geodetic is generally a helix, making a constant angle with the generating lines of the cylinder.

(*d*.) For an arbitrary conic surface,

$$S\nu\rho = 0, \qquad S\rho dUd\rho = 0;$$

integrate this differential equation, so as to deduce from it, $TV\rho Ud\rho =$ const.

Interpret this result; shew that the perpendicular from the vertex of the cone on the tangent to a given geodetie line is constant; this gives the rectilinear development.

When the cone is of the second degree, the same property is a particular case of a theorem respecting confocal surfaces.

(*e*.) For a surface of revolution,

$$S.\alpha\rho\nu = 0, \qquad S.\alpha\rho\, dUd\rho = 0;$$

integration gives,

$$\text{const.} = S.\alpha\rho Ud\rho = TV\alpha\rho SU(V\alpha\rho.d\rho);$$

the perpendicular distance of a point on a geodetic line from the axis of revolution varies inversely as the cosine of the angle under which the geodetic crosses a parallel (or circle) on the surface.

(*f.*) The differential equation, $S.a\rho dUd\rho = 0$, is satisfied not only by the geodetics, but also by the circles, on a surface of revolution; give the explanation of this fact of calculation, and shew that it arises from the coincidence between the normal plane to the circle and the plane of the meridian of the surface.

(*g.*) For any arbitrary surface, the equation of the geodetic may be thus transformed, $S.\nu d\rho d^2\rho = 0$; deduce this form, and shew that it expresses the normal property of the osculating plane.

(*h.*) If the element of the geodetic be constant, $dTd\rho = 0$, then the general equation formerly assigned may be reduced to $V.\nu d^2\rho = 0$.

Under the same condition, $d^2\rho = -\nu^{-1}Sd\nu d\rho$.

(*i.*) If the equation of a central surface of the second order be put under the form $f\rho = 1$, where the function f is scalar, and homogeneous of the second dimension, then the differential of that function is of the form $df\rho = 2S.\nu d\rho$, where the normal vector, $\nu = \phi\rho$, is a distributive function of ρ (homogeneous of the first dimension), $d\nu = d\phi\rho = \phi d\rho$.

This normal vector ν may be called the *vector of proximity* (namely, of the element of the surface to the centre); because its reciprocal, ν^{-1}, represents in length and in direction the perpendicular let fall from the centre on the tangent plane to the surface.

(*k.*) If we make $S\sigma\phi\rho = f(\sigma, \rho)$, *this* function f is commutative with respect to the *two* vectors on which it depends, $f(\rho, \sigma) = f(\sigma, \rho)$; it is also connected with the *former* function f, of a *single* vector ρ, by the relation, $f(\rho, \rho) = f\rho$: so that $f\rho = S\rho\phi\rho$.

$fd\rho = Sd\rho d\nu$; $dfd\rho = 2S.d\nu d^2\rho$; for a geodetic, with constant element,

$$\frac{dfd\rho}{2fd\rho} + S\frac{d\nu}{\nu} = 0;$$

this equation is immediately integrable, and gives const. $= T\nu\sqrt{(fUd\rho)}$ = reciprocal of Joachimstal's pro duct, PD.

(*l.*) If we give the name of "Didonia" to the curve (discussed by Delaunay) which, on a given surface and with a given perimeter, contains the greatest area, then for such a Didonian curve we have by quaternions the formula, $\int S.Uvd\rho\delta\rho + c\delta\int Td\rho = 0$,

where c is an arbitrary constant.

Derive hence the differential equation of the second order, equivalent (through the constant c) to one of the third order, $c^{-1}d\rho = V.UvdUd\rho$.

Geodetics are, therefore, that limiting case of Didonias for which the constant c is infinite.

On a plane, the Didonia is a circle, of which the equation, obtained by integration from the general form, is

$$\rho = \varpi + cUvd\rho,$$

ϖ being vector of centre, and c being radius of circle.

(*m.*) Operating by $S.Ud\rho$, the general differential equation of the Didonia takes easily the following forms·

$$c^{-1}Td\rho = S(Uvd\rho.dUd\rho);$$

$$c^{-1}Td\rho^2 = S(Uvd\rho.d^2\rho);$$

$$c^{-1}Td\rho^3 = S.Uvd\rho d^2\rho;$$

$$c^{-1} = S\frac{d^2\rho d\rho^{-2}}{Uvd\rho}.$$

(*n.*) The vector ω, of the centre of the osculating circle to a curve in space, of which the element $Td\rho$ is constant, has for expression,

$$\omega = \rho + \frac{d\rho^2}{d^2\rho}$$

Hence for the general Didonia,

$$c^{-1} = S\frac{(\omega-\rho)^{-1}}{Uvd\rho};$$

$$T(\rho-\omega) = cSU\frac{\rho-\omega}{vd\rho}.$$

(*o.*) Hence, the radius of curvature of any one Didonia varies, in general, proportionally to the cosine of the inclination of the osculating plane of the curve to the tangent plane of the surface.

And hence, by Meusnier's theorem, the difference of the squares of the curvatures of curve and surface is constant; the curvature of the surface meaning here the reciprocal of the radius of the sphere which osculates in the reduction of the element of the Didonia.

(*p.*) In general, for any curve on any surface, if ξ denote the vector of the intersection of the axis of the element (or the axis of the circle osculating to the curve) with the tangent plane to the surface, then

$$\xi = \rho + \frac{\nu\, d\rho^3}{S.\nu\, d\rho\, d^2\rho}.$$

Hence, for the general Didonia, with the same signification of the symbols,

$$\xi = \rho - cU\nu\, d\rho;$$

and the constant c expresses the length of the interval $\rho - \xi$, intercepted on the tangent plane, between the point of the curve and the axis of the osculating circle.

(*q*.) If, then, a sphere be described, which shall have its centre on the tangent plane, and shall contain the osculating circle, the radius of this sphere shall always be equal to c.

(*r*.) The recent expression for ξ, combined with the first form of the general differential equation of the Didonia, gives

$$d\xi - -cVdU\nu U d\rho; \qquad V\nu d\xi = 0.$$

(*s*.) Hence, or from the geometrical signification of the constant c, the known property may be proved, that if a developable surface be circumscribed about the arbitrary surface, so as to touch it along a Didonia, and if this developable be then unfolded into a plane, the curve will at the same time be flattened (generally) into a circular arc, with radius $= c$.

24. Find the condition that the equation

$$S\rho\, (\phi + f)^{-1} \rho = 1$$

may give three real values of f for any given value of ρ. If f be a function of a scalar parameter ξ, shew how to find the form of this function in order that we may have

$$-\nabla^2 \xi = \frac{d^2\xi}{dx^2} + \frac{d^2\xi}{dy^2} + \frac{d^2\xi}{dz^2} = 0.$$

Prove that the following is the relation between f and ξ,

$$c\xi = \int \frac{df}{\sqrt{(g_1+f)(g_2+f)(g_3+f)}} = \int \frac{df}{\sqrt{m_f}}$$

in the notation of § 147.

25. Shew, after Hamilton, that the proof of Dupin's theorem, that "each member of one of three series of orthogonal surfaces cuts each member of each of the other series along its lines of curvature," may be expressed in quaternion notation as follows:

If $\quad S\nu d\rho = 0, \qquad S\nu' d\rho = 0, \qquad S.\nu\nu' d\rho = 0$
be integrable, and if

$$S\nu\nu' = 0, \quad \text{then} \quad V\nu' d\rho = 0, \quad \text{makes} \quad S.\nu\nu' d\nu = 0.$$

Or, as follows,

If $\quad S\nu\nabla\nu = 0, \quad S\nu'\nabla\nu' = 0, \quad S\nu''\nabla\nu'' = 0, \quad$ and $\quad V.\nu\nu'\nu'' = 0,$

then
$$S.\nu''(S\nu'\nabla.\nu) = 0,$$

where
$$\nabla = i\frac{d}{dx} + j\frac{d}{dy} + k\frac{d}{dz}.$$

26. Shew that the equation

$$V\alpha\rho = \rho V\beta\rho$$

represents the line of intersection of a cylinder and cone, of the second order, which have β as a common generating line.

27. Two spheres are described, with centres at A, B, where $\overline{OA} = \alpha$, $\overline{OB} = \beta$, and radii a, b. Any line, OPQ, drawn from the origin, cuts them in P, Q respectively. Shew that the equation of the locus of intersection of AP, BQ has the form

$$V(\alpha + aU(\rho - \alpha))(\beta + bU(\rho - \beta)) = 0.$$

Shew that this involves $\quad S.\alpha\beta\rho = 0,$
and therefore that the left side is a scalar multiple of $V.\alpha\beta$, so that the locus is a plane curve.

Also shew that in the particular case

$$V\alpha\beta = 0,$$

the locus is the surface formed by the revolution of a Cartesian oval about its axis.

CHAPTER X.

KINEMATICS.

336.] WHEN a point's vector, ρ, is a function of the time t, we have seen (§ 36) that its vector-velocity is expressed by $\frac{d\rho}{dt}$ or, in Newton's notation, by $\dot{\rho}$.

That is, if

$$\rho = \phi t$$

be the equation of an orbit, *containing* (as the reader may see) *not merely the form of the orbit, but the law of its description also*, then

$$\dot{\rho} = \phi' t$$

gives at once the form of the *Hodograph* and the law of its description.

This shews immediately that the *vector-acceleration of a point's motion*,

$$\frac{d^2\rho}{dt^2} \text{ or } \ddot{\rho},$$

is the vector-velocity in the hodograph. Thus the fundamental properties of the hodograph are proved almost intuitively.

337.] Changing the independent variable, we have

$$\dot{\rho} = \frac{d\rho}{ds}\frac{ds}{dt} = v\rho',$$

if we employ the dash, as before, to denote $\frac{d}{ds}$.

This merely shews, in another form, that $\dot{\rho}$ expresses the velocity in magnitude and direction. But a second differentiation gives

$$\ddot{\rho} = \dot{v}\rho' + v^2\rho''.$$

This shews that the vector-acceleration can be resolved into two components, the first, $\dot{v}\rho'$, being in the direction of motion and equal in magnitude to the acceleration of the velocity, $\dot{v}$ or $\frac{dv}{dt}$; the second, $v^2\rho''$, being in the direction of the radius of absolute curvature, and having for its amount the square of the velocity multiplied by the curvature.

[It is scarcely conceivable that this important fundamental proposition, of which no simple analytical proof seems to have been obtained by Cartesian methods, can be proved more elegantly than by the process just given.]

338.] If the motion be in a plane curve, we may write the equation as follows, so as to introduce the usual polar cöordinates, r and θ,

$$\rho = r\alpha^{\frac{2\theta}{\pi}}\beta,$$

where α is a unit-vector perpendicular to, β a unit-vector in, the plane of the curve.

Here, of course, r and θ may be considered as connected by one scalar equation; or better, each may be looked on as a function of t. By differentiation we get

$$\dot{\rho} = \dot{r}\alpha^{\frac{2\theta}{\pi}}\beta + r\dot{\theta}\alpha\alpha^{\frac{2\theta}{\pi}}\beta,$$

which shews at once that $\dot{r}$ is the velocity along, $r\dot{\theta}$ that perpendicular to, the radius vector. Again,

$$\ddot{\rho} = (\ddot{r} - r\dot{\theta}^2)\alpha^{\frac{2\theta}{\pi}}\beta + (2\dot{r}\dot{\theta} + r\ddot{\theta})\alpha\alpha^{\frac{2\theta}{\pi}}\beta,$$

which gives, by inspection, the components of acceleration along, and perpendicular to, the radius vector.

339.] For *uniform acceleration in a constant direction*, we have at once,

$$\ddot{\rho} = \alpha.$$

Whence

$$\dot{\rho} = \alpha t + \beta,$$

where β is the vector-velocity at epoch. This shews that the hodograph is a straight line described uniformly.

Also

$$\rho = \frac{\alpha t^2}{2} + \beta t,$$

no constant being added if the origin be assumed to be the position of the moving point at epoch.

Since the resolved parts of ρ, parallel to β and α, vary respectively as the first and second powers of t, the curve is evidently a parabola (§ 31 (f)).

But we may easily deduce from the equation the following result,

$$T(\rho + \tfrac{1}{2}\beta\alpha^{-1}\beta) = -SU\alpha\left(\rho + \frac{\beta^2}{2}\alpha^{-1}\right),$$

the equation of a paraboloid of revolution, whose axis is α. Also

$$S.\alpha\beta\rho = 0,$$

and therefore the distance of any point in the path from the point $-\frac{1}{2}\beta\alpha^{-1}\beta$ is equal to its distance from the line whose equation is

$$\rho = -\frac{\beta^2}{2}\alpha^{-1} + x\alpha V\alpha\beta.$$

Thus we recognise the focus and directrix property.

340.] That the moving point may reach a point γ we must have, for some real value of t,

$$\gamma = \frac{\alpha}{2}t^2 + \beta t.$$

Now suppose $T\beta$, the velocity of projection, to be given $= v$, and, for shortness, write ϖ for $U\beta$.

Then
$$\gamma = \frac{\alpha}{2}t^2 + vt\varpi.$$

Since
$$T\varpi = 1,$$

we have
$$\frac{T\alpha^2 t^4}{4} - (v^2 - S\alpha\gamma)t^2 + T\gamma^2 = 0.$$

The values of t^2 are *real* if

$$(v^2 - S\alpha\gamma)^2 - T\alpha^2 T\gamma^2$$

is positive. Now, as $T\alpha T\gamma$ is never less than $S\alpha\gamma$, it is evident that $v^2 - S\alpha\gamma$ must always be positive if the roots are possible. Hence, when they are possible, both values of t^2 are *positive*. Thus we have *four* values of t which satisfy the conditions, and it is easy to see that since, disregarding the signs, they are equal two and two, each pair refer to the same path, but *described in opposite directions* between the origin and the extremity of γ. There are therefore, if any, in general two parabolas which satisfy the conditions. The directions of projection are (of course) given by the corresponding values of ϖ.

341.] The envelop of all the trajectories possible with a given velocity, evidently corresponds to

$$(v^2 - S\alpha\gamma)^2 - T\alpha^2 T\gamma^2 = 0,$$

for then γ is the vector of intersection of two indefinitely close paths in the same vertical plane.

Now
$$v^2 - S\alpha\gamma = T\alpha T\gamma$$

is evidently the equation of a paraboloid of revolution of which the origin is the focus, the axis parallel to α, and the directrix plane at a distance $\frac{v^2}{T\alpha}$.

All the ordinary problems connected with parabolic motion are easily solved by means of the above formulae. Some, however, are even more easily treated by assuming a horizontal unit-vector in the plane of motion, and expressing β in terms of it and α. But this must be left to the student.

342.] For *acceleration directed to or from a fixed point*, we have, taking that point as origin, and putting P for the magnitude of the central acceleration,

$$\ddot{\rho} = PU\rho.$$

Whence, at once, $V\rho\ddot{\rho} = 0.$

Integrating, $V\rho\dot{\rho} = \gamma =$ a constant vector.

The interpretation of this simple formula is—*first*, ρ and $\dot{\rho}$ are in a plane perpendicular to γ, hence the path is in a plane (of course passing through the origin); *second*, the area of the triangle, two of whose sides are ρ and $\dot{\rho}$ is constant.

[It is scarcely possible to imagine that a more simple proof than this can be given of the fundamental facts, that a central orbit is a plane curve, and that equal areas are described by the radius vector in equal times.]

343.] When the *law of acceleration to or from the origin is that of the inverse square of the distance*, we have

$$P = \frac{\mu}{T\rho^2},$$

where μ is *negative* if the acceleration be directed *to* the origin.

Hence $\ddot{\rho} - \frac{\mu U\rho}{T\rho^2}$

The following beautiful method of integration is due to Hamilton. (See Chapter IV.)

Generally, $\frac{dU\rho}{dt} = -\frac{U\rho . V\rho\dot{\rho}}{T\rho^2} = -\frac{U\rho . \gamma}{T\rho^2},$

therefore $\ddot{\rho}\gamma = -\mu\frac{dU\rho}{dt},$

and $\dot{\rho}\gamma = \epsilon - \mu U\rho,$

where ϵ is a constant vector, perpendicular to γ, because

$$S\gamma\dot{\rho} = 0.$$

Hence, in this case, we have for the hodograph,

$$\dot{\rho} = \epsilon\gamma^{-1} - \mu U\rho . \gamma^{-1}$$

Of the two parts of this expression, which are both vectors, the first is constant, and the second is constant in length. Hence the locus of the extremity of $\dot{\rho}$ is a circle in a plane perpendicular to γ (i.e. parallel to the plane of the orbit), whose radius is $\frac{\mu}{T\gamma}$, and whose centre is at the extremity of the vector $\epsilon\gamma^{-1}$.

[This equation contains the whole theory of the *Circular Hodo-*

graph. Its consequences are developed at length in Hamilton's *Elements.*]

344.] We may write the equations of this circle in the form

$$T(\dot{\rho}-\epsilon\gamma^{-1}) = \frac{\mu}{T\gamma},$$

(a sphere), and $$S\gamma\dot{\rho} = 0$$

(a plane through the origin, and through the centre of the sphere).

The equation of the orbit is found by operating by $V.\rho$ upon that of the hodograph. We thus obtain

$$\gamma = V.\rho\epsilon\gamma^{-1} + \mu T\rho\gamma^{-1},$$

or $$\gamma^2 = S\epsilon\rho + \mu T\rho,$$

or $$\mu T\rho = S\epsilon\,(\gamma^2\epsilon^{-1} - \rho);$$

in which last form we at once recognise the focus and directrix property. This is in fact the equation of a conicoid of revolution about its principal axis (ϵ), and the origin is one of the foci. The orbit is found by combining it with the equation of its plane,

$$S\gamma\rho = 0.$$

We see at once that $\gamma^2\epsilon^{-1}$ is the vector distance of the directrix from the focus; and similarly that the eccentricity is $\frac{T\epsilon}{\mu}$, and the major axis $\frac{-2\mu\gamma^2}{\mu^2+\epsilon^2}$.

345.] To take a simpler case: *let the acceleration vary as the distance from the origin.*

Then $$\ddot{\rho} = \pm m^2\rho,$$

the upper or lower sign being used according as the acceleration is *from* or *to* the centre.

This is $$\left(\frac{d^2}{dt^2} + m^2\right)\rho = 0.$$

Hence $$\rho = \alpha\varepsilon^{mt} + \beta\varepsilon^{-mt};$$

or $$\rho = \alpha\cos mt + \beta\sin mt,$$

where α and β are arbitrary, but constant, vectors; and ε is the base of Napier's logarithms.

The first is the equation of a hyperbola (§ 31, k) of which α and β are the directions of the asymptotes; the second, that of an ellipse of which α and β are semi-conjugate diameters.

Since $$\dot{\rho} = m\,\{\alpha\varepsilon^{mt} - \beta\varepsilon^{-mt}\},$$

or $$-\,m\,\{-\alpha\sin mt + \beta\cos mt\},$$

the hodograph is again a hyperbola or ellipse. But in the first case it is, if we neglect the change of dimensions indicated by the scalar factor m, conjugate to the orbit; in the case of the ellipse it is similar and similarly situated.

346.] Again, *let the acceleration be as the inverse third power of the distance*, we have

$$\ddot{\rho} = \frac{\mu U\rho}{T\rho^3}.$$

Of course, we have, as usual,

$$V\rho\dot{\rho} = \gamma.$$

Also, operating by $S.\dot{\rho}$,

$$S\dot{\rho}\ddot{\rho} = \frac{\mu S\rho\dot{\rho}}{T\rho^4},$$

of which the integral is

$$\dot{\rho}^2 = C - \frac{\mu}{\rho^2},$$

the equation of energy.

Again, $$S\rho\ddot{\rho} = \frac{\mu}{\rho^2}$$

Hence $$S\rho\ddot{\rho} + \dot{\rho}^2 = C,$$

or $$S\rho\dot{\rho} = Ct,$$

no constant being added if we reckon the time from the passage through the apse, where $S\rho\dot{\rho} = 0$.

We have, therefore, by a second integration,

$$\rho^2 = Ct^2 + C'. \qquad (1)$$

[To determine C', remark that

$$\rho\dot{\rho} = Ct + \gamma,$$

or $$\rho^2\dot{\rho}^2 = C^2t^2 - \gamma^2$$

But $$\rho^2\dot{\rho}^2 = C\rho^2 - \mu \quad \text{(by the equation of energy)},$$

$$= C^2t^2 + CC' - \mu, \quad \text{by (1)}.$$

Hence $$CC' = \mu - \gamma^2.]$$

To complete the solution, we have, by § 133,

$$V\frac{\dot{\rho}}{\rho} = \frac{dU\rho}{dt}(U\rho)^{-1} = \frac{d}{dt}\log\frac{U\rho}{\beta},$$

where β is a unit-vector in the plane of the orbit.

But $$V\frac{\dot{\rho}}{\rho} = -\frac{\gamma}{\rho^2}.$$

Hence $$\log\frac{U\rho}{\beta} = -\gamma\int\frac{dt}{Ct^2 + C'}.$$

The elimination of t between this equation and (1) gives $T\rho$ in terms of $U\rho$, or the required equation of the path.

We may remark that if θ be the ordinary polar angle in the orbit,

$$\log\frac{U\rho}{\beta} = \theta U\gamma.$$

Hence we have $\theta = -T\gamma \int \frac{dt}{Ct^2 + C'}$,

and $r^2 = -(Ct^2 + C')$,

from which the ordinary equations of Cotes' spirals can be at once found. [See Tait and Steele's *Dynamics of a Particle*, third edition, Appendix (A).]

347.] *To find the conditions that a given curve may be the hodograph corresponding to a central orbit.*

If ϖ be its vector, *given as a function of the time*, $\int \varpi dt$ is that of the orbit; hence the requisite conditions are given by

$$V \varpi \int \varpi dt = \gamma,$$

where γ is a constant vector.

We may transform this into other shapes more resembling the Cartesian ones.

Thus $$V \dot{\varpi} \int \varpi dt = 0,$$

and $$V \ddot{\varpi} \int \varpi dt + V \dot{\varpi} \varpi = 0.$$

From the first $$\int \varpi dt = x \dot{\varpi}$$

and therefore $$x V \varpi \dot{\varpi} = \gamma,$$

or the curve is *plane*. And

$$x V \ddot{\varpi} \dot{\varpi} + V \dot{\varpi} \varpi = 0;$$

or eliminating x, $$\gamma V \dot{\varpi} \ddot{\varpi} = -(V \varpi \dot{\varpi})^2.$$

Now if v' be the velocity in the hodograph, R' its radius of curvature, p' the perpendicular on the tangent; this equation gives at once $$hv' = R'p'^2,$$

which agrees with known results.

348.] *The equation of an epitrochoid or hypotrochoid*, referred to the centre of the fixed circle, is evidently

$$\rho = a i^{\frac{2\omega t}{\pi}} \alpha + b i^{\frac{2\omega_1 t}{\pi}} \alpha,$$

where α is a unit-vector in the plane of the curve and i another perpendicular to it. Here ω and ω_1 are the angular velocities in the two circles, and t is the time elapsed since the tracing point and the centres of the two circles were in one straight line.

Hence, for the length of an arc of such a curve,

$$s = \int T \dot{\rho} dt = \int dt \sqrt{\{\omega^2 a^2 + 2\omega\omega_1 ab \cos(\omega - \omega_1) t + \omega_1^2 b^2\}},$$

$$= \int dt \sqrt{\left\{(\omega a \mp \omega_1 b)^2 \pm 4\omega\omega_1 ab \left|\begin{matrix}\cos^2 \\ \sin^2\end{matrix}\right| \frac{\omega - \omega_1}{2} t\right\}},$$

which is, of course, an elliptic function.

But when the curve becomes an epicycloid or a hypocycloid, $\omega a + \omega_1 b = 0$, and

$$s = 2\sqrt{(\pm\omega\omega_1 ab)}\int dt \left\{\begin{matrix}\cos\\ \sin\end{matrix}\right\} \frac{\omega-\omega_1}{2} t,$$

which can be expressed in finite terms, as was first shewn by Newton in the *Principia*.

The hodograph is another curve of the same class, whose equation is

$$\dot{\rho} = i\left(a\omega i^{\frac{2\omega t}{\pi}} a + b\omega_1 i^{\frac{2\omega_1 t}{\pi}} a\right);$$

and the acceleration is denoted in magnitude and direction by the vector

$$\ddot{\rho} = -a\omega^2 i^{\frac{2\omega t}{\pi}} a - b\omega_1^2 i^{\frac{2\omega_1 t}{\pi}} a.$$

Of course the equations of the common *Cycloid* and *Trochoid* may be easily deduced from these forms by making a indefinitely great and ω indefinitely small, but the product $a\omega$ finite; and transferring the origin to the point $\rho = a a$.

349.] Let i be the normal-vector to any plane.

Let ϖ and ρ be the vectors of any two points in a rigid plate in contact with the plane.

After any small displacement of the rigid plate in its plane, let $d\varpi$ and $d\rho$ be the increments of ϖ and ρ.

Then $Sid\varpi = 0$, $Sid\rho = 0$; and, since $T(\varpi-\rho)$ is constant,

$$S(\varpi-\rho)(d\varpi-d\rho) = 0.$$

And we may evidently assume

$$d\rho = \omega i(\rho-\tau),$$

$$d\varpi = \omega i(\varpi-\tau);$$

where of course τ is the vector of some point in the plane, to a rotation ω about which the displacement is therefore equivalent.

Eliminating it, we have

$$\omega i = \frac{d(\varpi-\rho)}{\varpi-\rho},$$

which gives ω, and thence τ is at once found.

For any other point σ in the plane figure

$$Sid\sigma = 0,$$

$$S(\rho-\sigma)(d\rho-d\sigma) = 0. \quad \text{Hence } d\rho-d\sigma = \omega_1 i(\rho-\sigma).$$

$$S(\sigma-\varpi)(d\varpi-d\sigma) = 0. \quad \text{Hence } d\sigma-d\varpi = \omega_2 i(\sigma-\varpi).$$

From which, at once, $\omega_1 = \omega_2 = \omega$, and

$$d\sigma = \omega i(\sigma-\tau),$$

or this point also is displaced by a rotation ω about an axis through the extremity of τ and parallel to i.

350.] In the case of a rigid body moving about a fixed point let ϖ, ρ, σ denote the vectors of any three points of the body; the fixed point being origin.

Then ϖ^2, ρ^2, σ^2 are constant, and so are $S\varpi\rho$, $S\rho\sigma$, and $S\sigma\varpi$.

After any small displacement we have, for ϖ and ρ,

$$\left.\begin{aligned} S\varpi d\varpi &= 0, \\ S\rho d\rho &= 0, \\ S\varpi d\rho + S\rho d\varpi &= 0. \end{aligned}\right\} \quad\quad (1)$$

Now these three equations are satisfied by

$$d\varpi = V\alpha\varpi, \qquad d\rho = V\alpha\rho,$$

where α is *any* vector whatever. But if $d\varpi$ and $d\rho$ are *given*, then

$$Vd\varpi\, d\rho = V.V\alpha\varpi V\alpha\rho = \alpha S.\alpha\rho\varpi.$$

Operate by $S.V\varpi\rho$, and remember (1),

$$S^2\varpi d\rho = S^2\rho d\varpi = S^2.\alpha\rho\varpi.$$

Hence
$$\alpha = \frac{Vd\varpi d\rho}{S\varpi d\rho} = \frac{Vd\rho d\varpi}{S\rho d\varpi}. \quad\quad (2)$$

Now consider σ,
$$\left.\begin{aligned} S\sigma d\sigma &= 0, \\ S\rho d\sigma &= -S\sigma d\rho, \\ S\varpi d\sigma &= -S\sigma d\varpi. \end{aligned}\right\}$$

$d\sigma = V\alpha\sigma$ satisfies them all, by (2), and we have thus the proposition that *any small displacement of a rigid body about a fixed point is equivalent to a rotation.*

351.] *To represent the rotation of a rigid body about a given axis, through a given finite angle.*

Let α be a unit-vector in the direction of the axis, ρ the vector of any point in the body with reference to a fixed point in the axis, and θ the angle of rotation.

Then
$$\begin{aligned} \rho &= \alpha^{-1}S\alpha\rho + \alpha^{-1}V\alpha\rho, \\ &= -\alpha S\alpha\rho - \alpha V\alpha\rho. \end{aligned}$$

The rotation leaves, of course, the first part unaffected, but the second evidently becomes

$$-\alpha^{\frac{2\theta}{\pi}}\alpha V\alpha\rho,$$

or
$$\alpha V\alpha\rho\cos\theta + V\alpha\rho\sin\theta.$$

Hence ρ becomes

$$\begin{aligned} \rho_1 &= -\alpha S\alpha\rho - \alpha V\alpha\rho\cos\theta + V\alpha\rho\sin\theta, \\ &= \left(\cos\frac{\theta}{2} + \alpha\sin\frac{\theta}{2}\right)\rho\left(\cos\frac{\theta}{2} - \alpha\sin\frac{\theta}{2}\right), \\ &= \alpha^{\frac{\theta}{\pi}}\rho\alpha^{-\frac{\theta}{\pi}}. \end{aligned}$$

352.] Hence *to compound two rotations about axes which meet*, we may evidently write, as the effect of an additional rotation ϕ about the unit-vector β,

$$\rho_2 = \beta^{\frac{\phi}{\pi}} \rho_1 \beta^{-\frac{\phi}{\pi}}$$

Hence $$\rho_2 = \beta^{\frac{\phi}{\pi}} \alpha^{\frac{\theta}{\pi}} \rho \alpha^{-\frac{\theta}{\pi}} \beta^{-\frac{\phi}{\pi}}$$

If the β-rotation had been first, and then the α-rotation, we should have had

$$\rho'_2 = \alpha^{\frac{\theta}{\pi}} \beta^{\frac{\phi}{\pi}} \rho \beta^{-\frac{\phi}{\pi}} \alpha^{-\frac{\theta}{\pi}},$$

and the non-commutative property of quaternion multiplication shews that we have *not*, in general,

$$\rho'_2 = \rho_2.$$

If α, β, γ be radii of the unit sphere to the corners of a spherical triangle whose angles are $\frac{\theta}{2}, \frac{\phi}{2}, \frac{\psi}{2}$, we know that

$$\gamma^{\frac{\psi}{\pi}} \beta^{\frac{\phi}{\pi}} \alpha^{\frac{\theta}{\pi}} = -1. \quad \text{(Hamilton, \textit{Lectures}, p. 267.)}$$

Hence $$\beta^{\frac{\phi}{\pi}} \alpha^{\frac{\theta}{\pi}} = -\gamma^{-\frac{\psi}{\pi}},$$

and we may write $$\rho_2 = \gamma^{-\frac{\psi}{\pi}} \rho \gamma^{\frac{\psi}{\pi}},$$

or, *successive rotations about radii to two corners of a spherical triangle, and through angles double of those of the triangle, are equivalent to a single rotation about the radius to the third corner, and through an angle double of the exterior angle of the triangle.*

Thus any number of successive *finite* rotations may be compounded into a single rotation about a definite axis.

353.] When the rotations are indefinitely small, the effect of one is, by § 351, $$\rho_1 = \rho + \mathfrak{a} V\alpha\rho,$$

and for the two, neglecting products of small quantities,

$$\rho_2 = \rho + \mathfrak{a} V\alpha\rho + \mathfrak{b} V\beta\rho,$$

$\mathfrak{a}$ and $\mathfrak{b}$ representing the angles of rotation about the unit-vectors α and β respectively.

But this is equivalent to

$$\rho_2 = \rho + T(\mathfrak{a}\alpha + \mathfrak{b}\beta) VU(\mathfrak{a}\alpha + \mathfrak{b}\beta)\rho,$$

representing a rotation through an angle $T(\mathfrak{a}\alpha + \mathfrak{b}\beta)$, about the unit-vector $U(\mathfrak{a}\alpha + \mathfrak{b}\beta)$. Now the latter is the *direction*, and the former the *length*, of the diagonal of the parallelogram whose sides are $\mathfrak{a}\alpha$ and $\mathfrak{b}\beta$.

We may write these results more simply, by putting α for $\mathfrak{a}\alpha$, β for $\mathfrak{b}\beta$, where α and β are now no longer unit-vectors, but represent by their versors the *axes*, and by their tensors the *angles* (small), of rotation.

Thus
$$\rho_1 = \rho + V\alpha\rho,$$
$$\rho_2 = \rho + V\alpha\rho + V\beta\rho,$$
$$= \rho + V(\alpha+\beta)\rho.$$

354.] The general theorem, of which a few preceding sections illustrate special cases, is this:

By a rotation, about the axis of q, through double the angle of q, the quaternion r becomes the quaternion qrq^{-1}. Its tensor and angle remain unchanged, its plane or axis alone varies.

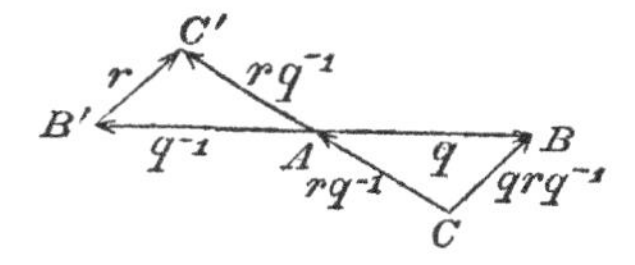

A glance at the figure is sufficient for the proof, if we note that of course $T.qrq^{-1} = Tr$, and therefore that we need consider the *versor* parts only. Let Q be the pole of q,

$$AB = q, \quad AB' = q^{-1}, \quad B'C' = r.$$

Join $C'A$, and make $AC = C'A$. Join CB.

Then $\overset{\frown}{CB}$ is qrq^{-1}, its arc CB is evidently equal in length to that of r, $B'C'$; and its plane (making the same angle with $B'B$ that that of $B'C'$ does) has evidently been made to revolve about Q, the pole of q, through double the angle of q.

If r be a vector, $= \rho$, then $q\rho q^{-1}$ (which is also a vector) is the result of a rotation through double the angle of q about the axis of q. Hence, as Hamilton has expressed it, if B represent a rigid system, or assemblage of vectors,
$$qBq^{-1}$$
is its new position after rotating through double the angle of q about the axis of q.

355.] To compound such rotations, we have
$$r.qBq^{-1}.r^{-1} = rq.B.(rq)^{-1}.$$

To cause rotation through an angle t-fold the double of the angle of q we write
$$q^t Bq^{-t}.$$

To reverse the direction of this rotation write $q^{-t}Bq^t$.

To *translate* the body B without rotation, each point of it moving through the vector α, we write $\alpha + B$.

To produce rotation of the translated body about the same axis, and through the same angle, as before,
$$q(\alpha + B)q^{-1}.$$

Had we rotated first, and then translated, we should have had
$$\alpha + qBq^{-1}.$$

The obvious discrepance between these last results might perhaps be useful to those who do not believe in the Moon's rotation, but to such men quaternions are unintelligible.

356.] *Given the instantaneous axis in terms of the time, it is required to find the single rotation which will bring the body from any initial position to its position at a given time.*

If α be the initial vector of a point of the body, ϖ the value of the same at time t, and q the required quaternion, we have

$$\varpi = q\alpha q^{-1}. \qquad (1)$$

Differentiating with respect to t, this gives

$$\begin{aligned}\dot{\varpi} &= \dot{q}\alpha q^{-1} - q\alpha q^{-1}\dot{q}q^{-1},\\ &= \dot{q}q^{-1}.q\alpha q^{-1} - q\alpha q^{-1}.\dot{q}q^{-1},\\ &= 2V.(V\dot{q}q^{-1}.q\alpha q^{-1}).\end{aligned}$$

But $$\dot{\varpi} = V\epsilon\varpi = V.\epsilon q\alpha q^{-1}.$$

Hence, as $q\alpha q^{-1}$ may be any vector whatever in the displaced body, we must have $$\epsilon = 2V\dot{q}q^{-1}. \qquad (2)$$

This result may be stated in even a simpler form than (2), for we have always, whatever quaternion q may be,

$$V\dot{q}q^{-1} = .\frac{dUq}{dt}(Uq)^{-1},$$

and, therefore, if we suppose the tensor of q, which may have any value whatever, to be a constant (unity, for instance), we may write (2) in the form $$\epsilon q = 2\dot{q}. \qquad (3)$$

An immediate consequence, which will be of use to us later, is

$$q.q^{-1}\epsilon q = 2\dot{q}. \qquad (4)$$

357.] *To express* q *in terms of the usual angles* ψ, θ, ϕ.

Here the vectors i, j, k in the original position of the body correspond to $\overline{OA}$, $\overline{OB}$, $\overline{OC}$, respectively, at time t. The transposition is effected by—*first*, a rotation ψ about k; *second*, a rotation θ about the new position of the line originally coinciding with j; *third*, a rotation ϕ about the final position of the line at first coinciding with k.

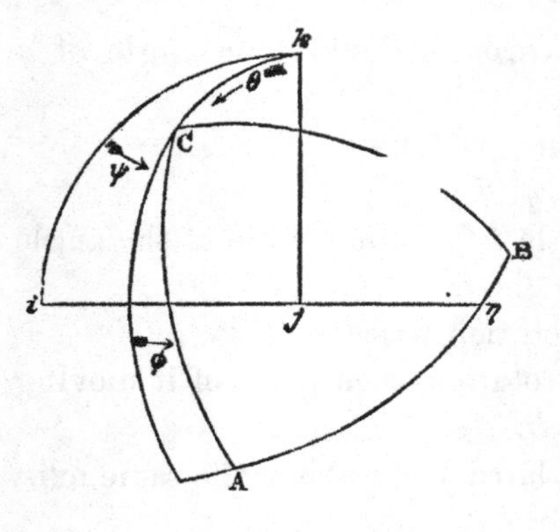

Let i, j, k be taken as the initial directions of the three vectors which at time t terminate at A, B, C respectively.

The rotation ψ about k has the operator

$$k^{\frac{\psi}{\pi}}(\ \ )k^{-\frac{\psi}{\pi}}.$$

This converts j into η, where

$$\eta = k^{\frac{\psi}{\pi}} j k^{-\frac{\psi}{\pi}} = j \cos\psi - i \sin\psi.$$

The body next rotates about η through an angle θ. This has the operator

$$\eta^{\frac{\theta}{\pi}}(\quad)\eta^{-\frac{\theta}{\pi}}.$$

It converts k into

$$\overline{OC} = \zeta = \eta^{\frac{\theta}{\pi}} k \eta^{-\frac{\theta}{\pi}} = \left(\cos\frac{\theta}{2} + \eta\sin\frac{\theta}{2}\right) k \left(\cos\frac{\theta}{2} - \eta\sin\frac{\theta}{2}\right)$$

$$= k\cos\theta + \sin\theta\,(i\cos\psi + j\sin\psi).$$

The body now turns through the angle ϕ about ζ, the operator being

$$\zeta^{\frac{\phi}{\pi}}(\quad)\zeta^{-\frac{\phi}{\pi}}.$$

Hence

$$q = \zeta^{\frac{\phi}{\pi}} \eta^{\frac{\theta}{\pi}} k^{\frac{\psi}{\pi}}$$

$$= \left(\cos\frac{\phi}{2} + \zeta\sin\frac{\phi}{2}\right)\left(\cos\frac{\theta}{2} + \eta\sin\frac{\theta}{2}\right)\left(\cos\frac{\psi}{2} + k\sin\frac{\psi}{2}\right)$$

$$= \left(\cos\frac{\phi}{2} + \zeta\sin\frac{\phi}{2}\right)\left[\cos\frac{\theta}{2}\cos\frac{\psi}{2} + k\cos\frac{\theta}{2}\sin\frac{\psi}{2}\right.$$

$$\left. + \sin\frac{\theta}{2}\cos\frac{\psi}{2}(j\cos\psi - i\sin\psi) + \sin\frac{\theta}{2}\sin\frac{\psi}{2}(i\cos\psi + j\sin\psi)\right]$$

$$= \left(\cos\frac{\phi}{2} + \zeta\sin\frac{\phi}{2}\right)\left[\cos\frac{\theta}{2}\cos\frac{\psi}{2} - i\sin\frac{\theta}{2}\sin\frac{\psi}{2} + j\sin\frac{\theta}{2}\cos\frac{\psi}{2} + k\cos\frac{\theta}{2}\sin\frac{\psi}{2}\right]$$

$$= \cos\frac{\phi}{2}\cos\frac{\theta}{2}\cos\frac{\psi}{2} + \sin\frac{\phi}{2}\sin\frac{\theta}{2}\sin\frac{\psi}{2}\sin\theta\cos\psi$$

$$- \sin\frac{\phi}{2}\sin\frac{\theta}{2}\cos\frac{\psi}{2}\sin\theta\sin\psi - \sin\frac{\phi}{2}\cos\frac{\theta}{2}\sin\frac{\psi}{2}\cos\theta$$

$$+ i\left(-\cos\frac{\phi}{2}\sin\frac{\theta}{2}\sin\frac{\psi}{2} + \sin\frac{\phi}{2}\cos\frac{\theta}{2}\cos\frac{\psi}{2}\sin\theta\cos\psi\right.$$

$$\left. \sin\frac{\phi}{2}\sin\frac{\theta}{2}\cos\frac{\psi}{2}\cos\theta + \sin\frac{\phi}{2}\cos\frac{\theta}{2}\sin\frac{\psi}{2}\sin\theta\sin\psi\right)$$

$$+ j\left(\cos\frac{\phi}{2}\sin\frac{\theta}{2}\cos\frac{\psi}{2} + \sin\frac{\phi}{2}\cos\frac{\theta}{2}\cos\frac{\psi}{2}\sin\theta\sin\psi\right.$$

$$\left. \sin\frac{\phi}{2}\sin\frac{\theta}{2}\sin\frac{\psi}{2}\cos\theta - \sin\frac{\phi}{2}\cos\frac{\theta}{2}\sin\frac{\psi}{2}\sin\theta\cos\psi\right)$$

$$+ k\left(\cos\frac{\phi}{2}\cos\frac{\theta}{2}\sin\frac{\psi}{2} + \sin\frac{\phi}{2}\cos\frac{\theta}{2}\cos\frac{\psi}{2}\cos\theta\right.$$

$$\left. + \sin\frac{\phi}{2}\sin\frac{\theta}{2}\sin\frac{\psi}{2}\sin\theta\sin\psi + \sin\frac{\phi}{2}\sin\frac{\theta}{2}\cos\frac{\psi}{2}\sin\theta\cos\psi\right)$$

$$= \cos\frac{\phi+\psi}{2}\cos\frac{\theta}{2} + i\sin\frac{\phi-\psi}{2}\sin\frac{\theta}{2} + j\cos\frac{\phi-\psi}{2}\sin\frac{\theta}{2} + k\sin\frac{\phi+\psi}{2}\cos\frac{\theta}{2},$$

which is, of course, essentially unsymmetrical.

358.] *To find the usual equations connecting ψ, θ, ϕ with the angular velocities about three rectangular axes fixed in the body.*

Having the value of q in last section in terms of the three angles, it may be useful to employ it, in conjunction with equation (3) of § 356, partly as a verification of that equation. Of course, this is an exceedingly roundabout process, and does not in the least resemble the simple one which is immediately suggested by quaternions.

We have $$2\dot{q} = \epsilon q = \{\omega_1 \overline{OA} + \omega_2 \overline{OB} + \omega_3 \overline{OC}\} q,$$
whence $$2q^{-1}\dot{q} = q^{-1}\{\omega_1 \overline{OA} + \omega_2 OB + \omega_3 \overline{OC}\} q,$$
or $$2\dot{q} = q(i\omega_1 + j\omega_2 + k\omega_3).$$

This breaks up into the four (equivalent to three independent) equations

$$2\frac{d}{dt}\left(\cos\frac{\phi+\psi}{2}\cos\frac{\theta}{2}\right)$$
$$= -\omega_1 \sin\frac{\phi-\psi}{2}\sin\frac{\theta}{2} - \omega_2\cos\frac{\phi-\psi}{2}\sin\frac{\theta}{2} - \omega_3\sin\frac{\phi+\psi}{2}\cos\frac{\theta}{2},$$
$$2\frac{d}{dt}\left(\sin\frac{\phi-\psi}{2}\sin\frac{\theta}{2}\right)$$
$$= \omega_1\cos\frac{\phi+\psi}{2}\cos\frac{\theta}{2} - \omega_2\sin\frac{\phi+\psi}{2}\cos\frac{\theta}{2} + \omega_3\cos\frac{\phi-\psi}{2}\sin\frac{\theta}{2},$$
$$2\frac{d}{dt}\left(\cos\frac{\phi-\psi}{2}\sin\frac{\theta}{2}\right)$$
$$= \omega_1\sin\frac{\phi+\psi}{2}\cos\frac{\theta}{2} + \omega_2\cos\frac{\phi+\psi}{2}\cos\frac{\theta}{2} - \omega_3\sin\frac{\phi-\psi}{2}\sin\frac{\theta}{2},$$
$$2\frac{d}{dt}\left(\sin\frac{\phi+\psi}{2}\cos\frac{\theta}{2}\right)$$
$$= -\omega_1\cos\frac{\phi-\psi}{2}\sin\frac{\theta}{2} + \omega_2\sin\frac{\phi-\psi}{2}\sin\frac{\theta}{2} + \omega_3\cos\frac{\phi+\psi}{2}\cos\frac{\theta}{2}.$$

From the second and third eliminate $\dot{\phi}-\dot{\psi}$, and we get by inspection
$$\cos\frac{\theta}{2}.\dot{\theta} = (\omega_1\sin\phi + \omega_2\cos\phi)\cos\frac{\theta}{2},$$
or
$$\dot{\theta} = \omega_1\sin\phi + \omega_2\cos\phi. \qquad (1)$$

Similarly, by eliminating $\dot{\theta}$ between the same two equations,
$$\sin\frac{\theta}{2}(\dot{\phi}-\dot{\psi}) = \omega_3\sin\frac{\theta}{2} + \omega_1\cos\phi\cos\frac{\theta}{2} - \omega_2\sin\phi\cos\frac{\theta}{2}.$$

And from the first and last of the group of four
$$\cos\frac{\theta}{2}(\dot{\phi}+\dot{\psi}) = \omega_3\cos\frac{\theta}{2} - \omega_1\cos\phi\sin\frac{\theta}{2} + \omega_2\sin\phi\sin\frac{\theta}{2}.$$

These last two equations give

$$\dot{\phi}+\dot{\psi}\cos\theta = \omega_3. \quad\ldots\ldots\ldots\ldots\ldots\ldots\ldots (2)$$

$$\dot{\phi}\cos\theta+\dot{\psi} = (-\omega_1\cos\phi+\omega_2\sin\phi)\sin\theta+\omega_3\cos\theta.$$

From the last two we have

$$\dot{\psi}\sin\theta = -\omega_1\cos\phi+\omega_2\sin\phi. \quad\ldots\ldots\ldots\ldots\ldots (3)$$

(1), (2), (3) are the forms in which the equations are usually given.

359.] *To deduce expressions for the direction-cosines of a set of rectangular axes in any position in terms of rational functions of three quantities only.*

Let α, β, γ be unit-vectors in the directions of these axes. Let q be, as in § 356, the requisite quaternion operator for turning the cöordinate axes into the position of this rectangular system. Then

$$q = w+xi+yj+zk,$$

where, as in § 356, we may write

$$1 = w^2+x^2+y^2+z^2.$$

Then we have $\quad q^{-1} = w-xi-yj+zk,$

and therefore

$$\alpha = qiq^{-1} = (wi-x-yk+zj)(w-xi-yj-zk)$$
$$- (w^2+x^2-y^2-z^2)i+2(wz+xy)j+2(xz-wy)k,$$

where the coefficients of i, j, k are the direction-cosines of α as required. A similar process gives by inspection those of β and γ.

As given by Cayley*, after Rodrigues, they have a slightly different and somewhat less simple form—to which, however, they are easily reduced by putting

$$w = \frac{x}{\lambda} = \frac{y}{\mu} - \frac{z}{\nu} - \frac{1}{\kappa^{\frac{1}{2}}}.$$

The geometrical interpretation of either set is obvious from the nature of quaternions. For (taking Cayley's notation) if θ be the angle of rotation : $\cos f$, $\cos g$, $\cos h$, the direction-cosines of the axis, we have

$$q = w+xi+yj+zk = \cos\frac{\theta}{2}+\sin\frac{\theta}{2}(i\cos f+j\cos g+k\cos h),$$

so that

$$w = \cos\frac{\theta}{2},$$

$$x = \sin\frac{\theta}{2}\cos f;$$

$$y = \sin\frac{\theta}{2}\cos g,$$

$$z = \sin\frac{\theta}{2}\cos h.$$

* *Camb. and Dub. Math. Journal.* Vol. i. (1846)

From these we pass at once to Rodrigues' subsidiary formulae,

$$\kappa = \frac{1}{w^2} = \sec^2\frac{\theta}{2},$$

$$\lambda = \frac{x}{w} = \tan\frac{\theta}{2}\cos f,$$

$$\&c. = \&c.$$

360.] By the definition of *Homogeneous Strain*, it is evident that if we take any three (non-coplanar) unit-vectors a, β, γ in an unstrained mass, they become after the strain other vectors, not neces sarily unit-vectors, a_1, β_1, γ_1.

Hence any other *given* vector, which of course may be thus expressed,
$$\rho = xa + y\beta + z\gamma,$$
becomes
$$\rho_1 = xa_1 + y\beta_1 + z\gamma_1,$$
and is therefore known if a_1, β_1, γ_1 be given.

More precisely
$$\rho S.a\beta\gamma = aS.\beta\gamma\rho + \beta S.\gamma a\rho + \gamma S.a\beta\rho$$
becomes
$$\rho_1 S.a\beta\gamma = \phi\rho S.a\beta\gamma = a_1 S.\beta\gamma\rho + \beta_1 S.\gamma a\rho + \gamma_1 S.a\beta\rho.$$

Thus the properties of ϕ, as in Chapter V, enable us to study with great simplicity strains or displacements in a solid or liquid.

For instance, *to find a vector whose direction is unchanged by the strain*, is to solve the equation
$$V\rho\phi\rho = 0, \quad \text{or} \quad \phi\rho = g\rho,$$
where g is a scalar unknown.

[This vector equation is equivalent to *three* simple equations, and contains only *three* unknown quantities; viz. *two* for the *direction* of ρ (the *tensor* does not enter, or, rather, is a factor of each side), and the unknown g.]

We have seen that every such equation leads to a cubic in g which may be written
$$g^3 - m_2 g^2 + m_1 g - m = 0,$$
where m_2, m_1, m are scalars depending in a known manner on the constant vectors involved in ϕ. This must have *one* real root, and may have *three*.

361.] For simplicity let us assume that a, β, γ form a rectangular system, then we may operate by $S.a$, $S.\beta$, and $S.\gamma$; and thus at once obtain the equation for g, in the form

$$\begin{vmatrix} Saa_1 + g, & Sa\beta_1, & Sa\gamma_1 \\ S\beta a_1, & S\beta\beta_1 + g, & S\beta\gamma_1 \\ S\gamma a_1, & S\gamma\beta_1, & S\gamma\gamma_1 + g \end{vmatrix} = 0. \quad \ldots\ldots\ldots (1)$$

To reduce this we have

$$\begin{vmatrix} S\alpha\alpha_1, & S\alpha\beta_1, & S\alpha\gamma_1 \\ S\beta\alpha_1, & S\beta\beta_1, & S\beta\gamma_1 \\ S\gamma\alpha_1, & S\gamma\beta_1, & S\gamma\gamma_1 \end{vmatrix}$$

$$= \frac{1}{S\alpha\alpha_1} \begin{vmatrix} S^2\alpha\alpha_1 + S^2\beta\alpha_1 + S^2\gamma\alpha_1, & \Sigma S\alpha\alpha_1 S\alpha\beta_1, & \Sigma S\alpha\alpha_1 S\alpha\gamma_1 \\ S\beta\alpha_1, & S\beta\beta_1, & S\beta\gamma_1 \\ S\gamma\alpha_1, & S\gamma\beta_1, & S\gamma\gamma_1 \end{vmatrix}$$

which, if the mass be rigid, becomes successively

$$\frac{1}{S\alpha\alpha_1} \begin{vmatrix} S\beta\beta_1, & S\beta\gamma_1 \\ S\gamma\beta_1, & S\gamma\gamma_1 \end{vmatrix} = \frac{1}{S\alpha\alpha_1} S\beta(\beta_1 S\gamma\gamma_1 - \gamma_1 S\gamma\beta_1)$$

$$= \frac{1}{S\alpha\alpha_1} S.\beta V\gamma V\gamma_1\beta_1 = -1.$$

Thus the equation becomes

$$-1 - g(S\alpha\alpha_1 + S\beta\beta_1 + S\gamma\gamma_1) + g^2(S\alpha\alpha_1 + S\beta\beta_1 + S\gamma\gamma_1) + g^3 = 0,$$

or $$(g-1)(g^2 + g(1 + S\alpha\alpha_1 + S\beta\beta_1 + S\gamma\gamma_1) + 1) = 0.$$

362.] If we take $T\rho = C$ we consider a portion of the mass initially spherical. This becomes of course

$$T\phi^{-1}\rho_1 = C,$$

an ellipsoid, in the strained state of the body.

Or if we consider a portion which is spherical after the strain, i.e

$$T\rho_1 = C,$$

its initial form was $$T\phi\rho = C,$$

another ellipsoid. The relation between these ellipsoids is obvious from their equations. (See § 311.)

In either case the axes of the ellipsoid correspond to a rectangular set of three diameters of the sphere (§ 257). But we must carefully separate the cases in which these corresponding lines in the two surfaces are, and are not, coincident. For, in the former case there is *pure* strain, in the latter the strain is accompanied by rotation. Here we have at once the distinction pointed out by Stokes* and Helmholtz† between the cases of fluid motion in which there is, or is not, a velocity-potential. In ordinary fluid motion the distortion is of the nature of a pure strain, i.e. is differentially non-rotational; while in vortex motion it is essentially accompanied by rotation. But the resultant of two pure strains is generally a strain accompanied by rotation. The question before us beautifully illustrates the properties of the linear and vector function.

* *Cambridge Phil Trans.* 1845.

† *Crelle,* vol. lv. 1857. See also *Phil Mag.* (Supplement) June 1867.

363.] *To find the criterion of a pure strain.* Take α, β, γ now as unit-vectors parallel to the axes of the strain-ellipsoid, they become after the strain $a\alpha$, $b\beta$, $c\gamma$.

Hence $$\rho_1 = \phi\rho = -a\alpha S\alpha\rho - b\beta S\beta\rho - c\gamma S\gamma\rho.$$

And we have, for the criterion of a pure strain, the property of the function ϕ, that it is *self-conjugate*, i. e.

$$S\rho\phi\sigma = S\sigma\phi\rho.$$

364.] *Two pure strains, in succession, generally give a strain accompanied by rotation.* For if ϕ, ψ represent the strains, since they are pure we have

$$\left.\begin{array}{l} S\rho\phi\sigma = S\sigma\phi\rho, \\ S\rho\psi\sigma = S\sigma\psi\rho. \end{array}\right\} \quad \ldots\ldots\ldots\ldots\ldots\ldots \quad (1)$$

But for the compound strain we have

$$\rho_1 = \chi\rho = \psi\phi\rho,$$

and we have *not* generally

$$S\rho\chi\sigma = S\sigma\chi\rho.$$

For $$S\rho\psi\phi\sigma = S\sigma\phi\psi\rho,$$

by (1), and $\psi\phi$ is not generally the same as $\phi\psi$ (See Ex. 7 to Chapter V.)

365.] The simplicity of this view of the question leads us to suppose that we may easily *separate the pure strain from the rotation in any case*, and exhibit the corresponding functions.

When the linear and vector function expressing a strain is self-conjugate the strain is pure. When not self-conjugate, it may be broken up into pure and rotational parts in various ways (analogous to the separation of a quaternion into the *sum* of a scalar and a vector part, or into the *product* of a tensor and a versor part), of which two are particularly noticeable. Denoting by a bar a self-conjugate function, we have thus either

$$\phi = \psi + V.\epsilon(\ \ ),$$

$$\phi = q\bar{\varpi}(\ \ )q^{-1}, \quad \text{or} \quad \phi = \varpi_1.q(\ \ )q^{-1},$$

where ϵ is a vector, and q a quaternion (which may obviously be regarded as a mere versor).

That this is possible is seen from the fact that ϕ involves nine independent constants, while ψ and $\bar{\varpi}$ each involve six, and ϵ and q each three. If ϕ' be the function conjugate to ϕ, we have

$$\phi' = \psi - V.\epsilon(\ \ ),$$

so that $$2\psi = \phi + \phi',$$

and $$2V.\epsilon(\ \ ) = \phi - \phi',$$

which completely determine the first decomposition. This is of course, perfectly well known in quaternions, but it does not seem to have been noticed as a theorem in the kinematics of strains that there is always one, and but one, mode of resolving a strain into the geometrical composition of the separate effects of (1) a *pure* strain, and (2) a rotation accompanied by uniform dilatation perpendicular to its axis, the dilatation being measured by (sec. $\theta - 1$) where θ is the angle of rotation.

In the second form (whose solution does not appear to have been attempted), we have

$$\phi = q\varpi(\quad)q^{-1},$$

where the pure strain precedes the rotation, and from this

$$\phi' = \bar{\varpi}.q^{-1}(\quad)q,$$

or in the conjugate strain the rotation (reversed) is followed by the pure strain. From these

$$\begin{aligned}\phi'\phi &= \varpi.q^{-1}(q\bar{\varpi}(\quad)q^{-1})\,q \\ &= \bar{\varpi}^2,\end{aligned}$$

and $\bar{\varpi}$ is to be found by the solution of a biquadratic equation *. It is evident, indeed, from the identical equation

$$S.\sigma\phi'\phi\rho = S.\rho\phi'\phi\sigma$$

that the operator $\phi'\phi$ is self-conjugate.

In the same way

$$\phi\phi'(\quad) = q\,\bar{\varpi}^2\,(q^{-1}(\quad)q)\,q^{-1},$$

or

$$q^{-1}(\phi\phi'\rho)\,q = \varpi^2\,(q^{-1}\rho q) = \phi'\phi\,(q^{-1}\rho q),$$

which shew the relations between $\phi\phi'$, $\phi'\phi$, and q.

To determine q we have

$$\phi\rho.q = q\varpi\rho$$

* Suppose the cubic in $\bar{\varpi}$ to be

$$\bar{\varpi}^3 + g\bar{\varpi}^2 + g_1\varpi + g_2 = 0,$$

write ω for $\phi'\phi$ in the given equation, and by its help this may be written as

$$(\bar{\varpi} + g)\,\omega + g_1\bar{\varpi} + g_2 = 0 = \bar{\varpi}\,(\omega + g_1) + g\omega + g_2.$$

Eliminating ϖ, we have

$$\omega^3 + (2g_1 - g^2)\,\omega^2 + (g_1^2 - 2gg_2)\,\omega - g_2^2 = 0.$$

This must agree with the (known) cubic in ω,

$$\omega^3 + m\omega^2 + m_1\omega + m_2 = 0,$$

suppose, so that by comparison of coefficients we have

$$2g_1 - g^2 = m, \qquad g_1^2 - 2gg_2 = m_1, \qquad g_2^2 = -m_2;$$

so that g_2 is known, and

$$g - \frac{g_1^2 - m_1}{2\sqrt{-m_2}},$$

where

$$2g_1 = m - \frac{(g_1^2 - m_1)^2}{4m_2}.$$

The values of the quantities g being found, $\bar{\varpi}$ is given in terms of ω by the equation above. (*Proc. R. S. E.*, 1870–71.)

whatever be ρ, so that

$$S.Vq(\phi-\varpi)\rho = 0,$$

or

$$S.\rho(\phi'-\varpi)Vq = 0,$$

which gives

$$(\phi'-\varpi)Vq = 0,$$

The former equation gives evidently

$$Vq \parallel V.(\phi-\varpi)\alpha(\phi-\varpi)\beta$$

whatever be α and β; and the rest of the solution follows at once. A similar process gives us the solution when the rotation precedes the pure strain.

366.] In general, if

$$\rho_1 = \phi\rho = -\alpha_1 S\alpha\rho - \beta_1 S\beta\rho - \gamma_1 S\gamma\rho,$$

the angle between any two lines, say ρ and σ, becomes in the altered state of the body

$$\cos^{-1}(-S.U\phi\rho U\phi\sigma).$$

The plane $S\zeta\rho = 0$ becomes (with the notation of § 144)

$$S\zeta\rho_1 = 0 = S\zeta\phi\rho = S\rho\phi'\zeta.$$

Hence the angle between the planes $S\zeta\rho = 0$, and $S\eta\rho = 0$, which is $\cos^{-1}(-S.U\zeta U\eta)$, becomes

$$\cos^{-1}(-S.U\phi'\zeta U\phi'\eta).$$

The *locus of lines equally elongated* is, of course,

$$T\phi U\rho = e,$$

or

$$T\phi\rho = eT\rho$$

a cone of the second order.

367.] In the case of a *Simple Shear*, we have, obviously,

$$\rho_1 = \phi\rho = \rho + \beta S\alpha\rho,$$

where

$$S\alpha\beta = 0.$$

The vectors which are unaltered in length are given by

$$T\rho_1 = T\rho,$$

or

$$2S\beta\rho S\alpha\rho + \beta^2 S^2\alpha\rho = 0,$$

which breaks up into

$$S.\alpha\rho = 0,$$

and

$$S\rho(2\beta + \beta^2\alpha) = 0.$$

The intersection of this plane with the plane of α, β is perpendicular to $2\beta + \beta^2\alpha$. Let it be $\alpha + x\beta$, then

$$S.(2\beta + \beta^2\alpha)(\alpha + x\beta) = 0,$$

i.e.

$$2x - 1 = 0.$$

Hence the intersection required is

$$\alpha + \frac{\beta}{2}.$$

For the axes of the strain, one is of course $\alpha\beta$, and the others are found by making $T\phi U\rho$ a maximum and minimum.

Let $$\rho = \alpha + x\beta,$$

then $$\rho_1 = \phi\rho = \alpha + x\beta - \beta,$$

and $$\frac{T\rho_1}{T\rho} = \text{max. or min.},$$

gives $$x^2 - x + \frac{1}{\beta^2} = 0,$$

from which the values of x are found.

Also, as a verification,

$$S.(\alpha + x_1\beta)(\alpha + x_2\beta) = -1 + \beta^2 x_1 x_2,$$

and should be $= 0$. It is so, since, by the equation,

$$x_1 x_2 = \frac{1}{\beta^2}.$$

Again

$$S\{\alpha + (x_1 - 1)\beta\}\{\alpha + (x_2 - 1)\beta\} = -1 + \beta^2\{x_1 x_2 - (x_1 + x_2) + 1\},$$

which ought also to be zero. And, in fact, $x_1 + x_2 = 1$ by the equation; so that this also is verified.

368.] We regret that our limits do not allow us to enter farther upon this very beautiful application.

But it may be interesting here, especially for the consideration of *any* continuous displacements of the particles of a mass, to introduce another of the extraordinary instruments of analysis which Hamilton has invented. Part of what is now to be given has been anticipated in last Chapter, but for continuity we commence afresh.

If $$F\rho = C \quad \ldots\ldots\ldots\ldots\ldots\ldots \quad (1)$$

be the equation of one of a system of surfaces, and if the differential of (1) be $$S\nu d\rho = 0, \quad \ldots\ldots\ldots\ldots\ldots\ldots \quad (2)$$

ν is a vector perpendicular to the surface, and *its length is inversely proportional to the normal distance between two consecutive surfaces.* In fact (2) shews that ν is perpendicular to $d\rho$, which is any tangent vector, thus proving the first assertion. Also, since in passing to a proximate surface we may write

$$S\nu\delta\rho = \delta C,$$

we see that $$F(\rho + \nu^{-1}\delta C) = C + \delta C.$$

This proves the latter assertion.

It is evident from the above that if (1) be an equipotential or an isothermal, surface, $-\nu$ *represents in direction and magnitude the force at any point or the flux of heat.* And we have seen (§ 317) that if

$$\nabla = i\frac{d}{dx} + j\frac{d}{dy} + k\frac{d}{dz},$$

giving $$\nabla^2 = -\frac{d^2}{dx^2} - \frac{d^2}{dy^2} - \frac{d^2}{dz^2},$$

then $$\nu = \nabla F\rho.$$

This is due to Hamilton (*Lectures on Quaternions*, p. 611).

369.] From this it follows that the effect of the vector operation ∇, upon any scalar function of the vector of a point, is to produce *the vector which represents in magnitude and direction the most rapid change in the value of the function.*

Let us next consider the effect of ∇ upon a *vector* function as

$$\sigma = i\xi + j\eta + k\zeta.$$

We have at once

$$\nabla\sigma = -\left(\frac{d\xi}{dx} + \frac{d\eta}{dy} + \frac{d\zeta}{dz}\right) - i\left(\frac{d\eta}{dz} - \frac{d\zeta}{dy}\right) - \&c.,$$

and in this semi-Cartesian form it is easy to see that:—

If σ represent a small vector displacement of a point situated at the extremity of the vector ρ (drawn from the origin)

$S\nabla\sigma$ represents the consequent cubical compression of the group of points in the vicinity of that considered, and

$V\nabla\sigma$ represents twice the vector axis of rotation of the same group of points.

Similarly $$S\sigma\nabla = -\left(\xi\frac{d}{dx} + \eta\frac{d}{dy} + \zeta\frac{d}{dz}\right) = -D_\sigma,$$

or is equivalent to total differentiation in virtue of our having passed from one end to the other of the vector σ.

370.] Suppose we fix our attention upon a group of points which originally filled a small sphere about the extremity of ρ as centre, whose equation referred to that point is

$$T\omega = e. \quad\ldots\ldots\ldots\ldots\ldots\ldots\ldots (1)$$

After displacement ρ becomes $\rho+\sigma$, and, by last section, $\rho+\omega$ becomes $\rho+\omega+\sigma-(S\omega\nabla)\sigma$. Hence the vector of the new surface which encloses the group of points (drawn from the extremity of $\rho+\sigma$) is $$\omega_1 = \omega - (S\omega\nabla)\sigma. \quad\ldots\ldots\ldots\ldots (2)$$

Hence ω is a homogeneous linear and vector function of ω_1; or

$$\omega = \phi\omega_1,$$

and therefore, by (1), $$T\phi\omega_1 = e,$$

the equation of the new surface, which is evidently a central surface of the second order, and therefore, of course, an ellipsoid.

We may solve (2) with great ease by approximation, if we remember that $T\sigma$ is very small, and therefore that in the small term we may put ω_1 for ω; i.e. omit squares of small quantities; thus

$$\omega = \omega_1 + (S\omega_1\nabla)\sigma.$$

371.] *If the small displacement of each point of a medium is in the direction of, and proportional to, the attraction exerted at that point by any system of material masses, the displacement is effected without rotation.*

For if $F\rho = C$ be the potential surface, we have $S\sigma d\rho$ a complete differential; i.e. in Cartesian coördinates

$$\xi\, dx + \eta\, dy + \zeta\, dz$$

is a differential of three independent variables. Hence the vector axis of rotation

$$i\left(\frac{d\zeta}{dy} - \frac{d\eta}{dz}\right) + \&c.,$$

vanishes by the vanishing of each of its constituents, or

$$V.\nabla\sigma = 0.$$

Conversely, *if there be no rotation, the displacements are in the direction of, and proportional to, the normal vectors to a series of surfaces.*

For $$0 = V.d\rho V.\nabla\sigma = (Sd\rho\nabla)\sigma - \nabla S\sigma d\rho,$$

where, in the last term, ∇ acts on σ alone.

Now, of the two terms on the right, the first is a complete differential, since it may be written $-D_{d\rho}\sigma$, and therefore the remaining term must be so.

Thus, in a distorted system, there is no compression if

$$S\nabla\sigma = 0,$$

and no rotation if $$V.\nabla\sigma = 0\cdot$$

and evidently merely transference if $\sigma = \alpha =$ a constant vector, which is one case of $$\nabla\sigma = 0.$$

In the important case of $$\sigma = e\nabla F\rho$$

there is evidently no rotation, since

$$\nabla\sigma = e\nabla^2 F\rho$$

is evidently a scalar. In this case, then, there are only translation and compression, and the latter is at each point proportional to the density of a distribution of matter, which would give the potential $F\rho$. For if r be such density, we have at once

$$\nabla^2 F\rho = 4\pi r^*.$$

372.] The *Moment of Inertia* of a body about a unit vector α as axis is evidently $$Mk^2 - -\Sigma m(V\alpha\rho)^2,$$

where ρ is the vector of the portion m of the mass, and the origin of ρ is in the axis.

* *Proc. R. S. E.*, 1862-3.

Hence if we take $kTa = e^2$, we have, as locus of the extremity of a,

$$Me^4 = -\Sigma m (Va\rho)^2 = MSa\phi a \text{ (suppose)},$$

the momental ellipsoid.

If ϖ be the vector of the centre of inertia, σ the vector of m with respect to it, we have $\rho = \varpi + \sigma$;

therefore

$$Mk^2 = -\Sigma m \{(Va\varpi)^2 + (Va\sigma)^2\}$$
$$= -M(Va\varpi)^2 + MSa\phi_1 a.$$

Now, for principal axes, k is max., min., or max.-min., with the condition

$$a^2 = -1.$$

Thus we have

$$Sa'(\varpi Va\varpi - \phi_1 a) = 0,$$
$$Sa'a = 0;$$

therefore $-\phi_1 a + \varpi Va\varpi = pa = k^2 a$ (by operating by Sa).

Hence

$$(\phi_1 + k^2 + \varpi^2)a = +\varpi Sa\varpi, \quad\quad (1)$$

determines the values of a, k^2 being found from the equation

$$S\varpi(\phi + k^2 + \varpi^2)^{-1}\varpi = 1. \quad\quad (2)$$

Now the normal to $S\sigma(\phi + k^2 + \varpi^2)^{-1}\sigma = 1$, (3)

at the point σ is $(\phi + k^2 + \varpi^2)^{-1}\sigma$.

But (3) passes through $-\varpi$, by (2), and *there* the normal is

$$(\phi + k^2 + \varpi^2)^{-1}\varpi,$$

which, by (1), is parallel to one of the required values of a. Thus we prove Binet's theorem that *the principal axes at any point are normals to the three surfaces, confocal with the momental ellipsoid, which pass through that point.*

EXAMPLES TO CHAPTER X.

1. Form, from kinematical principles, the equation of the cycloid; and employ it to prove the well-known elementary properties of the arc, tangent, radius of curvature, and evolute, of the curve.

2. Interpret, kinematically, the equation

$$\dot\rho = aU(\beta t - \rho),$$

where β is a given vector, and a a given scalar.

Shew that it represents a plane curve; and give it in an integrated form independent of t.

3. If we write $$\varpi = \beta t - \rho,$$
the equation in (2) becomes
$$\beta - \dot{\varpi} = aU\varpi.$$
Interpret this kinematically; and find an integal.

What is the nature of the step we have taken in transforming from the equation of (2) to that of the present question?

4. The motion of a point in a plane being given, refer it to

(*a*.) Fixed rectangular vectors in the plane.

(*b*.) Rectangular vectors in the plane, revolving uniformly about a fixed point.

(*c*.) Vectors, in the plane, revolving with different, but uniform, angular velocities.

(*d*.) The vector radius of a fixed circle, drawn to the point of contact of a tangent from the moving point.

In each case translate the result into Cartesian cöordinates.

5. Any point of a line of given length, whose extremities move in fixed lines in a given plane, describes an ellipse.

Shew how to find the centre, and axes, of this ellipse; and the angular velocity about the centre of the ellipse of the tracing point when the describing line rotates uniformly.

Transform this construction so as to shew that the ellipse is a hypotrochoid.

6. A point, A, moves uniformly round one circular section of a cone; find the angular velocity of the point, a, in which the generating line passing through A meets a subcontrary section about the centre of that section.

7. Solve, generally, the problem of finding the path by which a point will pass in the least time from one given point to another, the velocity at the point of space whose vector is ρ being expressed by the given scalar function $$f\rho.$$

Take also the following particular cases:—

(*a*.) $f\rho = a$ while $Sa\rho > 1$,
$f\rho = b$ while $Sa\rho < 1$.

(*b*.) $f\rho = Sa\rho$.

(*c*.) $f\rho = -\rho^2$. (Tait, *Trans. R. S. E.*, 1865.)

8. If, in the preceding question, $f\rho$ be such a function of $T\rho$ that any one swiftest path is a circle, every other such path is a circle, and all paths diverging from one point converge accurately in another. (Maxwell, *Cam. and Dub. Math. Journal*, IX. p. 9.)

9. Interpret, as results of the composition of successive conical rotations, the apparent truisms

$$\frac{\alpha}{\gamma}\frac{\gamma}{\beta}\frac{\beta}{\alpha} = 1,$$

and
$$\frac{\alpha}{\kappa}\frac{\kappa}{\iota}\frac{\iota}{\theta}\cdots\cdots\frac{\delta}{\gamma}\frac{\gamma}{\beta}\frac{\beta}{\alpha} = 1.$$

(Hamilton, *Lectures*, p. 334.)

10. Interpret, in the same way, the quaternion operators

$$q = (\delta\epsilon^{-1})^{\frac{1}{2}}(\epsilon\zeta^{-1})^{\frac{1}{2}}(\zeta\delta^{-1})^{\frac{1}{2}},$$

and
$$q = \left(\frac{\alpha}{\epsilon}\right)^{\frac{1}{2}}\left(\frac{\epsilon}{\delta}\right)^{\frac{1}{2}}\left(\frac{\delta}{\gamma}\right)^{\frac{1}{2}}\left(\frac{\gamma}{\beta}\right)^{\frac{1}{2}}\left(\frac{\beta}{\alpha}\right)^{\frac{1}{2}}. \quad (Ibid.)$$

11. Find the axis and angle of rotation by which one given rectangular set of unit-vectors α, β, γ is changed into another given set α_1, β_1, γ_1.

12. Shew that, if $\phi\rho = \rho + V\epsilon\rho$,
the linear and vector operation ϕ denotes rotation about the vector ϵ, together with uniform expansion in all directions perpendicular to it.

Prove this also by forming the operator which produces the expansion without the rotation, and that producing the rotation without the expansion; and finding their joint effect.

13. Express by quaternions the motion of a side of one right cone rolling uniformly upon another which is fixed, the vertices of the two being coincident.

14. Given the simultaneous angular velocities of a body about the principal axes through its centre of inertia, find the position of these axes in space at any assigned instant.

15. Find the linear and vector function, and also the quaternion operator, by which we may pass, in any simple crystal of the cubical system, from the normal to one given face to that to another. How can we use them to distinguish a series of faces belonging to the same zone?

16. Classify the simple forms of the cubical system by the properties of the linear and vector function, or of the quaternion operator.

17. Find the vector normal of a face which truncates symmetrically the edge formed by the intersection of two given faces.

18. Find the normals of a pair of faces symmetrically truncating the given edge.

19. Find the normal of a face which is equally inclined to three given faces.

20. Shew that the rhombic dodecahedron may be derived from the cube, or from the octahedron, by truncation of the edges.

21. Find the form whose faces replace, symmetrically, the edges of the rhombic dodecahedron.

22. Shew how the two kinds of hemihedral forms are indicated by the quaternion expressions.

23. Shew that the cube may be produced by truncating the edges of the regular tetrahedron.

24. Point out the modifications in the auxiliary vector function required in passing to the pyramidal and prismatic systems respectively.

25. In the rhombohedral system the auxiliary quaternion operator assumes a singularly simple form. Give this form, and point out the results indicated by it.

26. Shew that if the hodograph be a circle, and the acceleration be directed to a fixed point; the orbit must be a conic section, which is limited to being a circle if the acceleration follow any other law than that of gravity.

27. In the hodograph corresponding to acceleration $f(D)$ directed towards a fixed centre, the curvature is inversely as $D^2 f(D)$.

28. If two circular hodographs, having a common chord, which passes through, or tends towards, a common centre of force, be cut by any two common orthogonals, the sum of the two times of hodographically describing the two intercepted arcs (small or large) will be the same for the two hodographs. (Hamilton, *Elements*, p. 725.)

29. Employ the last theorem to prove, after Lambert, that the time of describing any arc of an elliptic orbit may be expressed in terms of the chord of the arc and the extreme radii vectores.

30. If $q(\ \ )q^{-1}$ be the operator which turns one set of rect angular unit-vectors α, β, γ into another set $\alpha_1, \beta_1, \gamma_1$, shew that there are three equations of the form

$$S\alpha\beta_1 - S\beta\alpha_1 = -\frac{4SqS\gamma q}{Tq^2}.$$

CHAPTER XI.

PHYSICAL APPLICATIONS.

373.] We propose to conclude the work by giving a few instances of the ready applicability of quaternions to questions of mathematical physics, upon which, even more than on the Geometrical or Kinematical applications, the real usefulness of the Calculus must mainly depend—except, of course, in the eyes of that section of mathematicians for whom Transversals and Anharmonic Pencils, &c. have a to us incomprehensible charm. Of course we cannot attempt to give examples in all branches of physics, nor even to carry very far our investigations in any one branch: this Chapter is not intended to teach Physics, but merely to shew by a few examples how expressly and naturally quaternions seem to be fitted for attacking the problems it presents.

We commence with a few general theorems in Dynamics—the formation of the equations of equilibrium and motion of a rigid system, some properties of the central axis, and the motion of a solid about its centre of inertia.

374.] When any forces act on a rigid body, the force β at the point whose vector is α, &c., then, if the body be slightly displaced, so that α becomes $\alpha + \delta\alpha$, the whole work done is

$$\Sigma S\beta\delta\alpha.$$

This must vanish if the forces are such as to maintain equilibrium. Hence *the condition of equilibrium of a rigid body is*

$$\Sigma S\beta\delta\alpha = 0.$$

For a displacement of translation $\delta\alpha$ is *any* constant vector, hence

$$\Sigma\beta = 0. \qquad\qquad (1)$$

For a rotation-displacement, we have by § 350, ϵ being the axis, and $T\epsilon$ being indefinitely small,

$$\delta\alpha = V\epsilon\alpha,$$

and $\quad \Sigma S.\beta V\epsilon\alpha = \Sigma S.\epsilon V\alpha\beta = S.\epsilon\Sigma(V\alpha\beta) = 0,$

whatever be ϵ, hence $\quad \Sigma.V\alpha\beta = 0. \quad\ldots\ldots\ldots\ldots\ldots\ldots (2)$

These equations, (1) and (2), are equivalent to the ordinary six equations of equilibrium.

375.] In general, for any set of forces, let

$$\Sigma\beta = \beta_1,$$

$$\Sigma.V\alpha\beta = \alpha_1,$$

it is required *to find the points for which the couple* α_1 *has its axis coincident with the resultant force* β_1. Let γ be the vector of such a point.

Then for it the axis of the couple is

$$\Sigma.V(\alpha-\gamma)\beta = \alpha_1 - V\gamma\beta_1,$$

and by condition $\quad x\beta_1 = \alpha_1 - V\gamma\beta_1.$

Operate by $S\beta_1$; therefore

$$x\beta_1^2 = S\alpha_1\beta_1,$$

and $\quad V\gamma\beta_1 = \alpha_1 - \beta_1^{-1}S\alpha_1\beta_1 = -\beta_1 V\alpha_1\beta_1^{-1},$

or $\quad \gamma = V\alpha_1\beta_1^{-1} + y\beta_1,$

a straight line (the *Central Axis*) parallel to the resultant force.

376.] *To find the points about which the couple is least.*

Here $\quad T(\alpha_1 - V\gamma\beta_1) = \text{minimum}.$

Therefore $\quad S.(\alpha_1 - V\gamma\beta_1)V\beta_1\gamma' = 0,$

where γ' is any vector whatever. It is useless to try $\gamma' = \beta_1$, but we may put it in succession equal to α_1 and $V\alpha_1\beta_1$. Thus

$$S.\gamma V.\beta_1 V\alpha_1\beta_1 = 0,$$

and $\quad (V\alpha_1\beta_1)^2 - \beta_1^2 S.\gamma V\alpha_1\beta_1 = 0.$

Hence $\quad \gamma = xV\alpha_1\beta_1 + y\beta_1,$

and by operating with $S.V\alpha_1\beta_1$, we get

$$\frac{1}{\beta_1^2}(V\alpha_1\beta_1)^2 = x(V\alpha_1\beta_1)^2,$$

or $\quad \gamma = V\alpha_1\beta_1^{-1} + y\beta_1,$

the same locus as in last section.

377.] The couple vanishes if

$$\alpha_1 - V\gamma\beta_1 = 0.$$

This necessitates $\quad S\alpha_1\beta_1 = 0,$

or the force must be *in* the plane of the couple. If this be the case,

$$\gamma = \alpha_1\beta_1^{-1} + x\beta_1,$$

still the central axis.

378.] To assign the values of forces ξ, ξ_1, to act at ϵ, ϵ_1, and be equivalent to the given system.

$$\xi + \xi_1 = \beta_1,$$

$$V\epsilon\xi + V\epsilon_1\xi_1 = a_1.$$

Hence $$V\epsilon\xi + V\epsilon_1(\beta_1 - \xi) = a_1,$$

and $$\xi = (\epsilon - \epsilon_1)^{-1}(a_1 - V\epsilon_1\beta_1) + x(\epsilon - \epsilon_1).$$

Similarly for ξ_1. The indefinite terms may be omitted, as they must evidently be equal and opposite. In fact they are any equal and opposite forces whatever acting in the line joining the given points.

379.] For the motion of a rigid system, we have of course

$$\Sigma S(m\ddot{a} - \beta)\delta a = 0,$$

by the general equation of Lagrange.

Suppose the displacements δa to correspond to a mere *translation*, then δa is *any* constant vector, hence

$$\Sigma(m\ddot{a} - \beta) = 0,$$

or, if a_1 be the vector of the centre of inertia, and therefore

$$a_1 \Sigma m = \Sigma m a,$$

we have at once $$\ddot{a}_1 \Sigma m - \Sigma\beta = 0,$$

and the centre of inertia moves as if the whole mass were concentrated in it, and acted upon by all the applied forces.

380.] Again, let the displacements δa correspond to a rotation about an axis ϵ, passing through the origin, then

$$\delta a = V\epsilon a,$$

it being assumed that $T\epsilon$ is indefinitely small.

Hence $$\Sigma S.\epsilon V a(m\ddot{a} - \beta) = 0,$$

for *all* values of ϵ, and therefore

$$\Sigma.Va(m\ddot{a} - \beta) = 0,$$

which contains the three remaining ordinary equations of motion.

Transfer the origin to the centre of inertia, i.e. put $a = a_1 + \varpi$, then our equation becomes

$$\Sigma V(a_1 + \varpi)(m\ddot{a}_1 + m\ddot{\varpi} - \beta) = 0.$$

Or, since $\Sigma m\varpi = 0$,

$$\Sigma V\varpi(m\ddot{\varpi} - \beta) + Va_1(\ddot{a}_1 \Sigma m - \Sigma\beta) = 0.$$

But $\ddot{a}_1 \Sigma m - \Sigma\beta = 0$, hence our equation is simply

$$\Sigma V\varpi(m\ddot{\varpi} - \beta) = 0.$$

Now $\Sigma V\varpi\beta$ is the couple, about the centre of inertia, produced by the applied forces; call it ϕ, then

$$\Sigma m V\varpi\ddot{\varpi} = \phi. \quad \text{.............................} \quad (1)$$

381.] Integrating once,

$$\Sigma m V \varpi \dot{\varpi} = \gamma + \int \phi \, dt. \qquad (2)$$

Again, as the motion considered is *relative* to the centre of inertia, it must be of the nature of rotation about some axis, in general variable. Let ϵ denote at once the direction of, and the angular velocity about, this axis. Then, evidently,

$$\dot{\varpi} = V \epsilon \varpi.$$

Hence, the last equation may be written

$$\Sigma m \varpi V \epsilon \varpi = \gamma + \int \phi dt.$$

Operating by $S.\epsilon$, we get

$$\Sigma m (V \epsilon \varpi)^2 = S \epsilon \gamma + S \epsilon \int \phi dt. \qquad (3)$$

But, by operating directly by $2\int S \dot{\epsilon} dt$ upon the equation (1), we get

$$\Sigma m (V \epsilon \varpi)^2 = -h^2 + 2 \int S \epsilon \phi dt. \qquad (4)$$

(2) and (4) contain the usual four integrals of the first order.

382.] When no forces act on the body, we have $\phi = 0$, and therefore

$$\Sigma m \varpi V \epsilon \varpi = \gamma, \qquad (5)$$

$$\Sigma m \dot{\varpi}^2 = \Sigma m (V \epsilon \varpi)^2 = -h^2, \qquad (6)$$

and, from (5) and (6),

$$S \epsilon \gamma = -h^2. \qquad (7)$$

One interpretation of (6) is, that the kinetic energy of rotation remains unchanged: another is, that the vector ϵ terminates in an ellipsoid whose centre is the origin, and which therefore assigns the angular velocity when the direction of the axis is given; (7) shews that the extremity of the instantaneous axis is always in a plane fixed in space.

Also, by (5), (7) is the equation of the tangent plane to (6) at the extremity of the vector ϵ. Hence the ellipsoid (6) *rolls* on the plane (7).

From (5) and (6), we have at once, as an equation which ϵ must satisfy,

$$\gamma^2 \Sigma . m (V \epsilon \varpi)^2 = -h^2 (\Sigma . m \varpi V \epsilon \varpi)^2.$$

This belongs to a cone of the second degree fixed in the body. Thus all the ordinary results regarding the motion of a rigid body under the action of no forces, the centre of inertia being fixed, are deduced almost intuitively: and the only difficulties to be met with in more complex properties of such motion are those of integration, which are inherent to the subject, and appear whatever analytical method is employed. (Hamilton, *Proc. R. I. A.* 1848.)

383.] Let a be the initial position of ϖ, q the quaternion by which the body can be at one step transferred from its initial position to its position at time t. Then

$$\varpi = q a q^{-1}$$

and Hamilton's equation (5) of last section becomes

$$\Sigma . mqaq^{-1}V . \epsilon qaq^{-1} = \gamma,$$

or

$$\Sigma . mq\,\{aS . aq^{-1}\epsilon q - q^{-1}\epsilon qa^2\}\,q^{-1} = \gamma .$$

Let

$$\phi\rho = \Sigma . m\,(aSa\rho - a^2\rho), \quad \ldots\ldots\ldots\ldots \quad (1)$$

where ϕ is a self-conjugate linear and vector function, whose constituent vectors are fixed in the body in its initial position. Then the previous equation may be written

$$q\phi\,(q^{-1}\epsilon q)\,q^{-1} = \gamma,$$

or

$$\phi\,(q^{-1}\epsilon q) = q^{-1}\gamma q .$$

For simplicity let us write

$$\left.\begin{array}{l} q^{-1}\epsilon q = \eta, \\ q^{-1}\gamma q = \zeta . \end{array}\right\} \quad \ldots\ldots\ldots\ldots \quad (2)$$

Then Hamilton's dynamical equation becomes simply

$$\phi\eta = \zeta . \quad \ldots\ldots\ldots\ldots \quad (3)$$

384.] It is easy to see what the new vectors η and ζ represent. For we may write (2) in the form

$$\left.\begin{array}{l} \epsilon = q\eta q^{-1}, \\ \gamma = q\zeta q^{-1}\,; \end{array}\right\} \quad \ldots\ldots\ldots\ldots \quad (2')$$

from which it is obvious that η is that vector in the initial position of the body which, at time t, becomes the instantaneous axis in the moving body. When no forces act, γ is constant, and ζ is the initial position of the vector which, at time t, is perpendicular to the invariable plane.

385.] The complete solution of the problem is contained in equations (2), (3) above, and (4) of § 356 *. Writing them again, we have

$$q\eta = 2\dot{q}, \quad \ldots\ldots\ldots\ldots \quad (4)$$

$$\gamma q = q\zeta, \quad \ldots\ldots\ldots\ldots \quad (2)$$

$$\phi\eta = \zeta . \quad \ldots\ldots\ldots\ldots \quad (3)$$

We have only to eliminate ζ and η, and we get

$$2\dot{q} = q\phi^{-1}(q^{-1}\gamma q), \quad \ldots\ldots\ldots\ldots \quad (5)$$

in which q is now the only unknown ; γ, if variable, being supposed known in terms of q and t. It is hardly conceivable that any simpler, or more easily interpretable, equation for q can be presented

* To these it is unnecessary to add

$$Tq = \text{constant},$$

as this constancy of Tq is proved by the *form* of (4). For, had Tq been variable, there must have been a quaternion in the place of the vector η. In fact,

$$\frac{d}{dt}(Tq)^2 = 2S . \dot{q}Kq = (Tq)^2 S\eta = 0 .$$

until symbols are devised far more comprehensive in their meaning than any we yet have.

386.] Before entering into considerations as to the integration of this equation, we may investigate some other consequences of the group of equations in § 385. Thus, for instance, differentiating (2), we have

$$\gamma q+\gamma q = \dot{q}\zeta+q\dot{\zeta},$$

and, eliminating $\dot{q}$ by means of (4),

$$\gamma q\eta + 2\dot{\gamma} q = q\eta\zeta + 2q\dot{\zeta},$$

whence
$$\dot{\zeta} = V\zeta\eta + q^{-1}\dot{\gamma}q\,;$$

which gives, in the case when no forces act, the forms

$$\dot{\zeta} = V\zeta\phi^{-1}\zeta, \quad \ldots\ldots\ldots\ldots \quad (6)$$

and
$$(\text{as } \dot{\zeta} = \phi\dot{\eta})$$

$$\phi\dot{\eta} = -V.\eta\phi\eta. \quad \ldots\ldots\ldots\ldots \quad (7)$$

To each of these the term $q^{-1}\dot{\gamma}q$, or $q^{-1}\psi q$, must be added on the right, if forces act.

387.] It is now desirable to examine the formation of the function ϕ. By its definition (1) we have

$$\phi\rho = \Sigma.m\,(\alpha S\alpha\rho - \alpha^2\rho),$$
$$= -\Sigma.m\alpha V\alpha\rho.$$

Hence
$$-S\rho\phi\rho = \Sigma.m\,(TV\alpha\rho)^2,$$

so that $-S\rho\phi\rho$ is the moment of inertia of the body about the vector ρ, multiplied by the square of the tensor of ρ. Thus the equation
$$S\rho\phi\rho = -h^2,$$

evidently belongs to an ellipsoid, of which the radii-vectores are inversely as the square roots of the moments of inertia about them; so that, if i, j, k be taken as unit-vectors in the directions of its axes respectively, we have

$$\left.\begin{aligned} Si\phi i &= -A, \\ Sj\phi j &= -B, \\ Sk\phi k &= -C, \end{aligned}\right\} \quad \ldots\ldots\ldots\ldots \quad (8)$$

A, B, C, being the principal moments of inertia. Consequently

$$\phi\rho = -\{AiSi\rho + BjSj\rho + CkSk\rho\}. \quad \ldots\ldots\ldots\ldots \quad (9)$$

Thus the equation (7) for η breaks up, if we put

$$\eta = i\omega_1 + j\omega_2 + k\omega_3,$$

into the three following scalar equations

$$\left.\begin{aligned} A\dot{\omega}_1 + (C-B)\,\omega_2\omega_3 &= 0, \\ B\dot{\omega}_2 + (A-C)\,\omega_3\omega_1 &= 0, \\ C\dot{\omega}_3 + (B-A)\,\omega_1\omega_2 &= 0, \end{aligned}\right\}$$

which are the same as those of Euler. Only, it is to be understood that the equations just written are not primarily to be considered as equations of rotation. They rather express, with reference to fixed axes in the initial position of the body, the motion of the extremity, $\omega_1, \omega_2, \omega_3$, of the vector corresponding to the instantaneous axis in the moving body. If, however, we consider $\omega_1, \omega_2, \omega_3$ as standing for their values in terms of w, x, y, z (§ 391 below), or any other cöordinates employed to refer the body to fixed axes, they *are* the equations of motion.

Similar remarks apply to the equation which determines ζ, for if we put

$$\zeta = i\varpi_1 + j\varpi_2 + k\varpi_3,$$

(6) may be reduced to three scalar equations of the form

$$\dot{\varpi}_1 - \left(\frac{1}{C} \quad \frac{1}{B}\right) \varpi_2 \varpi_3 = 0.$$

388.] Euler's equations in their usual form are easily deduced from what precedes. For, let

$$\phi\rho = q\phi(q^{-1}\rho q)q^{-1}$$

whatever be ρ; that is, let ϕ represent with reference to the moving principal axes what ϕ represents with reference to the principal axes in the initial position of the body, and we have

$$\begin{aligned}
\phi\dot{\epsilon} &= q\phi(q^{-1}\dot{\epsilon}q)q^{-1} = q\phi(\dot{\eta})q^{-1} \\
&= q\dot{\zeta}q^{-1} \qquad - qV(\zeta\phi^{-1}\zeta)q^{-1} \\
&= -qV(\eta\phi\eta)q^{-1} \\
&- -V.q\eta\phi(\eta)q^{-1} \\
&- -V.q\eta q^{-1}q\phi(q^{-}{}_1\epsilon q)q^{-1} \\
&- -V.\epsilon\phi\epsilon,
\end{aligned}$$

which is the required expression.

But perhaps the simplest mode of obtaining this equation is to start with Hamilton's unintegrated equation, which for the case of no forces is simply

$$\Sigma.mV\varpi\ddot{\varpi} = 0.$$

But from $\qquad \dot{\varpi} = V\epsilon\varpi$

we deduce $\qquad \ddot{\varpi} = V\epsilon\dot{\varpi} + V\dot{\epsilon}\varpi$

$$- \varpi\epsilon^2 - \epsilon S\epsilon\varpi + V\dot{\epsilon}\varpi,$$

so that $\qquad \Sigma.m(V\epsilon\varpi S\epsilon\varpi - \dot{\epsilon}\varpi^2 + \varpi S\dot{\epsilon}\varpi) = 0.$

If we look at equation (1), and remember that ϕ differs from ϕ simply in having ϖ substituted for a, we see that this may be written

$$V\epsilon\phi\epsilon + \phi\dot{\epsilon} = 0,$$

the equation before obtained. The first mode of arriving at it has been given because it leads to an interesting set of transformations, for which reason we append other two.

By (2) $\gamma = q\zeta q^{-1}$,

therefore $0 = \dot{q}q^{-1}.q\zeta q^{-1} + q\dot{\zeta}q^{-1} - q\zeta q^{-1}\dot{q}q^{-1}$,

or
$$q\dot{\zeta}q^{-1} = 2V.\gamma V\dot{q}q^{-1}$$
$$= V\gamma\epsilon.$$

But, by the beginning of this section, and by (5) of § 382, this is again the equation lately proved.

Perhaps, however, the following is neater. It occurs in Hamilton's *Elements*.

By (5) of § 382 $\phi\epsilon = \gamma$.

Hence
$$\phi\dot{\epsilon} = -\dot{\phi}\epsilon = -\Sigma.m(\dot{\varpi}V\epsilon\varpi + \varpi V\epsilon\dot{\varpi})$$
$$= -\Sigma.m\dot{\varpi}S\epsilon\varpi$$
$$= -V.\epsilon\Sigma.m\varpi S\epsilon\varpi$$
$$= -V\epsilon\phi\epsilon.$$

389.] However they are obtained, such equations as those of § 387 were shewn long ago by Euler to be integrable as follows.

Putting $2\int\omega_1\omega_2\omega_3 dt = s$,

we have $A\omega_1^2 = A\Omega_1^2 + (B-C)s$,

with other two equations of the same form. Hence

$$2\,dt = \frac{ds}{\left(\Omega_1^2 + \frac{B-C}{A}s\right)^{\frac{1}{2}}\left(\Omega_2^2 + \frac{C-A}{B}s\right)^{\frac{1}{2}}\left(\Omega_3^2 + \frac{A-B}{C}s\right)^{\frac{1}{2}}},$$

so that t is known in terms of s by an elliptic integral. Thus, finally, η or ζ may be expressed in terms of t; and in some of the succeeding investigations for q we shall suppose this to have been done. It is with this integration, or an equivalent one, that most writers on the farther development of the subject have commenced their investigations.

390.] By § 381, γ is evidently the vector moment of momentum of the rigid body; and the kinetic energy is

$$-\tfrac{1}{2}\Sigma.m\dot{\varpi}^2 = -\tfrac{1}{2}S\epsilon\gamma.$$

But $S\epsilon\gamma = S.q^{-1}\epsilon qq^{-1}\gamma q = S\eta\zeta$,

so that when no forces act

$$S\zeta\phi^{-1}\zeta = S\eta\phi\eta = -h^2.$$

But, by (2), we have also

$$T\zeta = T\gamma, \quad \text{or} \quad T\phi\eta = T\gamma,$$

so that we have, for the equations of the cones described in the initial position of the body by η and ζ, that is, for the cones described in the moving body by the instantaneous axis and by the perpendicular to the invariable plane,

$$h^2\zeta^2 + \gamma^2 S\zeta\phi^{-1}\zeta = 0,$$

$$h^2(\phi\eta)^2 + \gamma^2 S\eta\phi\eta = 0.$$

This is on the supposition that γ and h are constants. If forces act, these quantities are functions of t, and the equations of the cones then described in the body must be found by eliminating t between the respective equations. The final results to which such a process will lead must, of course, depend entirely upon the way in which t is involved in these equations, and therefore no general statement on the subject can be made.

391.] Recurring to our equations for the determination of q, and taking first the case of no forces, we see that, if we assume η to have been found (as in § 389) by means of elliptic integrals, we have to solve the equation

$$q\eta = 2\dot{q}\text{*},$$

that is, we have to integrate a system of four other differential equations harder than the first.

Putting, as in § 387, $\eta = i\omega_1 + j\omega_2 + k\omega_3$,

where ω_1, ω_2, ω_3 are supposed to be known functions of t, and

$$q = w + ix + jy + kz,$$

this system is $$\frac{1}{2}dt = \frac{dw}{W} = \frac{dx}{X} = \frac{dy}{Y} = \frac{dz}{Z},$$

* To get an idea of the nature of this equation, let us integrate it on the supposition that η is a *constant* vector. By differentiation and substitution, we get

$$2\ddot{q} = \dot{q}\eta = \tfrac{1}{2}\eta^2 q.$$

Hence

$$q = Q_1 \cos\frac{T\eta}{2}t + Q_2 \sin\frac{T\eta}{2}t.$$

Substituting in the given equation we have

$$T\eta\left(-Q_1\sin\frac{T\eta}{2}t + Q_2\cos\frac{T\eta}{2}t\right) = \left(Q_1\cos\frac{T\eta}{2}t + Q_2\sin\frac{T\eta}{2}t\right)\eta.$$

Hence

$$T\eta . Q_2 = Q_1\eta,$$

$$-T\eta . Q_1 = Q_2\eta,$$

which are virtually the same equation, and thus

$$q = Q_1\left(\cos\frac{T\eta}{2}t + U\eta\sin\frac{T\eta}{2}t\right)$$

$$= Q_1(U\eta)^{\frac{tT\eta}{\pi}}.$$

And the interpretation of $q(\quad)q^{-1}$ will obviously then be a rotation about η through the angle $tT\eta$, together with any other arbitrary rotation whatever. Thus any position whatever may be taken as the initial one of the body, and $Q_1(\quad)Q_1^{-1}$ brings it to its required position at time $t = 0$.

where
$$\begin{aligned} W &= -\omega_1 x - \omega_2 y - \omega_3 z, \\ X &= \quad \omega_1 w + \omega_3 y - \omega_2 z, \\ Y &= \quad \omega_2 w + \omega_1 z - \omega_3 x, \\ Z &= \quad \omega_3 w + \omega_2 x - \omega_1 y; \end{aligned}$$
or, as suggested by Cayley to bring out the skew symmetry,
$$\begin{aligned} X &= \quad . \quad \omega_3 y - \omega_2 z + \omega_1 w, \\ Y &= -\omega_3 x \quad . \quad + \omega_1 z + \omega_2 w, \\ Z &= \quad \omega_2 x - \omega_1 y \quad . \quad + \omega_3 w, \\ W &= -\omega_1 x - \omega_2 y - \omega_3 z \quad . \quad . \end{aligned}$$
Here, of course, one integral is
$$w^2 + x^2 + y^2 + z^2 = \text{constant}.$$

It may suffice thus to have alluded to a possible mode of solution, which, except for very simple values of η, involves very great difficulties. The quaternion solution, when η is of constant length and revolves uniformly in a right cone, will be given later.

392.] If, on the other hand, we eliminate η, we have to integrate
$$q\phi^{-1}(q^{-1}\gamma q) = 2\dot{q},$$
so that one integration theoretically suffices. But, in consequence of the present imperfect development of the quaternion calculus, the only known method of effecting this is to reduce the quaternion equation to a set of four ordinary differential equations of the first order. It may be interesting to form these equations.

Put
$$q = w + ix + jy + kz,$$
$$\gamma = ia + jb + kc,$$
then, by ordinary quaternion multiplication, we easily reduce the given equation to the following set:
$$\frac{dt}{2} = \frac{dw}{W} = \frac{dx}{X} = \frac{dy}{Y} = \frac{dz}{Z},$$
where
$$\begin{aligned} W &= -x\mathfrak{A} - y\mathfrak{B} - z\mathfrak{C} &\quad\text{or}\quad X &= \qquad\qquad y\mathfrak{C} - z\mathfrak{B} + w\mathfrak{A}, \\ X &= \quad w\mathfrak{A} + y\mathfrak{C} - z\mathfrak{B} & Y &= -x\mathfrak{C} \quad . \quad + z\mathfrak{A} + w\mathfrak{B}, \\ Y &= \quad w\mathfrak{B} + z\mathfrak{A} - x\mathfrak{C}, & Z &= \quad x\mathfrak{B} - y\mathfrak{A} \qquad + w\mathfrak{C}, \\ Z &= \quad w\mathfrak{C} + x\mathfrak{B} - y\mathfrak{A} & W &= -x\mathfrak{A} - y\mathfrak{B} - z\mathfrak{C} \quad . \quad , \end{aligned}$$
and
$$\mathfrak{A} = \frac{1}{A}[a(w^2 - x^2 - y^2 - z^2) + 2x(ax + by + cz) + 2w(bz - cy)],$$
$$\mathfrak{B} = \frac{1}{B}[b(w^2 - x^2 - y^2 - z^2) + 2y(ax + by + cz) + 2w(cx - az)],$$
$$\mathfrak{C} = \frac{1}{C}[c(w^2 - x^2 - y^2 - z^2) + 2z(ax + by + cz) + 2w(ay - bx)],$$

W, X, Y, Z are thus *homogeneous* functions of w, x, y, z of the third degree.

Perhaps the simplest way of obtaining these equations is to translate the group of § 385 into w, x, y, z at once, instead of using the equation from which ζ and η are eliminated.

We thus see that
$$\eta = i\mathfrak{A} + j\mathfrak{B} + k\mathfrak{C}.$$

One obvious integral of these equations ought to be
$$w^2 + x^2 + y^2 + z^2 = \text{constant},$$
which has been assumed all along. In fact, we see at once that
$$wW + xX + yY + zZ = 0$$
identically, which leads to the above integral.

These equations appear to be worthy of attention, partly because of the homogeneity of the denominators W, X, Y, Z, but particularly as they afford (what does not appear to have been sought) the means of solving this celebrated problem *at one step*, that is, without the previous integration of Euler's equations (§ 387).

A set of equations identical with these, but not in a homogeneous form (being expressed, in fact, in terms of κ, λ, μ, ν of § 359, instead of w, x, y, z), is given by Cayley (*Camb. and Dub. Math. Journal*, vol. i. 1846), and completely integrated (in the sense of being reduced to quadratures) by assuming Euler's equations to have been previously integrated. (Compare § 391.)

Cayley's method may be even more easily applied to the above equations than to his own; and I therefore leave this part of the development to the reader, who will at once see (as in § 391) that $\mathfrak{A}$, $\mathfrak{B}$, $\mathfrak{C}$ correspond to ω_1, ω_2, ω_3 of the η type, § 387.

393.] It may be well to notice, in connection with the formulae for direction cosines in § 359 above, that we may write

$$\mathfrak{A} - \frac{1}{A}[a(w^2 + x^2 - y^2 - z^2) + 2b(xy + wz) + 2c(xz - wy)],$$

$$\mathfrak{B} - \frac{1}{B}[2a(xy - wz) + b(w^2 - x^2 + y^2 - z^2) + 2c(yz + wx)],$$

$$\mathfrak{C} = \frac{1}{C}[2a(xz + wy) + 2b(yz - wx) + c(w^2 - x^2 - y^2 + z^2)].$$

These expressions may be considerably simplified by the usual assumption, that one of the fixed unit-vectors (i suppose) is perpendicular to the invariable plane, which amounts to assigning definitely the initial position of one line in the body; and which gives the relations
$$b = 0, \quad c = 0.$$

394.] When forces act, γ is variable, and the quantities a, b, c will in general involve all the variables w, x, y, z, t, so that the equations of last section become much more complicated. The type, however, remains the same if γ involves t only; if it involve q we must differentiate the equation, put in the form

$$\gamma = 2q\phi(q^{-1}\dot{q})q^{-1}$$

and we thus easily obtain the differential equation of the second order

$$\psi = 4V.\dot{q}\phi(q^{-1}\dot{q})q^{-1} + 2q\phi(V.q^{-1}\ddot{q})q^{-1};$$

if we recollect that, because $q^{-1}\dot{q}$ is a vector, we have

$$S.q^{-1}\ddot{q} = (q^{-1}\dot{q})^2.$$

Though remarkably simple, this formula, in the present state of the development of quaternions, must be looked on as intractable, except in certain very particular cases.

395.] Another mode of attacking the problem, at first sight entirely different from that in § 383, but in reality identical with it, is to seek the linear and vector function which expresses the *Homogeneous Strain* which the body must undergo to pass from its initial position to its position at time t.

Let

$$\varpi = \chi a,$$

a being (as in § 383) the initial position of a vector of the body, ϖ its position at time t. In this case χ is a linear and vector function. (See § 360.)

Then, obviously, we have, ϖ_1 being the vector of some other point, which had initially the value a_1,

$$S\varpi\varpi_1 = S.\chi a\chi a_1 = Saa_1,$$

(a particular case of which is

$$T\varpi = T\chi a = Ta)$$

and

$$V\varpi\varpi_1 = V.\chi a\chi a_1 = \chi Vaa_1.$$

These are necessary properties of the strain-function χ, depending on the fact that in the present application the system is rigid.

396.] The kinematical equation

$$\dot{\varpi} = V\epsilon\varpi$$

becomes

$$\dot{\chi} a = V.\epsilon\chi a$$

(the function $\dot{\chi}$ being formed from χ by the differentiation of its constituents with respect to t).

Hamilton's kinetic equation

$$\Sigma.m\varpi V\epsilon\varpi = \gamma,$$

becomes

$$\Sigma.m\chi aV.\epsilon\chi a = \gamma.$$

This may be written

$$\Sigma . m(\chi a S . \epsilon \chi a - \epsilon a^2) = \gamma,$$

or
$$\Sigma . m(a S . a \chi' \epsilon - \chi^{-1} \epsilon . a^2) = \chi\ ^{1}\gamma,$$

where χ' is the conjugate of χ.

But, because
$$S . \chi a \chi a_1 = S a a_1,$$

we have
$$S a a_1 = S . a \chi' \chi a_1,$$

whatever be a and a_1, so that

$$\chi' = \chi^{-1}$$

Hence
$$\Sigma . m(a S . a \chi^{-1} \epsilon - \chi^{-1} \epsilon . a^2) = \chi\ ^{1}\gamma,$$

or, by § 383,
$$\phi \chi^{-1} \epsilon = \chi^{-1} \gamma .$$

397.] Thus we have, as the analogues of the equations in §§ 383, 384,

$$\chi^{-1} \epsilon = \eta,$$
$$\chi^{-1} \gamma = \zeta,$$

and the former result
$$\dot{\chi} a = V . \epsilon \chi a$$

becomes
$$\dot{\chi} a = V . \chi \eta \chi a = \chi V \eta a .$$

This is our equation to determine χ, η being supposed known. To find η we may remark that

$$\phi \eta = \zeta,$$

and
$$\dot{\zeta} = \dot{\widehat{\chi^{-1}}} \gamma .$$

But
$$\chi \chi^{-1} a = a,$$

so that
$$\dot{\chi} \chi^{-1} a + \chi \dot{\widehat{\chi^{-1}}} a = 0 .$$

Hence
$$\dot{\zeta} = - \chi^{-1} \dot{\chi} \chi^{-1} \gamma$$
$$- - V . \eta \chi^{-1} \gamma = V \zeta \eta = V . \zeta \phi^{-1} \zeta,$$

or
$$\phi \dot{\eta} = - V \eta \phi \eta .$$

These are the equations we obtained before. Having found η from the last we have to find χ from the condition

$$\chi^{-1} \dot{\chi} a = V \eta a .$$

398.] We might, however, have eliminated η so as to obtain an equation containing χ alone, and corresponding to that of § 385. For this purpose we have

$$\eta = \phi^{-1} \zeta = \phi^{-1} \chi^{-1} \gamma,$$

so that, finally
$$\chi^{-1} \dot{\chi} a = V . \phi^{-1} \chi^{-1} \gamma a,$$

or
$$\dot{\widehat{\chi^{-1}}} a = V . \chi^{-1} a \phi^{-1} \chi^{-1} \gamma,$$

which may easily be formed from the preceding equation by putting $\chi^{-1} a$ for a, and attending to the value of $\dot{\widehat{\chi^{-1}}}$ given in last section.

399.] We have given this process, though really a disguised form of that in §§ 383, 385, and though the final equations to which it leads are not quite so easily attacked in the way of integration as those there arrived at, mainly to shew how free a use we can make of symbolic functional operators in quaternions without risk of error. It would be very interesting, however, to have the problem worked out afresh from this point of view by the help of the old analytical methods: as several new forms of long-known equations, and some useful transformations, would certainly be obtained.

400.] As a verification, let us now try to pass from the final equation, in χ alone, of § 398 to that of § 385 in q alone.

We have, obviously,

$$\varpi = qaq^{-}{}_{1} = \chi a,$$

which gives the relation between q and χ.

[It shews, for instance, that, as

$$S.\beta\chi a = S.a\chi'\beta,$$

while $$S.\beta\chi a = S.\beta q a q^{-1} = S.a q^{-1}\beta q,$$

we have $$\chi'\beta = q^{-1}\beta q,$$

and therefore that $$\chi\chi'\beta = q(q^{-1}\beta q)q^{-1} - \beta,$$

or $$\chi' = \chi^{-1}, \text{ as above.]}$$

Differentiating, we have

$$\dot{q}aq^{-}{}_{1} - qaq^{-1}\dot{q}q^{-1} - \dot{\chi}a.$$

Hence $$\chi^{-1}\dot{\chi}a = q^{-1}\dot{q}a - aq^{-1}\dot{q}$$
$$- 2V.V(q^{-1}\dot{q})a.$$

Also $$\phi^{-1}\chi^{-1}\gamma = \phi^{-1}(q^{-1}\gamma q),$$

so that the equation of § 398 becomes

$$2V.V(q^{-1}\dot{q})a = V.\phi^{-1}(q^{-1}\gamma q)a,$$

or, as a may have any value whatever,

$$2V.q^{-1}\dot{q} = \phi^{-1}(q^{-1}\gamma q),$$

which, if we put $$Tq = \text{constant}$$

as was originally assumed, may be written

$$2\dot{q} = q\phi^{-1}(q^{-1}\gamma q),$$

as in § 385.

401.] *To form the equation for Precession and Nutation.* Let σ be the vector, from the centre of inertia of the earth, to a particle m of its mass: and let ρ be the vector of the disturbing body, whose mass is M. The vector-couple produced is evidently

$$M\Sigma . mV.\sigma \frac{U(\rho-\sigma)}{T^2(\rho-\sigma)}$$

$$= M\Sigma . m \frac{V\sigma\rho}{T^3(\rho-\sigma)}$$

$$= M\Sigma . \frac{mV\sigma\rho}{T^3\rho} \frac{1}{\left(1 + \frac{2S\sigma\rho}{T^2\rho} + \frac{T^2\sigma}{T^2\rho}\right)^{\frac{3}{2}}}$$

$$= M\Sigma . \frac{mV\sigma\rho}{T^3\rho} \left(1 - \frac{3S\sigma\rho}{T^2\rho} + \&c.\right),$$

no farther terms being necessary, since $\frac{T\sigma}{T\rho}$ is always small in the actual cases presented in nature. But, because σ is measured from the centre of inertia,

$$\Sigma . m\sigma = 0.$$

Also, as in § 383,

$$\phi\rho = \Sigma . m(\sigma S\sigma\rho - \sigma^2\rho).$$

Thus the vector-couple required is

$$\frac{3M}{T^5\rho} V.\rho\phi\rho.$$

Referred to cöordinates moving with the body, ϕ becomes $\boldsymbol{\phi}$ as in § 388, and § 388 gives

$$\boldsymbol{\phi}\epsilon = \gamma = 3M \int \frac{V.\rho\boldsymbol{\phi}\rho}{T^5\rho} dt.$$

Simplifying the value of $\boldsymbol{\phi}$ by assuming that the earth has two principal axes of equal moment of inertia, we have

$$B\epsilon - (A-B)\alpha S\alpha\epsilon = \text{vector-constant} + 3M(A-B) \int \frac{V\alpha\rho S\alpha\rho}{T^5\rho} dt.$$

This gives

$$S\alpha\epsilon = \text{const.} = \Omega,$$

whence

$$\epsilon = -\Omega\alpha + \alpha\dot{\alpha},$$

so that, finally,

$$BV\alpha\ddot{\alpha} \quad A\Omega\dot{\alpha} = \frac{3M}{T^5\rho}(A-B)V\alpha\rho S\alpha\rho.$$

The most striking peculiarity of this equation is that the *form* of the solution is entirely changed, not modified as in ordinary cases of disturbed motion, according to the nature of the value of ρ.

Thus, when the right-hand side vanishes, we have an equation which, in the case of the earth, would represent the rolling of a cone fixed in the earth on one fixed in space, the angles of *both* being exceedingly small.

If ρ be finite, but constant, we have a case nearly the same as that of a top, the axis on the whole revolving conically about ρ.

But if we assume the expression

$$\rho = r\,(j \cos mt + k \sin mt),$$

(which represents a circular orbit described with uniform velocity,) a revolves on the whole conically about the vector i, perpendicular to the plane in which ρ lies. (*Trans. R. S. E.*, 1868–9.)

402.] *To form the equation of motion of a simple pendulum, taking account of the earth's rotation.* Let a be the vector (from the earth's centre) of the point of suspension, λ its inclination to the plane of the equator, a the earth's radius drawn to that point; and let the unit-vectors i, j, k be fixed in space, so that i is parallel to the earth's axis of rotation; then, if ω be the angular velocity of that rotation

$$a = a\left[i \sin \lambda + (j \cos \omega t + k \sin \omega t) \cos \lambda\right] \ldots \ldots \ldots \ldots (1)$$

This gives $\dot{a} = a\omega(-j \sin \omega t + k \cos \omega t) \cos \lambda$

$$= \omega V i a. \ldots \ldots \ldots \ldots (2)$$

Similarly $\ddot{a} = \omega V i \dot{a} = -\omega^2 (a - ai \sin \lambda) \ldots \ldots \ldots \ldots (3)$

403.] Let ρ be the vector of the bob m referred to the point of suspension, R the tension of the string, then if a_1 be the direction of pure gravity $m(\ddot{a} + \ddot{\rho}) = -mg U a_1 - R U \rho, \ldots \ldots \ldots \ldots (4)$

which may be written

$$V\rho\ddot{a} + V\rho\ddot{\rho} = \frac{g}{T a_1} V a_1 \rho. \ldots \ldots \ldots \ldots (5)$$

To this must be added, since r (the length of the string) is constant,

$$T\rho = r, \ldots \ldots \ldots \ldots (6),$$

and the equations of motion are complete.

404.] These two equations (5) and (6) contain every possible case of the motion, from the most infinitesimal oscillations to the most rapid rotation about the point of suspension, so that it is necessary to adapt different processes for their solution in different cases. We take here only the ordinary Foucault case, to the degree of approximation usually given.

405.] Here we neglect terms involving ω^2. Thus we write

$$\ddot{a} = 0,$$

and we write a for a_1, as the difference depends upon the ellipticity of the earth. Also, attending to this, we have

$$\rho = -\frac{r}{a} a + \varpi, \ldots \ldots \ldots \ldots (7)$$

where by (by (6)) $S a \varpi = 0, \ldots \ldots \ldots \ldots (8)$

and terms of the order ϖ^2 are neglected.

With (7), (5) becomes

$$-\frac{r}{a}Va\ddot{\varpi} = \frac{g}{a}Va\varpi;$$

so that, if we write $$\frac{g}{r} = n^2, \qquad\qquad (9)$$

we have $$Va(\ddot{\varpi}+n^2\varpi) = 0. \qquad\qquad (10)$$

Now, the two vectors $ai - a\sin\lambda$ and Via

have, as is easily seen, equal tensors; the first is parallel to the line drawn horizontally *northwards* from the point of suspension, the second horizontally *eastwards*.

Let, therefore, $$\varpi = x(ai - a\sin\lambda) + yVia, \qquad\qquad (11)$$

which (x and y being very small) is consistent with (6).

From this we have (employing (2) and (3), and omitting ω^2)

$$\dot{\varpi} = \dot{x}(ai - a\sin\lambda) + \dot{y}Via - x\omega\sin\lambda Via - y\omega(a - ai\sin\lambda),$$

$$\ddot{\varpi} = \ddot{x}(ai - a\sin\lambda) + \ddot{y}Via - 2\dot{x}\omega\sin\lambda Via - 2\dot{y}\omega(a - ai\sin\lambda).$$

With this (10) becomes

$$Va[\ddot{x}(ai - a\sin\lambda) + \ddot{y}Via - 2\dot{x}\omega\sin\lambda Via - 2\dot{y}\omega(a - ai\sin\lambda) + n^2x(ai - a\sin\lambda) + n^2yVia] = 0,$$

or, if we note that $V.aVia = a(ai - a\sin\lambda)$,

$$(-\ddot{x} - 2\dot{y}\omega\sin\lambda - n^2x)aVia + (\ddot{y} - 2\dot{x}\omega\sin\lambda + n^2y)a(ai - a\sin\lambda) = 0.$$

This gives at once $$\left.\begin{aligned}\ddot{x} + n^2x + 2\omega\dot{y}\sin\lambda &= 0,\\ \ddot{y} + n^2y - 2\omega\dot{x}\sin\lambda &= 0,\end{aligned}\right\} \qquad\qquad (12)$$

which are the equations usually obtained; and of which the solution is as follows:—

If we transform to a set of axes revolving in the horizontal plane at the point of suspension, the direction of motion being from the positive (northward) axis of x to the positive (eastward) axis of y, with angular velocity Ω, so that

$$\left.\begin{aligned}x &= \xi\cos\Omega t - \eta\sin\Omega t,\\ y &= \xi\sin\Omega t + \eta\cos\Omega t,\end{aligned}\right\} \qquad\qquad (13)$$

and omit the terms in Ω^2 and in $\omega\Omega$ (a process justified by the results, see equation (15)), we have

$$\left.\begin{aligned}(\ddot{\xi} + n^2\xi)\cos\Omega t - (\ddot{\eta} + n^2\eta)\sin\Omega t - 2\dot{y}(\Omega - \omega\sin\lambda) &= 0,\\ (\ddot{\xi} + n^2\xi)\sin\Omega t + (\ddot{\eta} + n^2\eta)\cos\Omega t + 2\dot{x}(\Omega - \omega\sin\lambda) &= 0.\end{aligned}\right\} \qquad (14)$$

So that, if we put $$\Omega = \omega\sin\lambda, \qquad\qquad (15)$$

we have simply $$\left.\begin{aligned}\ddot{\xi} + n^2\xi &= 0,\\ \ddot{\eta} + n^2\eta &= 0,\end{aligned}\right\} \qquad\qquad (16)$$

the usual equations of elliptic motion about a centre of force in the centre of the ellipse. (*Proc. R. S. E.*, 1869.)

406.] *To construct a reflecting surface from which rays, emitted from a point, shall after reflection diverge uniformly, but horizontally.*

Using the ordinary property of a reflecting surface, we easily obtain the equation

$$S.d\rho\left(\frac{\beta+aVa\rho}{\rho}\right)^{\frac{1}{2}}\rho = 0.$$

By Hamilton's grand *Theory of Systems of Rays*, we at once write down the second form

$$T\rho - T(\beta + aVa\rho) = \text{constant}.$$

The connection between these is easily shewn thus. Let ϖ and τ be any two vectors whose tensors are equal, then

$$\left(\frac{\tau+\varpi}{\tau}\right)^2 = 1+2\varpi\tau^{-1}+(\varpi\tau^{-1})^2$$
$$= 2\varpi\tau^{-1}(1+S\varpi\tau^{-1}),$$

whence, to a scalar factor *près*, we have

$$\left(\frac{\varpi}{\tau}\right)^{\frac{1}{2}} = \frac{\tau+\varpi}{\tau}.$$

Hence, putting $\varpi = U(\beta + aVa\rho)$ and $\tau = U\rho$, we have from the first equation above

$$S.d\rho[U\rho + U(\beta+aVa\rho)] = 0.$$

But $\qquad d(\beta+aVa\rho) = aVad\rho = -d\rho - aSad\rho,$

and $\qquad S.a(\beta+aVa\rho) = 0,$

so that we have finally

$$S.d\rho U\rho - S.d(\beta+aVa\rho)U(\beta+aVa\rho) = 0,$$

which is the differential of the second equation above. A curious particular case is a parabolic cylinder, as may be easily seen geometrically. The general surface has a parabolic section in the plane of a, β; and a hyperbolic section in the plane of β, $a\beta$.

It is easy to see that this is but a single case of a large class of integrable scalar functions, whose general type is

$$S.d\rho\left(\frac{\sigma-\rho}{\rho}\right)^{\frac{1}{2}}\rho = 0,$$

the equation of the reflecting surface; while

$$S(\sigma-\rho)d\sigma = 0$$

is the equation of the surface of the reflected wave: the integral of the former being, by the help of the latter, at once obtained in the form

$$T\rho \pm T(\sigma-\rho) = \text{constant}^*.$$

407.] We next take Fresnel's *Theory of Double Refraction*, but

* *Proc. R. S. E.*, 1870-71.

merely for the purpose of shewing how quaternions simplify the processes required, and in no way to discuss the plausibility of the physical assumptions.

Let $t\varpi$ be the vector displacement of a portion of the ether, with the condition
$$\varpi^2 = -1, \quad \ldots\ldots (1)$$
the force of restitution, on Fresnel's assumption, is
$$t(a^2 iSi\varpi + b^2 jSj\varpi + c^2 kSk\varpi) = t\phi\varpi,$$
using the notation of Chapter V. Here the function ϕ is obviously self-conjugate. a^2, b^2, c^2 are optical constants depending on the crystalline medium, and on the colour of the light, and may be considered as given.

Fresnel's second assumption is that the ether is incompressible, or that vibrations normal to a wave front are inadmissible. If, then, α be the unit normal to a plane wave in the crystal, we have of course
$$\alpha^2 = -1, \quad \ldots\ldots (2)$$
and
$$S\alpha\varpi = 0\,; \quad \ldots\ldots (3)$$
but, and in addition, we have
$$\varpi \ {}^1V\varpi\phi\varpi \parallel \alpha,$$
or
$$S.\alpha\varpi\phi\varpi = 0. \quad \ldots\ldots (4)$$
This equation (4) is the embodiment of Fresnel's second assumption, but it may evidently be read as meaning, *the normal to the front, the direction of vibration, and that of the force of restitution are in one plane.*

408.] Equations (3) and (4), if satisfied by ϖ, are also satisfied by $\varpi\alpha$, so that the plane (3) intersects the cone (4) in two lines at right angles to each other. That is, *for any given wave front there are two directions of vibration, and they are perpendicular to each other.*

409.] The square of the normal velocity of propagation of a plane wave is proportional to the ratio of the resolved part of the force of restitution in the direction of vibration, to the amount of displacement, hence
$$v^2 = S\varpi\phi\varpi.$$
Hence Fresnel's *Wave-surface* is the envelop of the plane
$$S\alpha\rho = \sqrt{S\varpi\phi\varpi}, \quad \ldots\ldots (5)$$
with the conditions
$$\varpi^2 = -1, \quad \ldots\ldots (1)$$
$$\alpha^2 = -1, \quad \ldots\ldots (2)$$
$$S\alpha\varpi = 0, \quad \ldots\ldots (3)$$
$$S.\alpha\varpi\phi\varpi = 0. \quad \ldots\ldots (4)$$

Formidable as this problem appears, it is easy enough. From (3) and (4) we get at once,

$$x\varpi = V.aVa\phi\varpi,$$

Hence, operating by $S.\varpi$,

$$-x = -S\varpi\phi\varpi = -v^2.$$

Therefore $$(\phi + v^2)\varpi = -aSa\phi\varpi,$$

and $$S.a(\phi + v^2)^{-1}a = 0. \qquad (6)$$

In passing, we may remark that *this equation gives the normal velo cities of the two rays whose fronts are perpendicular to* a. In Cartesian cöordinates it is the well-known equation

$$\frac{l^2}{a^2 - v^2} + \frac{m^2}{b^2 - v^2} + \frac{n^2}{c^2 - v^2} = 0.$$

By this elimination of ϖ, our equations are reduced to

$$S.a(\phi + v^2)^{-1}a = 0, \qquad (6)$$

$$v = -Sa\rho, \qquad (5)$$

$$a^2 = -1. \qquad (2)$$

They give at once, by § 309,

$$(\phi + v^2)^{-1}a + v\rho Sa(\phi + v^2)^{-2}a = ha.$$

Operating by $S.a$ we have

$$v^2 Sa(\phi + v^2)^{-2}a = h.$$

Substituting for h, and remarking that

$$Sa(\phi + v^2)^{-2}a = -T^2(\phi + v^2)^{-1}a,$$

because ϕ is self-conjugate, we have

$$v(\phi + v^2)^{-1}a = \frac{va - \rho}{\rho^2 + v^2}.$$

This gives at once, by rearrangement,

$$v(\phi + v^2)^{-1}a = (\phi - \rho^2)^{-1}\rho.$$

Hence $$(\phi - \rho^2)^{-1}\rho = \frac{va - \rho}{\rho^2 + v^2}.$$

Operating by $S.\rho$ on this equation we have

$$S\rho(\phi - \rho^2)^{-1}\rho = -1, \qquad (7)$$

which is the required equation.

[It will be a good exercise for the student to translate the last ten formulae into Cartesian cöordinates. He will thus reproduce almost exactly the steps by which Archibald Smith * first arrived at a simple and symmetrical mode of effecting the elimination. Yet, as we shall presently see, the above process is far from being the shortest and easiest to which quaternions conduct us.]

* *Cambridge Phil. Trans.*, 1835.

410.] The Cartesian form of the equation (7) is not the usual one. It is, of course,

$$\frac{x^2}{a^2-r^2}+\frac{y^2}{b^2-r^2}+\frac{z^2}{c^2-r^2}=-1.$$

But write (7) in the form

$$S.\rho\frac{\rho^2}{\phi-\rho^2}\rho=-\rho^2,$$

or

$$S.\rho\frac{\phi}{\phi-\rho^2}\rho=0,$$

and we have the usual expression

$$\frac{a^2x^2}{a^2-r^2}+\frac{b^2y^2}{b^2-r^2}+\frac{c^2z^2}{c^2-r^2}=0.$$

This last quaternion equation can also be put into either of the new forms

$$T\left(\frac{\phi}{\phi-\rho^2}\right)^{\frac{1}{2}}\rho=0,$$

or

$$T(\rho^{-2}-\phi^{-1})^{-\frac{1}{2}}\rho=0.$$

411.] By applying the results of §§ 171, 172 we may introduce a multitude of new forms. We must confine ourselves to the most simple; but the student may easily investigate others by a process precisely similar to that which follows.

Writing the equation of the wave as

$$S\rho(\phi^{-1}+g)^{-1}\rho=0,$$

where we have

$$g=-\rho^{\ 2},$$

we see that it may be changed to

$$S\rho(\phi^{-1}+h)^{-1}\rho=0,$$

if

$$mS\rho\phi\rho=gh\rho^2=-h.$$

Thus the new form is

$$S\rho(\phi^{-1}-mS\rho\phi\rho)^{-1}\rho=0. \quad\ldots\ldots\ldots\ldots\ldots\ldots (1)$$

Here $m=\frac{1}{a^2b^2c^2}$, $S\rho\phi\rho=a^2x^2+b^2y^2+c^2z^2$,

and the equation of the wave in Cartesian cöordinates is, putting

$$r_1^2=a^2x^2+b^2y^2+c^2z^2,$$

$$\frac{x^2}{b^2c^2-r_1^2}+\frac{y^2}{c^2a^2-r_1^2}+\frac{z^2}{a^2b^2-r_1^2}=0.$$

412.] By means of equation (1) of last section we may easily prove Plücker's Theorem. *The Wave-Surface is its own reciprocal with respect to the ellipsoid whose equation is*

$$S\rho\phi^{\frac{1}{2}}\rho=\frac{1}{\sqrt{m}}.$$

The equation of the plane of contact of tangents to this surface from the point whose vector is ρ is

$$S\varpi\phi^{\frac{1}{2}}\rho = \frac{1}{\sqrt{m}}$$

The reciprocal of this plane, with respect to the unit-sphere about the origin, has therefore a vector σ where

$$\sigma = \sqrt{m}\phi^{\frac{1}{2}}\rho.$$

Hence $$\rho = \frac{1}{\sqrt{m}}\phi^{-\frac{1}{2}}\sigma,$$

and when this is substituted in the equation of the wave we have for the reciprocal (with respect to the unit-sphere) of the reciprocal of the wave with respect to the above ellipsoid,

$$S.\sigma\left(\phi - \frac{1}{m}S\sigma\phi^{-1}\sigma\right)\sigma = 0.$$

This differs from the equation (1) of last section solely in having ϕ^{-1} instead of ϕ, and (consistently with this) $\frac{1}{m}$ instead of m. Hence it represents the index-surface. The required reciprocal of the wave with reference to the ellipsoid is therefore the wave itself.

413.] Hamilton has given a remarkably simple investigation of the form of the equation of the wave-surface, in his *Elements*, p. 736, which the reader may consult with advantage. The following is essentially the same, but several steps of the process, which a skilled analyst would not require to write down, are retained for the benefit of the learner.

Let $$S\mu\rho = -1 \qquad\qquad (1)$$

be the equation of any tangent plane to the wave, i.e. of any wave-front. Then μ is the vector of wave-slowness, and the normal velocity of propagation is therefore $\frac{1}{T\mu}$. Hence, if ϖ be the vector direction of displacement, $\mu^{-2}\varpi$ is the effective component of the force of restitution. Hence, $\phi\varpi$ denoting the whole force of restitution, we have $$\phi\varpi - \mu^{-2}\varpi \parallel \mu,$$

or $$\varpi \parallel (\phi - \mu^{-2})^{-1}\mu,$$

and, as ϖ is in the plane of the wave-front,

$$S\mu\varpi = 0,$$

or $$S\mu(\phi - \mu^{-2})^{-1}\mu = 0. \qquad\qquad (2)$$

This is, in reality, equation (6) of § 409. It appears here, however, as the equation of the *Index-Surface*, the polar reciprocal of the wave with respect to a unit-sphere about the origin. Of course the optical part of the problem is now solved, all that remains being the geometrical process of § 311.

414.] Equation (2) of last section may be at once transformed, by the process of § 410, into

$$S\mu(\mu^2-\phi^{-1})^{-1}\mu = 1.$$

Let us employ an auxiliary vector

$$\tau = (\mu^2-\phi^{-1})^{-1}\mu,$$

whence $\mu = (\mu^2-\phi^{-1})\tau$. .... (1)

The equation now becomes

$$S\mu\tau = 1, \quad \text{.............................. (2)}$$

or, by (1), $\mu^2\tau^2 - S\tau\phi^{-1}\tau = 1$. .... (3)

Differentiating (3), subtract its half from the result obtained by operating with $S.\tau$ on the differential of (1). The remainder is

$$\tau^2 S\mu d\mu - S\tau d\mu = 0.$$

But we have also (§ 311) $S\rho d\mu = 0$,

and therefore $x\rho = \mu\tau^2 - \tau$,

where x is a scalar.

This equation, with (2), shews that

$$S\tau\rho = 0. \quad \text{.............................. (4)}$$

Hence, operating on it by $S.\rho$, we have by (1) of last section

$$x\rho^2 = -\tau^2,$$

and therefore $\rho^{-1} = -\mu + \tau^{-1}$.

This gives $\rho^{-2} = \mu^2 - \tau^{-2}$.

Substituting from these equations in (1) above, it becomes

$$\tau^{-1} - \rho^{-1} = (\rho^{-2}+\tau^{-2}-\phi^{-1})\tau,$$

or $\tau = (\phi^{-1}-\rho^{-2})^{-1}\rho^{-1}$

Finally, we have for the required equation, by (4),

$$S\rho^{-1}(\phi^{-1}-\rho^{-2})^{-1}\rho^{-1} = 0,$$

or, by a transformation already employed,

$$S\rho(\phi-\rho^2)^{-1}\rho = -1.$$

415.] It may assist the student in the *practice* of quaternion analysis, which is our main object, if we give a few of these investigations by a somewhat varied process.

Thus, in § 407, let us write as in § 168,

$$a^2 iSi\varpi + b^2 jSj\varpi + c^2 kSk\varpi = \lambda' S\mu'\varpi + \mu' S\lambda'\varpi - p'\varpi.$$

We have, by the same processes as in § 407,

$$S.\varpi a\lambda' S\mu'\varpi + S.\varpi a\mu' S\lambda'\varpi = 0.$$

This may be written, *so far as the generating lines we require are concerned*,

$$S.\varpi a V.\lambda'\varpi\mu' = 0 = S.\varpi a\lambda'\varpi\mu',$$

since ϖa is a vector.

Or we may write

$$S.\mu' V.\varpi\lambda'\varpi a = 0 = S.\mu'\varpi\lambda'\varpi a. \quad \Bigg\} \ldots\ldots\ldots\ldots\ldots\ldots (1)$$

Equations (1) denote two cones of the second order which pass through the intersections of (3) and (4) of § 407. Hence their intersections are the directions of vibration.

416.] By (1) we have

$$S.\varpi\lambda'\varpi a\mu' = 0.$$

Hence $\varpi\lambda'\varpi$, a, μ' are coplanar; and, as ϖ is perpendicular to a, it is equally inclined to $V\lambda' a$ and $V\mu' a$.

For, if L, M, A be the projections of λ', μ', a on the unit sphere, BC the great circle whose pole is A, we are to find for the projections of the values of ϖ on the sphere points P and P', such that if LP be produced till

$$PQ = LP,$$

Q may lie on the great circle AM. Hence, evidently,

$$\widehat{CP} = \widehat{PB},$$

and $$\widehat{C'P'} = \widehat{P'B};$$

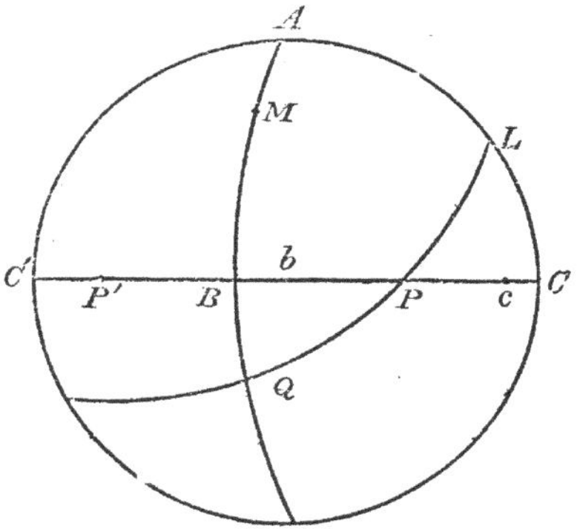

which proves the proposition, since the projections of $V\lambda' a$ and $V\mu' a$ on the sphere are points b and c in BC, distant by quadrants from C and B respectively.

417.] Or thus, $$S\varpi a = 0,$$

$$S.\varpi V.a\lambda'\varpi\mu' = 0,$$

therefore $$x\varpi = V.aV.a\lambda'\varpi\mu',$$

$$= -V.\lambda'\varpi\mu' - aSaV.\lambda'\varpi\mu'.$$

Hence $$(S\lambda'\mu' - x)\varpi = (\lambda' + aSa\lambda')S\mu'\varpi + (\mu' + aSa\mu')S\lambda'\varpi.$$

Operate by $S.\lambda'$, and we have

$$(x + S\lambda' aS\mu' a)S\lambda'\varpi = [\lambda'^2 a^2 - S^2\lambda' a]S\mu'\varpi$$
$$- S\mu'\varpi T^2 V\lambda' a.$$

Hence by symmetry,

$$\frac{S\mu'\varpi}{S\lambda'\varpi} T^2 V\lambda' a - \frac{S\lambda'\varpi}{S\mu'\varpi} T^2 V\mu' a,$$

or $$\frac{S\lambda'\varpi}{TV\lambda'a} \pm \frac{S\mu'\varpi}{TV\mu'a} = 0.$$

$$S\varpi\left(\frac{\lambda'}{TV\lambda'a} \pm \frac{\mu'}{TV\mu'a}\right) = 0,$$

and as $$S\varpi a = 0,$$

$$\varpi = U(UV\lambda'a + UV\mu'a).$$

418.] The optical interpretation of the common result of the last two sections is that *the planes of polarization of the two rays whose wave-fronts are parallel, bisect the angles contained by planes passing through the normal to the wave-front and the vectors* (optic axes) λ', μ'.

419.] As in § 409, the normal velocity is given by

$$v^2 = S\varpi\phi\varpi = 2\,S\lambda'\varpi S\mu'\varpi - p'\varpi^2$$

$$= p' \mp \frac{S^2.\lambda'\mu'a}{(T \mp S).V\lambda'aV\mu'a}.$$

[This transformation, effected by means of the value of ϖ in § 417, is left to the reader.]

Hence, if v_1, v_2 be the velocities of the two waves whose normal is a,

$$v_1^2 - v_2^2 = 2\,T.V\lambda'aV\mu'a$$

$$\propto \sin\widehat{\lambda'a}\sin\widehat{\mu'a}.$$

That is, *the difference of the squares of the velocities of the two waves varies as the product of the sines of the angles between the normal to the wave-front and the optic axes* (λ', μ').

420.] We have, obviously,

$$(T^2 - S^2).V\lambda'aV\mu'a = T^2V.V\lambda'aV\mu'a = S.^2\lambda'\mu'a.$$

Hence $$v^2 = p' \mp (T \pm S).V\lambda'aV\mu'a.$$

The equation of the index surface, for which

$$T\rho = \frac{1}{v}, \qquad U\rho = a,$$

is therefore $$1 = -p'\rho^2 \mp (T \pm S).V\lambda'\rho V\mu'\rho.$$

This will, of course, become the equation of the reciprocal of the index-surface, i.e. the wave-surface, if we put for the function ϕ its reciprocal: i. e. if in the values of λ', μ', p' we put $\frac{1}{a}$, $\frac{1}{b}$, $\frac{1}{c}$ for a, b, c respectively. We have then, and indeed it might have been deduced even more simply as a transformation of § 409 (7),

$$1 = -p\rho^2 \mp (T \pm S).V\lambda\rho V\mu\rho,$$

as another form of the equation of Fresnel's wave.

If we employ the ι, κ transformation of § 121, this may be written, as the student may easily prove, in the form

$$(\kappa^2-\iota^2)^2 = S^2(\iota-\kappa)\rho+(TV\iota\rho \mp TV\kappa\rho)^2.$$

421.] We may now, in furtherance of our object, which is to give varied examples of quaternions, not complete treatment of any one subject, proceed to deduce some of the properties of the wave-surface from the different forms of its equation which we have given.

422.] *Fresnel's construction of the wave by points.*

From § 273 (4) we see at once that the lengths of the principal semidiameters of the central section of the ellipsoid

$$S\rho\phi^{-1}\rho = 1,$$

by the plane $$S\alpha\rho = 0,$$

are determined by the equation

$$S.\alpha(\phi^{-1}-\rho^{-2})^{-1}\alpha = 0.$$

If these lengths be laid off along α, the central perpendicular to the cutting plane, their extremities lie on a surface for which $\alpha = U\rho$, and $T\rho$ has values determined by the equation.

Hence the equation of the locus is

$$S\rho(\phi^{-1}-\rho^{-2})^{-1}\rho = 0,$$

as in §§ 409, 414.

Of course the index-surface is derived from the reciprocal ellipsoid

$$S\rho\phi\rho = 1$$

by the same construction.

423.] Again, in the equation

$$1 = -p\rho^2 \mp (T \pm S).V\lambda\rho V\mu\rho,$$

suppose $V\lambda\rho = 0,$ or $V\mu\rho = 0,$

we obviously have

$$\rho = \pm\frac{U\lambda}{\sqrt{p}} \quad \text{or} \quad \rho = +\frac{U\mu}{\sqrt{p}},$$

and there are therefore four singular points.

To find the nature of the surface near these points put

$$\rho = \frac{U\lambda}{\sqrt{p}}+\varpi,$$

where $T\varpi$ is very small, and reject terms above the first order in $T\varpi$. The equation of the wave becomes, in the neighbourhood of the singular point,

$$2pS\lambda\varpi + S.\varpi V.\lambda V\lambda\mu = +T.V\lambda\varpi V\lambda\mu,$$

which belongs to a cone of the second order.

424.] From the similarity of its equation to that of the wave, it is obvious that the index-surface also has four conical cusps. As an infinite number of tangent planes can be drawn at such a point, the reciprocal surface must be capable of being touched by a plane at an infinite number of points; so that the wave-surface has four tangent planes which touch it along ridges.

To find their form, let us employ the last form of equation of the wave in § 420. If we put

$$TV\iota\rho = TV\kappa\rho, \quad \ldots\ldots\ldots\ldots\ldots\ldots \quad (1)$$

we have the equation of a cone of the second degree. It meets the wave at its intersections with the planes

$$S(\iota-\kappa)\rho = +(\kappa^2-\iota^2). \quad \ldots\ldots\ldots\ldots\ldots\ldots \quad (2)$$

Now the wave-surface is *touched* by these planes, because we cannot have the quantity on the first side of this equation greater in absolute magnitude than that on the second, so long as ρ satisfies the equation of the wave.

That the curves of contact are circles appears at once from (1) and (2), for they give in combination

$$\rho^2 = +S(\iota+\kappa)\rho, \quad \ldots\ldots\ldots\ldots\ldots\ldots \quad (3)$$

the equations of two spheres on which the curves in question are situated.

The diameter of this circular ridge is

$$TV.(\iota+\kappa)U(\iota-\kappa) = \frac{2TV\iota\kappa}{T(\iota-\kappa)} = \frac{1}{b}\sqrt{(a^2-b^2)(b^2-c^2)}.$$

[Simple as these processes are, the student will find on trial that the equation $\quad S\rho(\phi^{-1}-\rho^{-2})^{-1}\rho = 0,$
gives the results quite as simply. For we have only to examine the cases in which $-\rho^{-2}$ has the value of one of the roots of the symbolical cubic in ϕ^{-1}. In the present case $T\rho = b$ is the only one which requires to be studied.]

425.] By § 413, we see that the auxiliary vector of the succeeding section, viz.

$$\tau = (\mu^2-\phi^{-1})^{-1}\mu = (\phi^{-1}-\rho^{-2})^{-1}\rho^{-1},$$

is parallel to the direction of the force of restitution, $\phi\varpi$. Hence, as Hamilton has shewn, the equation of the wave, in the form

$$S\tau\rho = 0,$$

(4) of § 414, indicates that *the direction of the force of restitution is perpendicular to the ray.*

Again, as for any one versor of a vector of the wave there are two values of the tensor, which are found from the equation

$$S.U\rho(\phi^{-1}-\rho^{-2})^{-1}U\rho = 0,$$

we see by § 422 that *the lines of vibration for a given plane front are parallel to the axes of any section of the ellipsoid*

$$S.\rho\phi^{-1}\rho = 1$$

made by a plane parallel to the front; or to the tangents to the lines of curvature at a point where the tangent plane is parallel to the wave-front.

426.] Again, *a curve which is drawn on the wave-surface so as to touch at each point the corresponding line of vibration* has

$$\phi\, d\rho \parallel (\phi^{-1} - \rho^{-2})^{-1}\rho.$$

Hence $\quad S\phi\rho d\rho = 0, \quad$ or $\quad S\rho\phi\rho = C,$

so that such curves are the intersections of the wave with a series of ellipsoids concentric with it.

427.] For *curves cutting at right angles the lines of vibration* we have

$$d\rho \parallel V\rho\phi^{-1}(\phi^{-1} - \rho^{-2})^{-1}\rho$$
$$\parallel V\rho(\phi - \rho^2)^{-1}\rho.$$

Hence $\quad S\rho d\rho = 0, \quad$ or $\quad T\rho = C,$

so that the curves in question lie on concentric spheres.

They are also *spherical conics*, because where

$$T\rho = C$$

the equation of the wave becomes

$$S.\rho(\phi^{-1} + C^{-2})^{-1}\rho = 0,$$

the equation of a cyclic cone, whose vertex is at the common centre of the sphere and the wave-surface, and which cuts them in their curve of intersection. (*Quarterly Math. Journal*, 1859.)

428.] As another example we take the case of the action of electric currents on one another or on magnets; and the mutual action of permanent magnets.

A comparison between the processes we employ and those of Ampère (*Théorie des Phénomènes Electrodynamiques, &c.*, many of which are well given by Murphy in his *Electricity*) will at once shew how much is gained in simplicity and directness by the use of quaternions.

The same gain in simplicity will be noticed in the investigations of the mutual effects of permanent magnets, where the resultant forces and couples are at once introduced in their most natural and direct forms.

429.] Ampère's experimental laws may be stated as follows:

I. Equal and opposite currents in the same conductor produce equal and opposite effects on other conductors: whence it follows that an element of one current has no effect on an element of another which lies in the plane bisecting the former at right angles.

II. The effect of a conductor bent or twisted in any manner is equivalent to that of a straight one, provided that the two are traversed by equal currents, and the former *nearly* coincides with the latter.

III. No closed circuit can set in motion an element of a circular conductor about an axis through the centre of the circle and perpendicular to its plane.

IV. In similar systems traversed by equal currents the forces are equal.

To these we add the assumption that the action between two elements of currents is in the straight line joining them: and two others, viz. that the effect of any element of a current on another is directly as the product of the strengths of the currents, and of the lengths of the elements.

430.] Let there be two closed currents whose strengths are a and a_1; let α', α_1 be elements of these, α being the vector joining their middle points. Then the effect of α' on α_1 must, when resolved along α_1, be a complete differential with respect to α (i.e. with respect to the three independent variables involved in α), since the total resolved effect of the closed circuit of which α' is an element is zero by III.

Also by I, II, the effect is a function of $T\alpha$, $S\alpha\alpha'$, $S\alpha\alpha_1$, and $S\alpha'\alpha_1$, since these are sufficient to resolve α' and α_1 into elements parallel and perpendicular to each other and to α. Hence the mutual effect is
$$aa_1 U\alpha f(T\alpha, S\alpha\alpha', S\alpha\alpha_1, S\alpha'\alpha_1),$$
and the resolved effect parallel to α_1 is
$$aa_1 SU\alpha_1 U\alpha f.$$
Also, that action and reaction may be equal in absolute magnitude, f must be symmetrical in $S\alpha\alpha'$ and $S\alpha\alpha_1$. Again, α' (as differential of α) *can* enter *only to the first power*, and *must* appear in each term of f.

Hence
$$f = AS\alpha'\alpha_1 + BS\alpha\alpha' S\alpha\alpha_1.$$
But, by IV, this must be independent of the dimensions of the system. Hence A is of -2 and B of -4 dimensions in $T\alpha$. Therefore
$$\frac{1}{T\alpha}\{AS\alpha\alpha_1 S\alpha'\alpha_1 + BS\alpha\alpha' S^2\alpha\alpha_1\}$$
is a complete differential, with respect to α, if $d\alpha = \alpha'$. Let
$$A = \frac{C}{T\alpha^2},$$

where C is a constant depending on the units employed, therefore

$$d\frac{C}{2Ta^3} = \frac{B}{Ta} Saa',$$

or

$$B = \tfrac{3}{2}\frac{C}{Ta^4},$$

and the resolved effect

$$= \frac{Ca a_1}{2Ta_1} d\frac{S^2 a a_1}{Ta^3} = Ca a_1 \frac{Sa a_1}{Ta_1 Ta^5}(-a^2 Sa' a_1 + \tfrac{3}{2} Saa' Saa_1)$$

$$= Ca a_1 \frac{Sa a_1}{Ta_1 Ta^5}(S.Vaa' Vaa_1 + \tfrac{1}{2} Saa' Saa_1).$$

The factor in brackets is evidently proportional in the ordinary notation to $\sin\theta \sin\theta' \cos\omega - \frac{1}{2}\cos\theta\cos\theta'$.

431.] Thus the whole force is

$$\frac{Caa_1 a}{2Saa_1} d\frac{S^2 aa_1}{Ta^3} = \frac{Caa_1 a}{2Saa'} d_1 \frac{S^2 aa'}{Ta^3},$$

as we should expect, $d_1 a$ being $= a_1$. [This may easily be transformed into

$$-\frac{2Caa_1 Ua}{(Ta)^{\frac{1}{2}}} dd_1 (Ta)^{\frac{1}{2}},$$

which is the quaternion expression for Ampère's well-known form.]

432.] The whole effect on a_1 of the closed circuit, of which a' is an element, is therefore

$$\frac{Caa_1}{2}\int \frac{a}{Saa_1} d\frac{(Saa_1)^2}{Ta^3},$$

$$= \frac{Caa_1}{2}\left\{\frac{aSaa_1}{Ta^3} - V.a_1\int\frac{Vaa'}{Ta^3}\right\}$$

between proper limits. As the integrated part is the same at both limits, the effect is

$$\frac{Caa_1}{2} Va_1\beta, \text{ where } \beta = \int\frac{Vaa'}{Ta^3} = \int\frac{dUa}{a},$$

and depends on the form of the closed circuit.

433.] This vector β, which is of great importance in the whole theory of the effects of closed or indefinitely extended circuits, corresponds to the line which is called by Ampère "*directrice de l'action électrodynamique.*" It has a definite value at each point of space, independent of the existence of any other current.

Consider the circuit a polygon whose sides are indefinitely small; join its angular points with any assumed point, erect at the latter, perpendicular to the plane of each elementary triangle so formed, a vector whose length is $\frac{\omega}{r}$, where ω is the vertical angle of the triangle and r the length of one of the containing sides; the sum of such vectors is the "*directrice*" at the assumed point.

434.] The mere *form* of the result of § 432 shews at once that *if the element* α_1 *be turned about its middle point, the direction of the resultant action is confined to the plane whose normal is* β.

Suppose that the element α_1 is forced to remain perpendicular to some given vector δ, we have

$$S\alpha_1\delta = 0,$$

and the whole action in its plane of motion is proportional to

$$TV.\delta V\alpha_1\beta.$$

But
$$V.\delta V\alpha_1\beta = -\alpha_1 S\beta\delta.$$

Hence the action is evidently constant for all possible positions of α_1; or

The effect of any system of closed currents on an element of a conductor which is restricted to a given plane is (in that plane) independent of the direction of the element.

435.] Let the closed current be *plane* and *very small*. Let ϵ (where $T\epsilon = 1$) be its normal, and let γ be the vector of any point within it (as the centre of inertia of its area); the middle point of α_1 being the origin of vectors.

Let
$$\alpha = \gamma + \rho; \text{ therefore } \alpha' = \rho',$$

and
$$\beta = \int \frac{V\alpha\alpha'}{T\alpha^3} = \int \frac{V(\gamma+\rho)\rho'}{T(\gamma+\rho)^3}$$
$$= \frac{1}{T\gamma^3}\int V(\gamma+\rho)\rho' \left\{1 + \frac{3S\gamma\rho}{T\gamma^2}\right\}$$

to a sufficient approximation.

Now (between limits) $\int V\rho\rho' = 2A\epsilon$,

where A is the area of the closed circuit.

Also generally

$$\int V\gamma\rho' S\gamma\rho = \tfrac{1}{2}(S\gamma\rho V\gamma\rho + \gamma V.\gamma \int V\rho\rho')$$
$$= (\text{between limits}) \; A\gamma V\gamma\epsilon.$$

Hence for this case

$$\beta = \frac{A}{T\gamma^3}\left(2\epsilon + \frac{3\gamma V\gamma\epsilon}{T\gamma^2}\right)$$
$$= -\frac{A}{T\gamma^3}\left(\epsilon + \frac{3\gamma S\gamma\epsilon}{T\gamma^2}\right).$$

436.] If, instead of one small plane closed current, there be a series of such, of equal area, disposed regularly in a tubular form, let x be the distance between two consecutive currents measured along the axis of the tube; then, putting $\gamma' = x\epsilon$, we have for the whole effect of such a set of currents on α_1

$$\frac{CAaa_1}{2x} V_{\iota} a_1 \int \left(\frac{\gamma'}{T\gamma^3} + \frac{3\gamma S\gamma\gamma'}{T\gamma^5}\right)$$
$$= \frac{CAaa_1}{2x} \frac{Va_1\gamma}{T\gamma^3} \text{ (between proper limits).}$$

If the axis of the tubular arrangement be a closed curve this will evidently vanish. Hence *a closed solenoid exerts no influence on an element of a conductor. The same is evidently true if the solenoid be indefinite in both directions.*

If the axis extend to infinity in one direction, and γ_0 be the vector of the other extremity, the effect is

$$\frac{CAaa_1}{2x} \frac{Va_1\gamma_0}{T\gamma_0^3},$$

and is therefore *perpendicular to the element and to the line joining it with the extremity of the solenoid. It is evidently inversely as $T\gamma_0^2$ and directly as the sine of the angle contained between the direction of the element and that of the line joining the latter with the extremity of the solenoid. It is also inversely as x, and therefore directly as the number of currents in a unit of the axis of the solenoid.*

437.] To find the effect of the whole circuit whose element is a_1 on the extremity of the solenoid, we must change the sign of the above and put $a_1 = \gamma_0'$; therefore the effect is

$$-\frac{CAaa_1}{2x} \int \frac{V\gamma_0'\gamma_0}{T\gamma_0^3},$$

an integral of the species considered in § 432 whose value is easily assigned in particular cases.

438.] *Suppose the conductor to be straight, and indefinitely extended in both directions.*

Let $h\theta$ be the vector perpendicular to it from the extremity of the canal, and let the conductor be $\parallel \eta$, where $T\theta = T\eta = 1$.

Therefore $\quad \gamma_0 = h\theta + y\eta \quad$ (where y is a scalar),

$$V\gamma_0'\gamma_0 = hy' V\eta\theta,$$

and the integral in § 436 is

$$hV\eta\theta \int_{-\infty}^{+\infty} \frac{y'}{(h^2+y^2)^{\frac{3}{2}}} = \frac{2}{h} V\eta\theta.$$

The whole effect is therefore

$$\frac{CAaa_1}{xh} V\eta\theta,$$

and is thus *perpendicular to the plane passing through the conductor and the extremity of the canal, and varies inversely as the distance of the latter from the conductor.*

This is exactly the observed effect of an indefinite straight current on a magnetic pole, or particle of free magnetism.

439.] *Suppose the conductor to be circular, and the pole nearly in its axis.*

Let EPD be the conductor, AB its axis, and C the pole; BC perpendicular to AB, and small in comparison with $AE = h$ the radius of the circle.

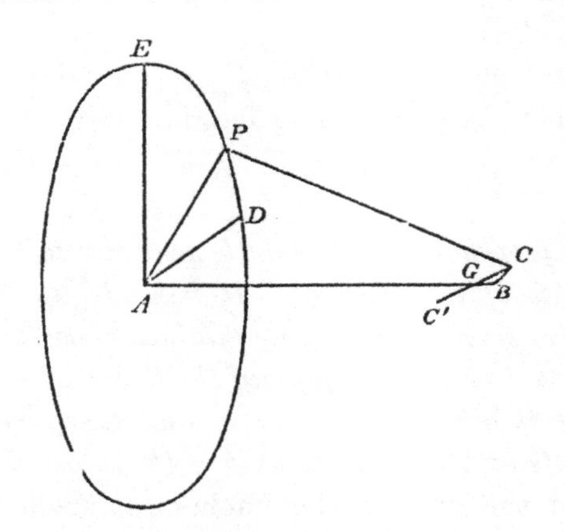

Let $\quad AB$ be $a_1 i, \quad BC = bk, \quad AP = h(jx + ky)$

where $$\left.\begin{matrix} x \\ y \end{matrix}\right\} = \left\{\begin{matrix} \cos \\ \sin \end{matrix}\right\} \angle EAP = \left\{\begin{matrix} \cos \\ \sin \end{matrix}\right\} \theta.$$

Then $$CP = \gamma = a_1 i + bk - h(jx + ky).$$

And the effect on $C \propto \int \frac{V\gamma\gamma'}{T\gamma^3}$,

$$\propto h \int \frac{\theta' \{(h - by)i + a_1 xj + a_1 yk\}}{(a_1^2 + b^2 + h^2 - 2bhy)^{\frac{3}{2}}},$$

where the integral extends to the whole circuit.

440.] Suppose in particular C to be one pole of a small magnet or solenoid CC' whose length is $2l$, and whose middle point is at G and distant a from the centre of the conductor.

Let $\angle CGB = \Delta$. Then evidently

$$a_1 = a + l \cos \Delta,$$
$$b = l \sin \Delta.$$

Also the effect on C becomes, if $a_1^2 + b^2 + h^2 = A^2$,

$$\frac{h}{A^3} \int \theta' \{(h - by)i + a_1 xj + a_1 yk\} \left\{1 + \frac{3hby}{A^2} + \frac{15}{2} \frac{h^2 b^2 y^2}{A^4} + \ldots\right\}$$
$$= \frac{\pi h^2}{A^3} \left(2i - \frac{3b^2 i}{A^2} + \frac{3a_1 bk}{A^2} + \frac{15}{2} \frac{h^2 b^2 i}{A^4} + \ldots\right),$$

since for the whole circuit

$$\int \theta' y^{2n} = 2\pi \frac{\lfloor 2n)}{2^{2n}(\lfloor n)^2},$$

$$\int \theta' y^{2n+1} = 0,$$

$$\int \theta' x y^m = 0.$$

If we suppose the centre of the magnet fixed, the vector axis of the couple produced by the action of the current on C is

$$lV.(i\cos\Delta + k\sin\Delta)\int\frac{V\gamma\gamma'}{T\gamma^3}$$

$$\propto \frac{\pi h^2 l \sin\Delta}{A^3} \cdot j\left\{2 - \frac{3b^2}{A^2} + \frac{15}{2}\frac{h^2 b^2}{A^4} - \frac{3a_1 b\cos\Delta}{A^2\sin\Delta}\right\}.$$

If A, &c. be now developed in powers of l, this at once becomes

$$\frac{\pi h^2 l\sin\Delta}{(a^2+h^2)^{\frac{3}{2}}} j\left\{2 - \frac{6al\cos\Delta}{a^2+h^2} + \frac{15a^2l^2\cos^2\Delta}{(a^2+h^2)^2} - \frac{3l^2}{a^2+h^2}\right.$$
$$\left. - \frac{3l^2\sin^2\Delta}{a^2+h^2} + \frac{15}{2}\frac{h^2l^2\sin^2\Delta}{(a^2+h^2)^2} - 3\frac{(a+l\cos\Delta)l\cos\Delta}{a^2+h^2}\left(1 - \frac{5al\cos\Delta}{a^2+h^2}\right)\right\}.$$

Putting $-l$ for l and changing the sign of the whole to get that for pole C', we have for the vector axis of the complete couple

$$\frac{4\pi h^2 l\sin\Delta}{(a^2+h^2)^{\frac{3}{2}}} j\left\{1 + \frac{3}{4}\frac{l^2(4a^2-h^2)(4-5\sin^2\Delta)}{(a^2+h^2)^2} + \&c.\right\},$$

which is almost exactly proportional to $\sin\Delta$ if $2a = h$ and l be small.

On this depends a modification of the tangent galvanometer. (Bravais, *Ann. de Chimie*, xxxviii. 309.)

441.] As before, the effect of an indefinite solenoid on a_1 is

$$\frac{CAaa_1}{2x}\frac{Va_1\gamma}{T\gamma^3}.$$

Now suppose a_1 to be an element of a small plane circuit, δ the vector of the centre of inertia of its area, the pole of the solenoid being origin.

Let $\gamma = \delta + \rho$, then $a_1 = \rho'$.

The whole effect is therefore

$$-\frac{CAaa_1}{2x}\int\frac{V(\delta+\rho)\rho'}{T(\delta+\rho)^3}$$

$$= \frac{CAA_1aa_1}{2xT\delta^3}\left(\epsilon_1 + \frac{3\delta S\delta\epsilon_1}{T\delta^2}\right),$$

where A_1 and ϵ_1 are, for the new circuit, what A and ϵ were for the former.

Let the new circuit also belong to an indefinite solenoid, and let δ_0 be the vector joining the poles of the two solenoids. Then the mutual effect is

$$\frac{CAA_1 aa_1}{2xx_1}\int\left(\frac{\delta'}{T\delta^3}+\frac{3\delta S\delta\delta'}{T\delta^5}\right)$$
$$=\frac{CAA_1 aa_1}{2xx_1}\frac{\delta_0}{(T\delta_0)^3}\propto\frac{U\delta_0}{(T\delta_0)^2},$$

which is exactly *the mutual effect of two magnetic poles. Two finite solenoids, therefore, act on each other exactly as two magnets, and the pole of an indefinite solenoid acts as a particle of free magnetism.*

442.] The mutual attraction of two indefinitely small plane closed circuits, whose normals are ϵ and ϵ_1, may evidently be deduced by twice differentiating the expression $\frac{U\delta}{T\delta^2}$ for the mutual action of the poles of two indefinite solenoids, making $d\delta$ in one differentiation $\parallel \epsilon$ and in the other $\parallel \epsilon_1$.

But it may also be calculated directly by a process which will give us in addition the couple impressed on one of the circuits by the other, supposing for simplicity the first to be *circular*

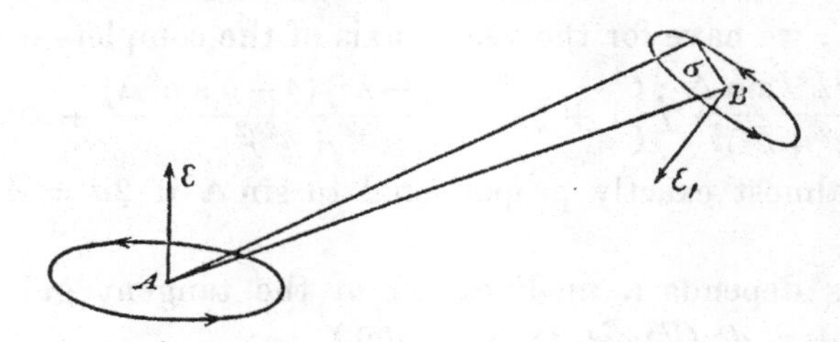

Let A and B be the centres of inertia of the areas of A and B, ϵ and ϵ_1 vectors normal to their planes, σ any vector radius of B, $AB=\beta$.

Then whole effect on σ', by §§ 432, 435,

$$\propto\frac{A}{T(\beta+\sigma)^3}V\sigma'\left\{\epsilon+\frac{3(\beta+\sigma)S(\beta+\sigma)\epsilon}{T(\beta+\sigma)^2}\right\},$$

$$\propto\frac{1}{T\beta^3}\left\{V\sigma'\epsilon\left(1+\frac{3S\beta\sigma}{T\beta^2}\right)+\frac{3V\sigma'\beta S\beta\epsilon}{T\beta^2}\left(1+\frac{5S\beta\sigma}{T\beta^2}\right)\right.$$
$$\left.+\frac{3V\sigma'\beta S\sigma\epsilon}{T\beta^2}+3\frac{V\sigma'\sigma S\beta\epsilon}{T\beta^2}\right\}.$$

But between proper limits,

$$\int V\sigma'\eta S\theta\sigma=-A_1 V.\eta V\theta\epsilon_1,$$

for generally $\int V\sigma'\eta S\theta\sigma=-\tfrac{1}{2}(V\eta\sigma S\theta\sigma+V.\eta V.\theta\int V\sigma\sigma')$.

Hence, after a reduction or two, we find that the whole force exerted by A on the centre of inertia of the area of B

$$\propto\frac{AA_1}{T\beta^5}\left\{\beta\left(S\epsilon\epsilon_1+\frac{5S\beta\epsilon S\beta\epsilon_1}{T\beta^2}\right)+\epsilon S\beta\epsilon_1+\epsilon_1 S\beta\epsilon\right\}.$$

This, as already observed, may be at once found by twice differentiating $\frac{U\beta}{T\beta^2}$. In the same way the vector moment, due to A, about the centre of inertia of B,

$$\propto \frac{A}{T\beta^3}\int V.\sigma\left(V\sigma'\epsilon + \frac{3V\sigma'\beta S\beta\epsilon}{T\beta^2}\right),$$

$$\propto -\frac{AA_1}{T\beta^3}\left(V\epsilon\epsilon_1 + \frac{3V\beta\epsilon_1 S\beta\epsilon}{T\beta^2}\right).$$

These expressions for the whole force of one small magnet on the centre of inertia of another, and the couple about the latter, seem to be the simplest that can be given. It is easy to deduce from them the ordinary forms. For instance, the whole resultant couple on the second magnet

$$\propto \frac{T\left(V\epsilon\epsilon_1 + \frac{3V\beta\epsilon_1 S\beta\epsilon}{T\beta^2}\right)}{T\beta^3},$$

may easily be shewn to coincide with that given by Ellis (*Camb. Math. Journal,* iv. 95), though it seems to lose in simplicity and capability of interpretation by such modifications.

443.] The above formulae shew that the whòle force exerted by one small magnet M, on the centre of inertia of another m, consists of four terms which are, in order,

1st. *In the line joining the magnets, and proportional to the cosine of their mutual inclination.*

2nd. *In the same line, and proportional to five times the product of the cosines of their respective inclinations to this line.*

3rd and 4th. *Parallel to* $\left\{\begin{matrix}\mathrm{m}\\ \mathrm{M}\end{matrix}\right\}$ *and proportional to the cosine of the inclination of* $\left\{\begin{matrix}\mathrm{M}\\ m\end{matrix}\right\}$ *to the joining line.*

All these forces are, in addition, inversely as the fourth power of the distance between the magnets.

For the couples about the centre of inertia of m we have

1st. *A couple whose axis is perpendicular to each magnet, and which is as the sine of their mutual inclination.*

2nd. *A couple whose axis is perpendicular to* m *and to the line joining the magnets, and whose moment is as three times the product of the sine of the inclination of* m, *and the cosine of the inclination of* M, *to the joining line.*

In addition these couples vary inversely as the third power of the distance between the magnets.

[These results afford a good example of what has been called the *internal* nature of the methods of quaternions, reducing, as they do at once, the forces and couples to others independent of any lines of reference, other than those necessarily belonging to the system under consideration. To shew their ready applicability, let us take a Theorem due to Gauss.]

444.] *If two small magnets be at right angles to each other, the moment of rotation of the first is approximately twice as great when the axis of the second passes through the centre of the first, as when the axis of the first passes through the centre of the second.*

In the first case $\quad \epsilon \parallel \beta \perp \epsilon_1;$

therefore moment $= \dfrac{C'}{T\beta^3} T(\epsilon\epsilon_1 - 3\epsilon\epsilon_1) = \dfrac{2C'}{T\beta^3} T\epsilon\epsilon_1.$

In the second $\quad \epsilon_1 \parallel \beta \perp \epsilon;$

therefore moment $= \dfrac{C'}{T\beta^3} T\epsilon\epsilon_1.$ Hence the theorem.

445.] Again, we may easily reproduce the results of § 442 if for the two small circuits we suppose two small magnets perpendicular to their planes to be substituted. β is then the vector joining the middle points of these magnets, and by changing the tensors we may take 2ϵ and $2\epsilon_1$ as the vector lengths of the magnets.

Hence evidently the mutual effect

$$\propto \frac{U}{T^2}(\beta+\epsilon-\epsilon_1) - \frac{U}{T^2}(\beta-\epsilon-\epsilon_1) + \frac{U}{T^2}(\beta-\epsilon+\epsilon_1) - \frac{U}{T^2}(\beta+\epsilon+\epsilon_1),$$

which is easily reducible to

$$-\frac{12}{T\beta^5}\left\{\beta\left(S\epsilon\epsilon_1 + \frac{5S\beta\epsilon S\beta\epsilon_1}{T\beta^2}\right) + \epsilon_1 S\beta\epsilon + \epsilon S\beta\epsilon_1\right\},$$

as before, if smaller terms be omitted.

If we operate with $V.\epsilon_1$ on the two first terms of the unreduced expression, and take the difference between this result and the same with the sign of ϵ_1 changed, we have the whole vector axis of the couple on the magnet $2\epsilon_1$, which is therefore, as before, seen to be proportional to

$$\frac{4}{T\beta^3}\left(V\epsilon_1\epsilon + \frac{3V\epsilon_1\beta S\beta\epsilon}{T\beta^2}\right).$$

446.] We might apply the foregoing formulae with great ease to other cases treated by Ampère, De Montferrand, &c.—or to two finite circular conductors as in Weber's Dynamometer—but in general the only difficulty is in the integration, which even in some of the simplest cases involves elliptic functions, &c., &c. (*Quarterly Math. Journal*, 1860.)

447.] Let $F(\gamma)$ be the potential of any system upon a unit particle at the extremity of γ.

$$F(\gamma) = C \quad \ldots\ldots\ldots\ldots\ldots\ldots \quad (1)$$

is the equation of a level surface.

Let the differential of (1) be

$$S\nu d\gamma = 0, \quad \ldots\ldots\ldots\ldots\ldots\ldots \quad (2)$$

then ν is a vector normal to (1), and is therefore the *direction* of the force.

But, passing to a proximate level surface, we have $S\nu\delta\gamma = \delta C$.

Make $\delta\gamma = x\nu$, then $-xT\nu^2 = \delta C$,

$$\text{or} \qquad -T\nu = \frac{\delta C}{T\delta\gamma}.$$

Hence ν expresses the force in *magnitude* also. (§ 368.)

Now by § 435 we have for the vector force exerted by a small plane closed circuit on a particle of free magnetism the expression

$$-\frac{A}{T\gamma^3}\left(\epsilon + \frac{3\gamma S\gamma\epsilon}{T\gamma^2}\right),$$

omitting the factors depending on the strength of the current and the strength of magnetism of the particle.

Hence the potential, by (2) and (1),

$$\begin{aligned}
&\propto A\int \frac{1}{T\gamma^3}\left(S\epsilon d\gamma + \frac{3S\gamma d\gamma S\gamma\epsilon}{T\gamma^2}\right),\\
&\propto \frac{AS\epsilon\gamma}{T\gamma^3},\\
&\propto \frac{\text{area of circuit projected perpendicular to } \gamma}{T\gamma^2},\\
&\propto \text{spherical opening subtended by circuit}
\end{aligned}$$

The constant is omitted in the integration, as the potential must evidently vanish for infinite values of $T\gamma$.

By means of Ampère's idea of breaking up a finite circuit into an indefinite number of indefinitely small ones, it is evident that the above result may be at once extended to the case of such a finite closed circuit.

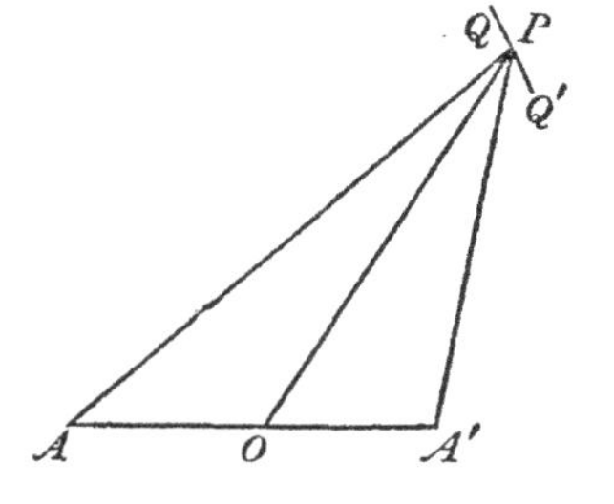

448.] Quaternions give a simple method of deducing the well-known property of the *Magnetic Curves*.

Let A, A' be two equal magnetic poles, whose vector distance, $2a$, is bisected in O, QQ' an indefinitely small magnet whose length is $2\rho'$, where $\rho = OP$. Then evidently, taking moments,

$$\frac{V(\rho+a)\rho'}{T(\rho+a)^3} = \pm \frac{V(\rho-a)\rho'}{T(\rho-a)^3},$$

where the upper or lower sign is to be taken according as the poles are like or unlike.

Operate by $S.Va\rho$,

$$\frac{Sa\rho'(\rho+a)^2 - Sa(\rho+a)S\rho'(\rho+a)}{T(\rho+a)^3} = \pm \text{ \{same with } -a\},$$

or $\quad S.aV\left(\frac{\rho'}{\rho+a}\right)U(\rho+a) = + \text{ \{same with } -a\},$

i.e. $\quad SadU(\rho+a) = + SadU(\rho-a),$

$$Sa\{U(\rho+a) \mp U(\rho-a)\} = \text{const.},$$

or $\quad \cos\angle OAP + \cos\angle OA'P = \text{const.},$

the property referred to.

If the poles be unequal, one of the terms to the left must be multiplied by the ratio of their strengths.

449.] If the vector of any point be denoted by

$$\rho = ix + jy + kz, \quad \ldots\ldots\ldots\ldots \quad (1)$$

there are many physically interesting and important transformations depending upon the effects of the quaternion operator

$$\nabla = i\frac{d}{dx} + j\frac{d}{dy} + k\frac{d}{dz} \quad \ldots\ldots\ldots\ldots \quad (2)$$

on various functions of ρ. When the function of ρ is a scalar, the effect of ∇ is to give the vector of most rapid increase. Its effect on a vector function is indicated briefly in § 369.

450.] We commence with one or two simple examples, which are not only interesting, but very useful in transformations.

$$\nabla\rho = \left(i\frac{d}{dx} + \&c.\right)(ix + \&c.) = -3, \quad \ldots\ldots\ldots\ldots \quad (3)$$

$$\nabla T\rho = \left(i\frac{d}{dx} + \&c.\right)(x^2+y^2+z^2)^{\frac{1}{2}} = \frac{ix+jy+kz}{(x^2+y^2+z^2)^{\frac{1}{2}}} = \frac{\rho}{T\rho} = U\rho, \quad (4)$$

$$\nabla(T\rho)^n = n(T\rho)^{n-1}\nabla T\rho = n(T\rho)^{n-2}\rho; \quad \ldots\ldots\ldots\ldots \quad (5)$$

and, of course, $\quad \nabla\frac{1}{(T\rho)^n} = -\frac{n\rho}{(T\rho)^{n+2}}; \quad \ldots\ldots\ldots\ldots \quad (5)^1$

whence, $\quad \nabla\frac{1}{T\rho} = -\frac{\rho}{T\rho^3} = -\frac{U\rho}{T\rho^2}, \quad \ldots\ldots\ldots\ldots \quad (6)$

and, of course, $\quad \nabla^2\frac{1}{T\rho} = -\nabla\frac{U\rho}{T\rho^2} = 0. \quad \ldots\ldots\ldots\ldots \quad (6)^1$

Also, $\quad \nabla\rho = -3 = T\rho\nabla U\rho + \nabla T\rho . U\rho = T\rho\nabla U\rho - 1,$

$$\therefore \quad \nabla U\rho = -\frac{2}{T\rho}. \quad \ldots\ldots\ldots\ldots \quad (7)$$

451.] By the help of the above results, of which (6) is especially useful (though obvious on other grounds), and (4) and (7) very remarkable, we may easily find the effect of ∇ upon more complex functions.

Thus, $$\nabla S\alpha\rho = -\nabla(\alpha x + \&c.) = -\alpha, \quad\quad (1)$$

$$\nabla V\alpha\rho = -\nabla V\rho\alpha = -\nabla(\rho\alpha - S\alpha\rho) = 3\alpha - \alpha = 2\alpha. \quad\quad (2)$$

Hence

$$\nabla\frac{V\alpha\rho}{T\rho^3} = \frac{2\alpha}{T\rho^3} - \frac{3\rho V\alpha\rho}{T\rho^5} = -\frac{2\alpha\rho^2 + 3\rho V\alpha\rho}{T\rho^5} = \frac{\alpha\rho^2 - 3\rho S\alpha\rho}{T\rho^5} \quad\quad (3)$$

Hence

$$S.\delta\rho\nabla\frac{V\alpha\rho}{T\rho^3} = \frac{\rho^2 S\alpha\delta\rho - 3S\alpha\rho S\rho\delta\rho}{T\rho^5} = -\frac{S\alpha\delta\rho}{T\rho^3} - \frac{3S\alpha\rho S\rho\delta\rho}{T\rho^5} = -\delta\frac{S\alpha\rho}{T\rho^3}. \quad (4)$$

This is a very useful transformation in various physical applications. By (6) it can be put in the sometimes more convenient form

$$S.\delta\rho\nabla\frac{V\alpha\rho}{T\rho^3} = \delta S.\alpha\nabla\frac{1}{T\rho}. \quad\quad (5)$$

And it is worthy of remark that, as may easily be seen, $-S$ may be put for V in the left-hand member of the equation.

452.] We have also

$$\nabla V.\beta\rho\gamma = \nabla\{\beta S\gamma\rho - \rho S\beta\gamma + \gamma S\beta\rho\} = -\gamma\beta + 3S\beta\gamma - \beta\gamma = S\beta\gamma. \quad (1)$$

Hence, if ϕ be any linear and vector function of the form

$$\phi\rho = \alpha + \Sigma V.\beta\rho\gamma + m\rho, \quad\quad (2)$$

i.e. a self-conjugate function with a constant vector added, then

$$\nabla\phi\rho = \Sigma S\beta\gamma - 3m = \text{scalar}. \quad\quad (3)$$

Hence, an integral of

$$\nabla\sigma = \text{scalar constant, is } \sigma = \phi\rho. \quad\quad (4)$$

If the constant value of $\nabla\sigma$ contain a vector part, there will be terms of the form $V\epsilon\rho$ in the expression for σ, which will then express a distortion accompanied by rotation. (§ 371.)

Also, a solution of $\nabla q = a$ (where q and a are quaternions) is

$$q = S\zeta\rho + V\epsilon\rho + \phi\rho.$$

It may be remarked also, as of considerable importance in physical applications, that, by (1) and (2) of § 451,

$$\nabla(S + \tfrac{1}{2}V)\alpha\rho = 0,$$

but we cannot here enter into details on this point.

453.] It would be easy to give many more of these transformations, which really present no difficulty; but it is sufficient to shew the ready applicability to physical questions of one or two of those already obtained; a property of great importance, as extensions of mathematical physics are far more valuable than mere analytical or geometrical theorems.

Thus, if σ be the vector-displacement of that point of a homogeneous elastic solid whose vector is ρ, we have, p being the consequent pressure produced,

$$\nabla p+\nabla^2\sigma = 0, \qquad (1)$$

whence $S\delta\rho\nabla^2\sigma = -S\delta\rho\nabla p = \delta p$, a complete differential. .... (2)

Also, generally, $\qquad p = kS\nabla\sigma,$

and if the solid be incompressible

$$S\nabla\sigma = 0. \qquad (3)$$

Thomson has shewn (*Camb. and Dub. Math. Journal*, ii. p. 62), that the forces produced by given distributions of matter, electricity, magnetism, or galvanic currents, can be represented at every point by displacements of such a solid producible by external forces. It may be useful to give his analysis, with some additions, in a quaternion form, to shew the insight gained by the simplicity of the present method.

454.] Thus, if $S\sigma\delta\rho = \delta\frac{1}{T\rho}$, we may write each equal to

$$-S\delta\rho\nabla\frac{1}{T\rho}.$$

This gives $\qquad \sigma = -\nabla\frac{1}{T\rho},$

the vector-force exerted by one particle of matter or free electricity on another. This value of σ evidently satisfies (2) and (3).

Again, if $\quad S.\delta\rho\nabla\sigma = \delta\frac{S\alpha\rho}{T\rho^3}$, either is equal to

$$-S.\delta\rho\nabla\frac{V\alpha\rho}{T\rho^3} \text{ by (4) of § 451.}$$

Here a particular case is

$$\sigma = -\frac{V\alpha\rho}{T\rho^3},$$

which is the vector-force exerted by an element α of a current upon a particle of magnetism at ρ. (§ 436.)

455.] Also, by § 451 (3),

$$\nabla\frac{V\alpha\rho}{T\rho^3} = \frac{\alpha\rho^2-3\rho S\alpha\rho}{T\rho^5},$$

and we see by §§ 435, 436 that this is the vector-force exerted by a small plane current at the origin (its plane being perpendicular to a) upon a magnetic particle, or pole of a solenoid, at ρ. This expression, being a pure vector, denotes an elementary rotation caused by the distortion of the solid, and it is evident that the above value of σ satisfies the equations (2), (3), and the distortion is therefore producible by external forces. Thus the effect of an element of a current on a magnetic particle is expressed directly by the displacement, while that of a small closed current or magnet is represented by the vector-axis of the rotation caused by the displacement.

456.] Again, let $$S\delta\rho\nabla^2\sigma = \delta\frac{Sa\rho}{T\rho^3}.$$

It is evident that σ satisfies (2), and that the right-hand side of the above equation may be written

$$-S.\delta\rho\nabla\frac{Va\rho}{T\rho^3}.$$

Hence a particular case is

$$\nabla\sigma = -\frac{Va\rho}{T\rho},$$

and this satisfies (3) also.

Hence the corresponding displacement is producible by external forces, and $\nabla\sigma$ is the rotation axis of the element at ρ, and is seen as before to represent the vector-force exerted on a particle of magnetism at ρ by an element a of a current at the origin.

457.] It is interesting to observe that a particular value of σ in this case is

$$\sigma = -\tfrac{1}{2}\nabla SaU\rho - \frac{a}{T\rho},$$

as may easily be proved by substitution.

Again, if $$S\delta\rho\sigma = -\delta\frac{Sa\rho}{T\rho^3},$$

we have evidently $$\sigma = \nabla\frac{Sa\rho}{T\rho^3}.$$

Now, as $\frac{Sa\rho}{T\rho^3}$ is the potential of a small magnet a, at the origin, on a particle of free magnetism at ρ, σ is the resultant magnetic force, and represents also a possible distortion of the elastic solid by external forces, since $\nabla\sigma = \nabla^2\sigma = 0$, and thus (2) and (3) are both satisfied.

458.] We conclude with some examples of quaternion integration of the kinds specially required for many important physical problems.

It may perhaps be useful to commence with a different form of definition of the operator ∇, as we shall thus, if we desire it, entirely avoid the use of ordinary Cartesian coördinates. For this purpose we write

$$S.\alpha\nabla = -d_\alpha,$$

where α is any unit-vector, the meaning of the right-hand operator (neglecting its sign) being the *rate of change of the function to which it is applied* per unit of length in the direction of the unit-vector α. If α be not a unit-vector we may treat it as a vector-velocity, and then the right-hand operator means the *rate of change per unit of time* due to the change of position.

Let α, β, γ be any rectangular system of unit-vectors, then by a fundamental quaternion transformation

$$\nabla = -\alpha S\alpha\nabla - \beta S\beta\nabla - \gamma S\gamma\nabla = \alpha d_\alpha + \beta d_\beta + \gamma d_\gamma,$$

which is identical with Hamilton's form so often given above. (*Lectures*, § 620.)

459.] This mode of viewing the subject enables us to see at once that the effect of applying ∇ to any scalar function of the position of a point is to give its *vector of most rapid increase.* Hence, when it is applied to a potential u, we have

$$\nabla u = \text{vector-force at } \rho.$$

If u be a velocity-potential, we obtain the velocity of the fluid element at ρ; and if u be the temperature of a conducting solid we obtain the flux of heat. Finally, whatever series of surfaces is represented by

$$u = C,$$

the vector ∇u is the normal at the point ρ, and its length is inversely as the normal distance at that point between two consecutive surfaces of the series.

Hence it is evident that

$$S.d\rho\nabla u = -du,$$

or, as it may be written,

$$-S.d\rho\nabla = d;$$

the left-hand member therefore expresses total differentiation in virtue of any arbitrary, but small, displacement $d\rho$.

460.] To interpret the operator $V.\alpha\nabla$ let us apply it to a potential function u. Then we easily see that u may be taken under the vector sign, and the expression

$$V(\alpha\nabla)u = V.\alpha\nabla u$$

denotes the vector-couple due to the force at ρ about a point whose relative vector is α.

Again, if σ be any vector function of ρ, we have by ordinary quaternion operations

$$V(a\nabla).\sigma = S.aV\nabla\sigma + aS\nabla\sigma - \nabla Sa\sigma.$$

The meaning of the third term (in which it is of course understood that ∇ operates on σ alone) is obvious from what precedes. It remains that we explain the other terms.

461.] These involve the very important quantities (not *operators* such as the expressions we have been hitherto considering),

$$S.\nabla\sigma \quad \text{and} \quad V.\nabla\sigma,$$

which form the basis of our investigations. Let us look upon σ as the displacement, or as the velocity, of a point situated at ρ, and consider the group of points situated near to that at ρ, as the quantities to be interpreted have reference to the deformation of the group.

462.] Let τ be the vector of one of the group relative to that situated at ρ. Then after a small interval of time t, the actual coördinates become $\rho + t\sigma$

and $\rho + \tau + t(\sigma - S(\tau\nabla)\sigma)$

by the definition of ∇ in § 458. Hence, if ϕ be the linear and vector function reprèsenting the deformation of the group, we have

$$\phi\tau = \tau - tS(\tau\nabla)\sigma.$$

The farther solution is rendered very simple by the fact that we may assume t to be so small that its square and higher powers may be neglected.

If ϕ' be the function conjugate to ϕ, we have

$$\phi'\tau = \tau - t\nabla S\tau\sigma.$$

Hence
$$\phi\tau = \tfrac{1}{2}(\phi + \phi')\tau + \tfrac{1}{2}(\phi - \phi')\tau$$
$$= \tau - \frac{t}{2}[S(\tau\nabla)\sigma + \nabla S\tau\sigma] - \frac{t}{2}V.\tau V\nabla\sigma.$$

The first three terms form a self-conjugate linear and vector function of τ, which we may denote for a moment by $\varpi\tau$. Hence

$$\phi\tau = \varpi\tau - \frac{t}{2}V.\tau V\nabla\sigma,$$

or, omitting t^2 as above,

$$\phi\tau = \varpi\tau - \frac{t}{2}V.\varpi\tau V\nabla\sigma.$$

Hence the deformation may be decomposed into (1) the pure strain ϖ, (2) the rotation $\frac{t}{2}V\nabla\sigma.$

Thus the *vector-axis of rotation of the group* is

$$\tfrac{1}{2}V\nabla\sigma.$$

If we were content to avail ourselves of the ordinary results of Cartesian investigations, we might at once have reached this con clusion by noticing that

$$V\nabla\sigma = i\left(\frac{d\zeta}{dy} - \frac{d\eta}{dz}\right) + j\left(\frac{d\xi}{dz} - \frac{d\zeta}{dx}\right) + k\left(\frac{d\eta}{dx} - \frac{d\xi}{dy}\right).$$

and remembering as in (§ 362) the formulae of Stokes and Helmholtz.

463.] In the same way, as

$$S\nabla\sigma = -\frac{d\xi}{dx} - \frac{d\eta}{dy} - \frac{d\zeta}{dz},$$

we recognise the *cubical compression* of the group of points considered. It would be easy to give this a more strictly quaternionic form by employing the definition of § 458. But, working with quaternions, we ought to obtain all our results by their help alone; so that we proceed to prove the above result by finding the volume of the ellipsoid into which an originally spherical group of points has been distorted in time t.

For this purpose, we refer again to the equation of deformation

$$\phi\tau = \tau - tS(\tau\nabla)\,\sigma,$$

and form the cubic in ϕ according to Hamilton's exquisite process. We easily obtain, remembering that t^2 is to be neglected*,

$$0 = \phi^3 - (3 - tS\nabla\sigma)\,\phi^2 + (3 - 2tS\nabla\sigma)\,\phi - (1 - tS\nabla\sigma),$$

or

$$0 = (\phi - 1)^2\,(\phi - 1 + tS\nabla\sigma).$$

The roots of this equation are the ratios of the diameters of the ellipsoid whose directions are unchanged to that of the sphere. Hence the volume is increased by the factor

$$1 - tS\nabla\sigma,$$

from which the truth of the preceding statement is manifest.

* Thus, in Hamilton's notation, λ, μ, ν being any three non-coplanar vectors, and m, m_1, m_2 the coefficients of the cubic,

$$\begin{aligned}
-mS\,\lambda\mu\nu &= S.\phi'\lambda\phi'\mu\phi'\nu \\
&= S.(\lambda - t\nabla S\lambda\sigma)(\mu - t\nabla S\mu\sigma)(\nu - t\nabla S\nu\sigma) \\
&= S.(\lambda - t\nabla S\lambda\sigma)(V\mu\nu - tV\mu\nabla S\nu\sigma + tV\nu\nabla S\mu\sigma) \\
&= S.\lambda\mu\nu - t\,[S.\mu\nu\nabla S\lambda\sigma + S.\nu\lambda\nabla S\mu\sigma + S.\lambda\mu\nabla S\nu\sigma] \\
&= S.\lambda\mu\nu - tS.[\lambda S.\mu\nu\nabla + \mu S.\nu\lambda\nabla + \nu S.\lambda\mu\nabla]\,\sigma \\
&= S.\lambda\mu\nu - tS.\lambda\mu\nu S\nabla\sigma. \\
m_1 S.\lambda\mu\nu &= S.\lambda\phi'\mu\phi'\nu + S.\mu\phi'\nu\phi'\lambda + S.\nu\phi'\lambda\phi'\mu \\
&= S.\lambda\,(V\mu\nu - tV\mu\nabla S\nu\sigma + tV\nu\nabla S\mu\sigma) + \&c. \\
&= S.\lambda\mu\nu - tS.\lambda\mu\nabla S\nu\sigma - tS.\nu\lambda\nabla S\mu\sigma + \&c. \\
&= 3S.\lambda\mu\nu - 2tS\nabla\sigma S.\lambda\mu\nu. \\
-m_2 S.\lambda\mu\nu &= S.\lambda\mu\phi'\nu + S.\mu\nu\phi'\lambda + S.\nu\lambda\phi'\mu \\
&= S.\lambda\mu\nu - tS.\lambda\mu\nabla S\nu\sigma + \&c. \\
&= 3S.\lambda\mu\nu - tS\nabla\sigma S.\lambda\mu\nu
\end{aligned}$$

464.] As the process in last section depends essentially on the use of a non-conjugate vector function, with which the reader is less likely to be acquainted than with the more usually employed forms, I add another investigation.

Let $$\varpi = \phi\tau = \tau - tS(\tau\nabla)\sigma.$$

Then $$\tau = \phi^{-1}\varpi = \varpi + tS(\varpi\nabla)\sigma.$$

Hence since if, before distortion, the group formed a sphere of radius 1, we have $$T\tau = 1,$$
the equation of the ellipsoid is
$$T(\varpi + tS(\varpi\nabla)\sigma) = 1,$$
or $$\varpi^2 + 2tS\varpi\nabla S\varpi\sigma = -1.$$

This may be written
$$S.\varpi\chi\varpi = S.\varpi(\varpi + t\nabla S\varpi\sigma + tS(\varpi\nabla)\sigma) = -1,$$
where χ is now self-conjugate.

Hamilton has shewn that the reciprocal of the product of the squares of the semiaxes is
$$-S.\chi i\chi j\chi k,$$
whatever rectangular system of unit-vectors is denoted by i, j, k.

Substituting the value of χ, we have
$$S.(i + t\nabla Si\sigma + tS(i\nabla)\sigma)(j + \&c.)(k + \&c.)$$
$$= -S.(i + t\nabla Si\sigma + tS(i\nabla)\sigma)(i + 2tiS\nabla\sigma - tS(i\nabla)\sigma - t\nabla Si\sigma)$$
$$= 1 + 2tS\nabla\sigma$$

The ratio of volumes of the ellipsoid and sphere is therefore, as before,
$$\frac{1}{\sqrt{1 + 2tS\nabla\sigma}} = 1 - tS\nabla\sigma.$$

465.] In what follows we have constantly to deal with integrals extended over a closed surface, compared with others taken through the space enclosed by such a surface; or with integrals over a limited surface, compared with others taken round its bounding curve. The notation employed is as follows. If Q per unit of length, of surface, or of volume, at the point ρ, Q being any quaternion, be the quantity to be summed, these sums will be denoted by
$$\iint Q\,ds \quad \text{and} \quad \iiint Q\,ds,$$
when comparing integrals over a closed surface with others through the enclosed space; and by
$$\iint Q\,ds \quad \text{and} \quad \int QTd\rho,$$
when comparing integrals over an unclosed surface with others round its boundary. No ambiguity is likely to arise from the double use of
$$\iint Q\,ds,$$

for its meaning in any case will be obvious from the integral with which it is compared.

466.] We have just shewn that, if σ be the vector displacement of a point originally situated at

$$\rho = ix + jy + kz,$$

then

$$S.\nabla\sigma$$

expresses the increase of density of aggregation of the points of the system caused by the displacement.

467.] Suppose, now, space to be uniformly filled with points, and a closed surface Σ to be drawn, through which the points can freely move when displaced.

Then it is clear that the increase of number of points within the space Σ, caused by a displacement, may be obtained by either of two processes—by taking account of the increase of density at all points within Σ, or by estimating the excess of those which pass inwards through the surface over those which pass outwards. These are the principles usually employed (for a mere element of volume) in forming the so-called 'Equation of Continuity.'

Let ν be the normal to Σ at the point ρ, drawn outwards, then we have at once (by equating the two different expressions of the same quantity above explained) the equation

$$\iiint S.\nabla\sigma ds = \iint S.\sigma U\nu ds,$$

which is our fundamental equation so long as we deal with triple integrals.

468.] As a first and very simple example of its use, suppose σ to represent the vector force exerted upon a unit particle at ρ (of ordinary matter, electricity, or magnetism) by any distribution of attracting matter, electricity, or magnetism partly outside, partly inside Σ. Then, if P be the potential at ρ,

$$\sigma = \nabla P,$$

and if r be the density of the attracting matter, &c., at ρ,

$$\nabla\sigma = \nabla^2 P = 4\pi r$$

by Poisson's extension of Laplace's equation.

Substituting in the fundamental equation, we have

$$4\pi\iiint r ds = 4\pi M = \iint S.\nabla P U\nu ds,$$

where M denotes the whole quantity of matter, &c., inside Σ. This is a well-known theorem.

469.] Let P and P_1 be any scalar functions of ρ, we can of course find the distribution of matter, &c., requisite to make either of them the potential at ρ; for, if the necessary densities be r and r_1 respectively, we have as before

$$\nabla^2 P = 4\pi r, \qquad \nabla^2 P_1 = 4\pi r_1.$$

Now $$\nabla(P\nabla P_1) = \nabla P \nabla P_1 + P\nabla^2 P_1,$$

Hence, if in the above formula we put

$$\sigma = P\nabla P_1,$$

we obtain

$$\iiint S.\nabla P \nabla P_1 ds = -\iiint P\nabla^2 P_1 ds + \iint PS.\nabla P_1 U\nu ds,.$$
$$= -\iiint P_1 \nabla^2 P ds + \iint P_1 S.\nabla P U\nu ds,$$

which are the common forms of Green's Theorem. Sir W. Thomson's extension of it follows at once from the same proof.

470.] If P_1 be a many-valued function, but ∇P_1 single-valued, and if Σ be a multiply-connected* space, the above expressions require a modification which was first shewn to be necessary by Helmholtz, and first supplied by Thomson. For simplicity, suppose Σ to be doubly-connected (as a ring or endless rod, whether knotted or not). Then if it be cut through by a surface s, it will become simply-connected, but the surface-integrals have to be increased by terms depending upon the portions thus added to the whole surface. In the first form of Green's Theorem, just given, the only term altered is the last: and it is obvious that if p_1 be the increase of P_1 after a complete circuit of the ring, the portion to be added to the right-hand side of the equation is

$$p_1 \iint S.\nabla P U\nu ds,$$

taken over the cutting surface only. Similar modifications are easily seen to be produced by each additional complexity in the space Σ.

471.] The immediate consequences of Green's theorem are well known, so that I take only one instance.

Let P and P_1 be the potentials of one and the same distribution of matter, and let none of it be within Σ. Then we have

$$\iiint (\nabla P)^2 ds = \iint PS.\nabla P U\nu ds,$$

so that if ∇P is zero all over the surface of Σ, it is zero all through the interior, i.e., the potential is constant inside Σ. If P be the velocity-potential in the irrotational motion of an incompressible fluid, this equation shews that there can be no such motion of the

* Called by Helmholtz, after Riemann, *mehrfach zusammenhängend*. In translating Helmholtz's paper (*Phil. Mag.* 1867) I used the above as an English equivalent. Sir W. Thomson in his great paper on *Vortex Motion* (*Trans. R. S. E.* 1868) uses the expression "multiply-continuous."

fluid unless there is a normal motion at some part of the bounding surface, so long at least as Σ is simply-connected.

Again, if Σ is an equipotential surface,

$$\iiint (\nabla P)^2 ds = P \iint S.\nabla P U\nu ds = P \iiint \nabla^2 P ds$$

by the fundamental theorem. But there is by hypothesis no matter inside Σ, so this shews that the potential is constant throughout the interior. Thus there can be no equipotential surface, not including some of the attracting matter, within which the potential can change. Thus it cannot have a maximum or minimum value at points unoccupied by matter.

472.] If, in the fundamental theorem, we suppose

$$\sigma = \nabla \tau,$$

which imposes the condition that

$$S.\nabla\sigma = 0,$$

i.e., that the σ displacement is effected without condensation, it becomes

$$\iint S.\nabla\tau U\nu ds = \iiint S.\nabla^2\tau ds = 0.$$

Suppose any closed curve to be traced on the surface Σ, dividing it into two parts. This equation shews that the surface-integral is the same for both parts, the difference of sign being due to the fact that the normal is drawn in opposite directions on the two parts. Hence we see that, with the above limitation of the value of σ, the double integral is the same for all surfaces bounded by a given closed curve. It must therefore be expressible by a single integral taken round the curve. The value of this integral will presently be determined.

473.] The theorem of § 467 may be written

$$\iiint \nabla^2 P ds = \iint S.U\nu\nabla P ds = \iint S(U\nu\nabla) P ds.$$

From this we conclude at once that if

$$\sigma = iP + jP_1 + kP_2,$$

(which may, of course, represent any vector whatever) we have

$$\iiint \nabla^2 \sigma ds = \iint S(U\nu\nabla)\sigma ds,$$

or, if

$$\nabla^2\sigma = \tau,$$

$$\iiint \tau ds = \iint S(U\nu\nabla^{-1})\tau ds.$$

This gives us the means of representing, by a surface-integral, a vector-integral taken through a definite space. We have already seen how to do the same for a scalar-integral—so that we can now express in this way, subject, however, to an ambiguity presently to be mentioned, the general integral

$$\iiint q ds,$$

where q is any quaternion whatever. It is evident that it is only in certain classes of cases that we can *exvect* a perfectly definite expression of such a volume-integral in terms of a surface-integral.

474.] In the above formula for a vector-integral there may present itself an ambiguity introduced by the inverse operation

$$\nabla^{-1}$$

to which we must devote a few words. The assumption

$$\nabla^2\sigma = \tau$$

is tantamount to saying that, as the constituents of σ are the potentials of certain distributions of matter, &c., those of τ are the corresponding densities each multiplied by 4π.

If, therefore, τ be given throughout the space enclosed by Σ, σ is given by this equation *so far only* as it depends upon the distribution within Σ, and must be completed by an arbitrary vector depending on *three* potentials of mutually independent distributions exterior to Σ.

But, if σ be given, τ is perfectly definite; and as

$$\nabla\sigma = \nabla^{-1}\tau,$$

the value of ∇^{-1} is also completely defined. These remarks must be carefully attended to in using the theorem above: since they involve as particular cases of their application many curious theorems in Fluid Motion, &c.

475.] As a particular case, the equation

$$V\nabla\sigma = 0$$

of course gives $\nabla\sigma = u$, a scalar.

Now, if v be the potential of a distribution whose density is u, we have $\nabla^2 v = 4\pi u$.

We know that this equation gives one, and but one, definite value for v, so that there is no ambiguity in

$$v = 4\pi\nabla^{-2}u,$$

and therefore $\sigma = \frac{1}{4\pi}\nabla v$ is also determinate.

476.] This shews the nature of the arbitrary term which must be introduced into the solution of the equation

$$V\nabla\sigma = \tau.$$

To solve this equation is (§ 462) to find the displacement of any one of a group of points when the consequent rotation is given.

Here $S\nabla\tau = S.\nabla V\nabla\sigma = S\nabla^2\sigma = 0\cdot$

so that, omitting the arbitrary term (§ 475), we have

$$\nabla^2\sigma = \nabla\tau,$$

and each constituent of σ is, as above, determinate.

Thomson * has put the solution in a form which may be written

$$\sigma = \tfrac{1}{3}\int V\tau\, d\rho + \nabla u,$$

if we understand by $\int (\ )d\rho$ integrating the term in dx as if y and z were constants, &c. Bearing this in mind, we have as verification,

$$V\nabla\sigma = \tfrac{1}{3}\Sigma Vi\left\{V\tau i + \int V\frac{d\tau}{dx}d\rho\right\}$$

$$= \tfrac{1}{3}\left\{2\tau + \Sigma\int\frac{d\tau}{dx}dx + \Sigma\int d\rho Si\frac{d\tau}{dx}\right\}$$

$$= \tfrac{1}{3}\{3\tau + \int d\rho S\nabla\tau\} = \tau.$$

477.] We now come to relations between the results of integra tion extended over a non-closed surface and round its boundary.

Let σ be any vector function of the position of a point. The line-integral whose value we seek as a fundamental theorem is

$$\int S.\sigma d\tau,$$

where τ is the vector of any point in a small closed curve, drawn from a point within it, and in its plane.

Let σ_0 be the value of σ at the origin of τ, then

$$\sigma = \sigma_0 - S(\tau\nabla)\sigma_0,$$

so that $$\int S.\sigma d\tau = \int S.(\sigma_0 - S(\tau\nabla)\sigma_0)d\tau.$$

But $$\int d\tau = 0,$$

because the curve is closed; and (Tait on *Electro-Dynamics*, § 13 *Quarterly Math. Journal*, Jan. 1860) we have generally

$$\int S.\tau\nabla S.\sigma_0 d\tau = \tfrac{1}{2}S.\nabla(\tau S\sigma_0\tau - \sigma_0\int V.\tau d\tau).$$

Here the integrated part vanishes for a closed circuit, and

$$\tfrac{1}{2}\int V.\tau d\tau = ds\, Uv,$$

where ds is the area of the small closed curve, and Uv is a unit-vector perpendicular to its plane. Hence

$$\int S.\sigma_0 d\tau = S.\nabla\sigma_0 Uv.ds.$$

Now, any finite portion of a surface may be broken up into small elements such as we have just treated, and the sign only of the integral along each portion of a bounding curve is changed when we go round it in the opposite direction. Hence, just as Ampère did with electric currents, substituting for a finite closed circuit a network of an infinite number of infinitely small ones, in each contiguous pair of which the common boundary is described by equal currents in opposite directions, we have for a finite unclosed surface $$\int S.\sigma d\rho = \iint S.\nabla\sigma Uv.ds.$$

There is no difficulty in extending this result to cases in which the

* *Electrostatics and Magnetism*, § 521, or *Phil. Trans.*, 1852.

bounding curve consists of detached ovals, or possesses multiple points. This theorem seems to have been first given by Stokes (*Smith's Prize Exam.* 1854), in the form

$$\int(\alpha dx+\beta dy+\gamma dz)$$
$$=\iint ds\left(l\left(\frac{d\gamma}{dy}-\frac{d\beta}{dz}\right)+m\left(\frac{d\alpha}{dz}-\frac{d\gamma}{dx}\right)+n\left(\frac{d\beta}{dx}-\frac{d\alpha}{dy}\right)\right).$$

It solves the problem suggested by the result of § 472 above.

478.] If σ represent the vector force acting on a particle of matter at ρ, $-S.\sigma d\rho$ represents the work done while the particle is displaced along $d\rho$, so that the single integral

$$\int S.\sigma d\rho$$

of last section, taken with a negative sign, represents the work done during a complete cycle. When this integral vanishes it is evident that, if the path be divided into any two parts, the work spent during the particle's motion through one part is equal to that gained in the other. Hence the system of forces must be conservative, i. e., must do the same amount of work for all paths having the same extremities.

But the equivalent double integral must also vanish. Hence a conservative system is such that

$$\iint ds S.\nabla\sigma U\nu = 0,$$

whatever be the form of the finite portion of surface of which ds is an element. Hence, as $\nabla\sigma$ has a fixed value at each point of space, while $U\nu$ may be altered at will, we must have

$$V\nabla\sigma = 0,$$

or $$\nabla\sigma = \text{scalar}.$$

If we call X, Y, Z the component forces parallel to rectangular axes, this extremely simple equation is equivalent to the well-known conditions

$$\frac{dX}{dy}-\frac{dY}{dx}=0,\quad \frac{dY}{dz}-\frac{dZ}{dy}=0,\quad \frac{dZ}{dx}-\frac{dX}{dz}=0.$$

Returning to the quaternion form, as far less complex, we see that

$$\nabla\sigma = \text{scalar} = 4\pi r, \text{ suppose},$$

implies that $$\sigma=\nabla P,$$

where P is a scalar such that

$$\nabla^2 P = 4\pi r;$$

that is, P is the potential of a distribution of matter, magnetism, or statical electricity, of volume-density r.

Hence, for a non-closed path, under conservative forces

$$\begin{aligned} -\int S.\sigma d\rho &= -\int S.\nabla P d\rho \\ &= -\int S(d\rho\nabla)P \\ &= \int d_{d\rho}P = \int dP \\ &= P_1 - P_0, \end{aligned}$$

depending solely on the values of P at the extremities of the path.

479.] A vector theorem, which is of great use, and which corresponds to the Scalar theorem of § 473, may easily be obtained. Thus, with the notation already employed,

$$\begin{aligned} \int V.\sigma d\tau &= \int V(\sigma_0 - S(\tau\nabla)\sigma_0)d\tau, \\ &= -\int S(\tau\nabla)V.\sigma_0 d\tau. \end{aligned}$$

Now $\quad V(V.\nabla V.\tau d\tau)\sigma_0 = -S(\tau\nabla)V.\sigma_0 d\tau - S(d\tau\nabla)V\tau\sigma_0,$

and $\quad d(S(\tau\nabla)V\sigma_0\tau) \;=\; S(\tau\nabla)V.\sigma_0 d\tau + S(d\tau\nabla)V\sigma_0\tau.$

Subtracting, and omitting the term which is the same at both limits, we have $\quad \int V.\sigma d\tau = -V.(V.U\nu\nabla)\sigma_0 ds.$

Extended as above to any closed curve, this takes at once the form

$$\int V.\sigma d\rho = -\iint ds V.(V.U\nu\nabla)\sigma.$$

Of course, in many cases of the attempted representation of a quaternion surface-integral by another taken round its bounding curve, we are met by ambiguities as in the case of the space-integral, § 474 : but their origin, both analytically and physically, is in general obvious.

480.] If P be any scalar function of ρ, we have (by the process of § 477, above)

$$\begin{aligned} \int P d\tau &= \int (P_0 - S(\tau\nabla)P_0)d\tau \\ &= -\int S.\tau\nabla P_0.d\tau. \end{aligned}$$

But $\quad V.\nabla V.\tau d\tau = d\tau S.\tau\nabla - \tau S.d\tau\nabla,$

and $\quad d(\tau S\tau\nabla) = d\tau S.\tau\nabla + \tau S.d\tau\nabla.$

These give

$$\int P d\tau = -\tfrac{1}{2}(\tau S\tau\nabla - V.V\tau d\tau\nabla)P_0 = ds V.U\nu\nabla P_0,$$

Hence, for a closed curve of any form, we have

$$\int P d\rho = \iint ds V.U\nu\nabla P,$$

from which the theorems of §§ 477, 479 may easily be deduced.

481.] Commencing afresh with the fundamental integral

$$\iiint S\nabla\sigma ds = \iint S.\sigma U\nu ds,$$

put $\quad \sigma = u\beta,$

and we have $\quad \iiint S\beta\nabla u ds = \iint u S.\beta U\nu ds;$

from which at once $\iiint \nabla u\, ds = \iint u U\nu\, ds,$ .... (1)

or $\iiint \nabla \tau\, ds = \iint U\nu . \tau\, ds.$ .... (2)

Putting $u_1 \tau$ for τ, and taking the scalar, we have

$$\iiint (S\tau \nabla u_1 + u_1 S\nabla \tau)\, ds = \iint u_1 S\tau U\nu\, ds,$$

whence $\iiint (S(\tau\nabla)\sigma + \sigma S.\nabla\tau)\, ds = \iint \sigma S\tau U\nu\, ds.$ .... (3)

482.] As one example of the important results derived from these simple formulae, take the following, viz.:—

$$\iint V.(V\sigma U\nu)\tau\, ds = \iint \sigma S\tau U\nu\, ds - \iint U\nu S\sigma\tau\, ds,$$

where by (3) and (1) we see that the right-hand member may be written

$$= \iiint (S(\tau\nabla)\sigma + \sigma S\nabla\tau - \nabla S\sigma\tau)\, ds$$

$$= -\iiint V.V(\nabla\sigma)\tau\, ds. \qquad (4)$$

This, and similar formulae, are easily applied to find the potential and vector-force due to various distributions of magnetism. To shew how this is introduced, we briefly sketch the mode of expressing the potential of a distribution.

483.] Let σ be the vector expressing the direction and intensity of magnetisation, per unit of volume, at the element ds. Then if the magnet be placed in a field of magnetic force whose potential is u, we have for its potential energy

$$E = -\iiint S\sigma \nabla u\, ds$$

$$= \iiint u S\nabla\sigma\, ds - \iint u S\sigma U\nu\, ds.$$

This shews at once that the magnetism may be resolved into a volume-density $S(\nabla\sigma)$, and a surface-density $-S\sigma U\nu$. Hence, for a solenoidal distribution, $S.\nabla\sigma = 0.$

What Thomson has called a lamellar distribution (*Phil. Trans.* 1852), obviously requires that

$$S\sigma\, d\rho$$

be integrable without a factor; i. e., that

$$V\nabla\sigma = 0.$$

A complex lamellar distribution requires that the same expression be integrable by the aid of a factor. If this be u, we have at once

$$V\nabla(u\sigma) = 0,$$

or

$$S.\sigma\nabla\sigma = 0.$$

With these preliminaries we see at once that (4) may be written

$$\iint V.(V\sigma U\nu)\tau\, ds = -\iiint V.\tau V\nabla\sigma\, ds - \iiint V.\sigma\nabla\tau\, ds + \iiint S\sigma\nabla.\tau\, ds.$$

Now, if $\tau = \nabla\left(\frac{1}{r}\right),$

where r is the distance between any external point and the element

ds, the last term on the right is the vector-force exerted by the magnet on a unit-pole placed at the point. The second term on the right vanishes by Laplace's equation, and the first vanishes as above if the distribution of magnetism be lamellar, thus giving Thomson's result in the form of a surface integral.

484.] An application may be made of similar transformations to Ampère's *Directrice de l'action électrodynamique*, which, § 432 above, is the vector-integral

$$\int \frac{V\rho d\rho}{T\rho^3},$$

where $d\rho$ is an element of a closed circuit, and the integration extends round the circuit. This may be written

$$-\int V.(d\rho\nabla)\frac{1}{r},$$

so that its value as a surface integral is

$$\iint S(U\nu\nabla)\nabla\frac{1}{r}ds - \iint U\nu\nabla^2\frac{1}{r}ds.$$

Of this the last term vanishes, unless the origin is in, or infinitely near to, the surface over which the double integration extends. The value of the first term is seen (by what precedes) to be the vector-force due to uniform normal magnetisation of the same surface.

485.] Also, since $\nabla U\rho = \frac{2}{T\rho}$,

we obtain at once

$$-2\iiint \frac{ds}{T\rho} = \iint S.U\rho U\nu ds,$$

whence, by differentiation, or by putting $\rho + \alpha$ for ρ, and expanding in ascending powers of $T\alpha$ (both of which tacitly assume that the origin is external to the space integrated through, i. e., that $T\rho$ nowhere vanishes), we have

$$-2\iiint \frac{dsU\rho}{T\rho^2} = \iint \frac{V.U\rho V.U\nu U\rho}{T\rho}ds = 2\iint \frac{U\nu ds}{T\rho};$$

and this, again, involves

$$\iint \frac{U\nu ds}{T\rho} = \iint \frac{U\rho}{T\rho} S.U\nu U\rho ds.$$

486.] The interpretation of these, and of more complex formulae of a similar kind, leads to many curious theorems in attraction and in potentials. Thus, from (1) of § 481, we have

$$\iiint \frac{\nabla t}{T\rho}ds - \iiint \frac{tU\rho}{T\rho^2}ds = \iint \frac{tU\nu}{T\rho}ds,$$

which gives the attraction of a mass of density t in terms of the potentials of volume distributions and surface distributions. Putting

$$\sigma = it_1 + jt_2 + kt_3,$$

this becomes

$$\iiint \frac{\nabla \sigma ds}{T\rho} - \iiint \frac{U\rho . \sigma ds}{T\rho^2} = \iint \frac{U\nu . \sigma ds}{T\rho}.$$

By putting $\sigma = \rho$, and taking the scalar, we recover a formula given above; and by taking the vector we have

$$V \iint U\nu U\rho \, ds = 0.$$

This may be easily verified from the formula

$$\int P d\rho = V \iint U\nu . \nabla P ds,$$

by remembering that $\nabla T\rho = U\rho$.

Again if, in the fundamental integral, we put

$$\sigma = tU\rho,$$

we have $$\iiint \frac{S(\rho\nabla)t}{T\rho} ds - 2\iiint \frac{t ds}{T\rho} = \iint t S . U\nu U\rho ds.$$

487.] As another application, let us consider briefly the *Stress-function in an elastic solid.*

At any point of a strained body let λ be the vector stress per unit of area perpendicular to i, μ and ν the same for planes perpendicular to j and k respectively.

Then, by considering an indefinitely small tetrahedron, we have for the stress per unit of area perpendicular to a unit-vector ω the expression $$\lambda S i \omega + \mu S j \omega + \nu S k \omega = -\phi \omega,$$
so that the stress across any plane is represented by *a linear and vector function* of the unit normal to the plane.

But if we consider the equilibrium, as regards rotation, of an infinitely small parallelepiped whose edges are parallel to i, j, k respectively, we have (supposing there are no molecular couples)

$$V(i\lambda + j\mu + k\nu) = 0,$$

or $$\Sigma V i \phi i = 0,$$

or $$V . \nabla \phi \rho = 0.$$

This shews (§ 173) that in this case ϕ is *self-conjugate*, or, in other words, involves not nine distinct constants but only six.

488.] Consider next the equilibrium, as regards translation, of any portion of the solid filling a simply-connected closed space. Let u be the potential of the external forces. Then the condition is obviously $$\iint \phi (U\nu) ds + \iiint ds \nabla u = 0,$$
where ν is the normal vector of the element of surface ds. Here

the double integral extends over the whole boundary of the closed space, and the triple integral throughout the whole interior.

To reduce this to a form to which the method of § 467 is directly applicable, operate by $S.\alpha$ where α is any constant vector whatever, and we have $\iint S.\phi\alpha U\nu ds + \iiint ds S\alpha\nabla u = 0$ by taking advantage of the self-conjugateness of ϕ. This may be written $\iiint ds(S.\nabla\phi\alpha + S.\alpha\nabla u) = 0,$ and, as the limits of integration may be any whatever,

$$S.\nabla\phi\alpha + S.\alpha\nabla u = 0. \quad\text{..........................}\quad (1)$$

This is the required equation, the indeterminateness of α rendering it equivalent to *three* scalar conditions.

There are various modes of expressing this without the α. Thus, if Δ be used for ∇ when the constituents of ϕ are considered, we may write

$$\nabla u = -S\nabla\Delta.\phi\rho.$$

In integrating this expression through a given space, we must remark that ∇ and ρ are merely artificial symbols of construction, and therefore are not to be looked on as variables in the integral.

489.] As a verification, it may be well to shew that from this equation we can get the condition of equilibrium, as regards *rotation*, of a simply connected portion of the body, which can be written by inspection as

$$\iint V.\rho\phi(U\nu)ds + \iiint V.\rho\nabla u ds = 0.$$

This is easily done as follows: (1) gives

$$S.\nabla\phi\sigma + S.\sigma\nabla u = 0,$$

if, and only if, σ satisfy the condition

$$S.\phi(\nabla)\sigma = 0.$$

Now this condition is satisfied if

$$\sigma = V\alpha\rho$$

where α is any constant vector. For

$$S.\phi(\nabla)V\alpha\rho = -S.\alpha V\phi(\nabla)\rho$$
$$= S.\alpha V\nabla\phi\rho = 0.$$

Hence $$\iiint ds(S.\nabla\phi V\alpha\rho + S.\alpha\rho\nabla u) = 0,$$

or $$\iint ds S.\alpha\rho\phi U\nu + \iiint ds S.\alpha\rho\nabla u = 0.$$

Multiplying by α, and adding the results obtained by making α in succession each of three rectangular vectors, we obtain the required equation.

490.] Suppose σ to be the displacement of a point originally at ρ, then the work done by the stress on any simply connected portion of the solid is obviously

$$W = \iint S.\phi(U\nu)\sigma ds,$$

because $\phi(Uv)$ is the vector force overcome per unit of area on the element ds. This is easily transformed to

$$W = \iiint S.\nabla\phi\sigma\, ds.$$

491.] In this case obviously the strain-function is

$$\chi(\varpi) = \varpi - S.(\varpi\nabla)\sigma.$$

Now if the strain be a mere rotation, in which case

$$S.\chi\varpi\chi\tau - S.\varpi\tau = 0,$$

whatever be the vectors ϖ and τ, no work is done by the stress. Hence the expression for the work done by the stress must vanish if these conditions are fulfilled.

Again, it is easily seen that when the strain is infinitely small the work must be a homogeneous function of the second degree of these critical quantities; for, if it exist, it is essentially positive. Hence, even when finite, the work on unit-volume may be expressed as $w = \Sigma.(S.\chi\epsilon\chi\epsilon' - S\epsilon\epsilon')(S.\chi\eta\chi\eta' - S\eta\eta')$,
where ϵ, ϵ', η, η', which are in general functions of σ, become constant vectors if the stress is indefinitely small. When this is the case it is easy to see that, whatever be the number of terms under Σ, w involves twenty-one separate and independent constants only; viz. the coefficients of the homogeneous products of the second order of the six values of form

$$S.\chi\varpi\chi\tau - S\varpi\tau$$

for the values i, j, k of ϖ or τ.

Supposing the strain to be indefinitely small, we have for the variation of w, the expression

$$\delta w = \Sigma(S.\delta\chi\epsilon\chi\epsilon' + S.\delta\chi\epsilon'\chi\epsilon)(S.\chi\eta\chi\eta' - S\eta\eta')$$
$$+ \Sigma(S.\chi\epsilon\chi\epsilon' - S\epsilon\epsilon')(S.\delta\chi\eta\chi\eta' + S.\delta\chi\eta'\chi\eta).$$

Now, by the first equation, we have

$$\delta\chi\varpi = -S(\varpi\nabla)\delta\sigma.$$

Hence, writing the result for one of the factors only, the variation of the whole work done by straining a mass is

$$\delta W = \delta\iiint w\, ds = \iiint \delta w\, ds$$
$$= -\Sigma\iiint ds(S.\chi\eta\chi\eta' - S\eta\eta')\{S.\chi\epsilon' S.(\epsilon\nabla)\delta\sigma + S.\chi\epsilon S(\epsilon'\nabla)\delta\sigma\}.$$

Now, if we have at the limits

$$\delta\sigma = 0,$$

i.e. if the surface of the mass is altered in a *given* way, we have obviously,

$$\iiint ds\, S.\varpi S(\epsilon\nabla)\delta\sigma = -\iiint ds\, S.\delta\sigma S(\epsilon\nabla)\varpi.$$

Hence

$$\delta W = \Sigma \iiint ds\, S.\delta\sigma \lfloor S(\epsilon\nabla)\{\chi\epsilon'(S.\chi\eta\chi\eta' - S\eta\eta')\} + S(\epsilon'\nabla)\{\chi\epsilon(S.\chi\eta\chi\eta' - S\eta\eta')\}\rfloor.$$

Now any arbitrary change in σ will in general *increase* the amount of work done, so that we have

$$0 = \Sigma[S(\epsilon\nabla)\{\chi\epsilon'(S.\chi\eta\chi\eta' - S\eta\eta')\} + S(\epsilon'\nabla)\{\chi\epsilon(S.\chi\eta\chi\eta' - S\eta\eta')\}],$$

which is our equation for the determination of σ, as the constants $\epsilon, \epsilon', \eta, \eta'$ are dependent solely on the elastic properties of the sub stance distorted, and may therefore be considered as known; while χ essentially involves σ.

492.] Since the algebraic operator

$$\epsilon^{h\frac{d}{dx}},$$

when applied to any function of x, simply changes x into $x+h$, it is obvious that if σ be a vector not acted on by

$$\nabla = i\frac{d}{dx} + j\frac{d}{dy} + k\frac{d}{dz},$$

we have

$$\epsilon^{-S\sigma\nabla} f(\rho) = f(\rho+\sigma),$$

whatever function f may be. From this it is easy to deduce Taylor's theorem in one important quaternion form.

If Δ bear to the constituents of σ the same relation as ∇ bears to those of ρ, and if f and F be any two functions which satisfy the commutative law in multiplication, this theorem takes the curious form

$$\epsilon^{-S\Delta\nabla} f(\rho)\, F(\sigma) = f(\rho+\Delta)\, F(\sigma) = F(\sigma+\nabla) f(\rho);$$

of which a particular case is

$$\epsilon^{\frac{d^2}{dx\,dy}} f(x)\, F(y) = f\left(x+\frac{d}{dy}\right) F(y) = F\left(y+\frac{d}{dx}\right) f(x).$$

The modifications which the general expression undergoes, when f and F are not commutative, are easily seen.

If one of these be an inverse function, such as, for instance, may occur in the solution of a linear differential equation, these theorems of course do not give the arbitrary part of the integral, but they often materially aid in the determination of the rest.

Other theorems, involving operators such as $\epsilon^{S\rho\nabla}$, $\epsilon^{S.\alpha\rho\nabla}$, &c., &c. are easily deduced, and all have numerous applications.

493.] But there are among them results which appear startling from the excessively free use made of the separation of symbols. Of these one is quite sufficient to shew their general nature.

Let P be any scalar function of ρ. It is required to find the difference between the value of P at ρ, and its *mean* value throughout a very small sphere, of radius r and volume v, which has the extremity of ρ as centre.

From what is said above, it is easy to see that we have the following expression for the required result:—

$$\frac{1}{v}\iiint(\epsilon^{-S\sigma\nabla}-1)P\,ds,$$

where σ is the vector joining the centre of the sphere with the ele ment of volume ds, and the integration (which relates to σ and ds alone) extends through the whole volume of the sphere. Expanding the exponential, we may write this expression in the form

$$-\frac{1}{v}\iiint\left(S\sigma\nabla-\frac{1}{2}(S\sigma\nabla)^2+\ldots\right)P\,ds$$

$$=-\frac{1}{v}S.\nabla P\iiint\sigma\,ds+\frac{1}{2v}\iiint(S\sigma\nabla)^2P\,ds-\&c.$$

higher terms being omitted on account of the smallness of r, the limit of $T\sigma$.

Now, symmetry shews at once that

$$\iiint\sigma\,ds=0.$$

Also, whatever constant vector be denoted by α,

$$\iiint(S\alpha\sigma)^2ds=-\alpha^2\iiint(S\sigma U\alpha)^2ds.$$

Since the integration extends throughout a sphere, it is obvious that the integral on the right is half of what we may call the moment of inertia of the volume about a diameter. Hence

$$\iiint(S\sigma U\alpha)^2ds-\frac{vr^2}{5}\cdot$$

If we now write ∇ for α, *as the integration does not refer to* ∇, we have by the foregoing results (neglecting higher powers of r)

$$\frac{1}{v}\iiint(\epsilon^{-S\sigma\nabla}-1)P\,ds=-\frac{r^2}{10}\nabla^2P,$$

which is the expression given by Clerk-Maxwell*. Although, for simplicity, P has here been supposed a scalar, it is obvious that in the result above it may at once be written as a quaternion.

494.] If ρ be the vector of the element ds, where the surface density is $f\rho$, the potential at σ is

$$\iint ds\,f\rho\,FT(\rho-\sigma),$$

F being the potential function, which may have any form whatever.

By the preceding, § 492, this may be transformed into

$$\iint ds\,f\rho\,\epsilon^{S\sigma\nabla}FT\rho\,;$$

* *London Math. Soc. Proc.*, vol. iii, no. 34, 1871.

or, far more conveniently for the integration, into

$$\iint ds f\rho \epsilon^{S\rho\Delta} FT\sigma,$$

where Δ depends on the constituents of σ in the same manner as ∇ depends on those of ρ.

A still farther simplification may be introduced by using a vector σ_0, which is finally to be made zero, along with its corresponding operator Δ_0, for the above expression then becomes

$$\iint ds\, \epsilon^{S\rho(\Delta-\Delta_0)} f\sigma_0 FT\sigma,$$

where ρ appears in a comparatively manageable form. It is obvious that, so far, our formulae might be made applicable to any distribution. We now restrict them to a superficial one.

495.] Integration of this last *form* can always be easily effected in the case of a surface of revolution, the origin being a point in the axis. For the expression, so far as the integration is concerned, can in that case be exhibited as a single integral

$$\int_p^q dx\, \phi x \epsilon^{ax},$$

where ϕ may be any scalar function, and x depends on the cosine of the inclination of ρ to the axis. And

$$\int_p^q dx\, \phi x \epsilon^{ax} = \phi\left(\frac{d}{da}\right) \cdot \frac{\epsilon^{qa} - \epsilon^{pa}}{a}.$$

As the interpretation of the general results is a little troublesome, let us take the case of a spherical shell, the origin being the centre and the density unity, which, while simple, sufficiently illustrates the proposed mode of treating the subject.

We easily see that in the above simple case, α being any constant vector whatever, and a being the radius of the sphere,

$$\iint ds\, \epsilon^{S\alpha\rho} = 2\pi a \int_{-a}^{+a} \epsilon^{xT\alpha} dx = \frac{2\pi a}{T\alpha} (\epsilon^{aT\alpha} - \epsilon^{-aT\alpha}).$$

Now, it appears that *we are at liberty to treat Δ as α has just been treated.* It is necessary, therefore, to find the effects of such operators as $T\Delta$, $\epsilon^{aT\Delta}$, &c., which seem to be novel, upon a scalar function of $T\sigma$; or $\mathfrak{T}$, as we may for the present call it.

Now $$(T\Delta)^2 F = -\Delta^2 F = F'' + \frac{2F'}{\mathfrak{T}},$$

whence it is easy to guess at a particular form of $T\Delta$. To be sure that it is the only one, assume

$$T\Delta = \phi \frac{d}{d\mathfrak{T}} + \psi,$$

where ϕ and ψ are scalar functions of $\mathfrak{T}$ to be found. This gives

$$(T\Delta)^2F = \left(\phi \frac{d}{d\mathfrak{T}} + \psi\right)(\phi F' + \psi F)$$

$$= \phi^2 F'' + (\phi\phi' + \psi\phi + \phi\psi)F' + (\phi\psi' + \psi^2)F.$$

Comparing, we have

$$\phi^2 = 1,$$

$$\phi\phi' + \psi\phi + \phi\psi = \frac{2}{\mathfrak{T}};$$

$$\phi\psi' + \psi^2 = 0.$$

From the first, $\qquad \phi = \pm 1,$

whence the second gives $\qquad \psi = \pm \frac{1}{\mathfrak{T}};$

the signs of ϕ and ψ being alike. The third is satisfied identically.

That is $\qquad + T\Delta = \frac{d}{d\mathfrak{T}} + \frac{1}{\mathfrak{T}}\cdot$

Also, an easy induction shews that

$$+ (T\Delta)^n = \left(\frac{d}{d\mathfrak{T}}\right)^n + \frac{n}{\mathfrak{T}}\left(\frac{d}{d\mathfrak{T}}\right)^{n-1}$$

Hence we have at once

$$\epsilon^{aT\Delta} = 1 + a\left(\frac{d}{d\mathfrak{T}} + \frac{1}{\mathfrak{T}}\right) + \ldots \pm \frac{a^n}{1.2\ldots n}\left[\left(\frac{d}{d\mathfrak{T}}\right)^n + \frac{n}{\mathfrak{T}}\left(\frac{d}{d\mathfrak{T}}\right)^{n-1}\right] + \&c.$$

$$= \epsilon^{\pm a \frac{d}{d\mathfrak{T}}} + \frac{a}{\mathfrak{T}}\epsilon^{\pm a\frac{d}{d\mathfrak{T}}},$$

by the help of which we easily arrive at the well-known results. This we leave to the student*.

496.] As an elementary example of the use of ∇ in connection with the Calculus of Variations, let us consider the expression

$$A = \int QTd\rho,$$

where $Td\rho$ is an element of a finite arc along which the integration extends, and Q is in general a scalar function of ρ and constants.

We have

$$\delta A = \int (\delta Q T d\rho + Q\delta T d\rho)$$

$$= \int (\delta Q T d\rho - QS.Ud\rho d\delta\rho)$$

$$= -[QSUd\rho\delta\rho] + \int (\delta Q Td\rho + S.\delta\rho d(QUd\rho)),$$

where the portion in square brackets refers to the limits only, and gives the terminal conditions. The remaining portion may easily be put in the form

$$S\int \delta\rho(d(QUd\rho) - \nabla Q.Td\rho).$$

* Proc. R. S. E., 1871-2.

If the curve is to be determined by the condition that the varia tion of A shall vanish, we must have, as $\delta\rho$ may have any direction,

$$d(QUd\rho) - \nabla Q.Td\rho = 0,$$

or, with the notation of Chap. IX,

$$\frac{d}{ds}(Q\rho') - \nabla Q = 0.$$

This simple equation shews that

(1) The osculating plane of the sought curve contains the vector ∇Q.

(2) The curvature at any point is inversely as Q, and directly as the component of ∇Q parallel to the radius of absolute curvature.

497.] As a first application, suppose A to represent the *action* of a particle moving freely under a system of forces which have a potential; so that

$$Q = T\dot{\rho},$$

and

$$\dot{\rho}^2 = 2(P - H),$$

where P is the potential, H the energy constant.

These give

$$T\dot{\rho}\nabla T\dot{\rho} = Q\nabla Q = -\nabla P,$$

and

$$Q\rho' = \dot{\rho},$$

so that the equation above becomes simply

$$\ddot{\rho} + \nabla P = 0,$$

which is obviously true.

498.] If we look to the superior limit only, the first expression for δA becomes in the present case

$$-[T\dot{\rho}SUd\rho\,\delta\rho] = -S\dot{\rho}\,\delta\rho.$$

If we suppose a variation of the constant H, we get the following term from the unintegrated part

$$t\delta H.$$

Hence we have at once Hamilton's equations of *varying action* in the forms

$$\nabla A = \dot{\rho}$$

and

$$\frac{dA}{dH} = t.$$

The first of these gives, by the help of the condition above,

$$(\nabla A)^2 = 2(P - H),$$

the well-known partial differential equation of the first order and second degree.

499.] To shew that, if A be any solution whatever of this equation, the vector ∇A represents the velocity in a free path capable of being described under the action of the given system of forces, we have

$$\frac{d}{dt}\dot{\rho} = \ddot{\rho} = -\nabla P = -\tfrac{1}{2}\nabla(\nabla A)^2$$
$$= -S(\nabla A.\nabla)\nabla A.$$

But $$\frac{d}{dt}\cdot\nabla A = -S(\dot{\rho}\nabla)\nabla A.$$

A comparison shews at once that the equality

$$\nabla A = \dot{\rho}$$

is consistent with each of these vector equations.

500.] Again, if ∂ refer to the constants only,

$$\tfrac{1}{2}\partial(\nabla A)^2 = S.\nabla A\partial\nabla A = -\partial H$$

by the differential equation.

But we have also $$\frac{\partial A}{\partial H} = t,$$

which gives $$\frac{d}{dt}(\partial A) = -S(\dot{\rho}\nabla)\partial A = \partial H.$$

These two expressions for ∂H again agree in giving

$$\nabla A = \dot{\rho},$$

and thus shew that the differential coefficients of A with regard to the two constants of integration must, themselves, be constants. We thus have the equations of two surfaces whose intersection determines the path.

501.] Let us suppose next that A represents the *time* of passage, so that the brachistochrone is required. Here we have

$$Q = \frac{1}{T\dot{\rho}},$$

the other condition being as in § 497, and we have

$$\frac{d}{dt}\dot{\rho}^{-1} - \dot{\rho}^{-2}\nabla P = 0,$$

which may be reduced to the symmetrical form

$$\ddot{\rho} + \dot{\rho}^{-1}\nabla P\dot{\rho} = 0.$$

It is very instructive to compare this equation with that of the free path as above, § 497.

The application of Hamilton's method may be easily made, as in the preceding example. (Tait, *Trans. R. S. E.*, 1865.)

502.] As a particular case, let us suppose gravity to be the only force, then $$\nabla P = \alpha,$$

a constant vector, so that

$$\frac{d}{dt}\dot{\rho}^{-1} - \dot{\rho}^{-2}\alpha = 0.$$

The form of this equation suggests the assumption

$$\dot{\rho}^{-1} = \beta - p\alpha \tan qt,$$

where p and q are scalars and

$$S\alpha\beta = 0.$$

Substituting, we get

$$-pq \sec^2 qt + (-\beta^2 - p^2 \alpha^2 \tan^2 qt) = 0,$$

which gives $$pq = T^2\beta = p^2 T^2 \alpha.$$

Now let $$p\beta^{-1}\alpha = \gamma;$$

this must be a unit-vector perpendicular to α and β, so that

$$\dot{\rho}^{-1} = \frac{\beta}{\cos qt}(\cos qt - \gamma \sin qt),$$

whence $$\dot{\rho} = \cos qt (\cos qt + \gamma \sin qt)\beta^{-1}$$

(which may be verified at once by multiplication).

Finally, taking the origin so that the constant of integration may vanish, we have

$$2\rho\beta = t + \frac{1}{2q}(\sin 2qt - \gamma \cos 2qt),$$

which is obviously the equation of a cycloid referred to its vertex. The tangent at the vertex is parallel to β, and the axis of symmetry to α.

503.] In the case of a chain hanging under the action of given forces $$Q = Pr,$$

where P is the potential, r the mass of unit-length.

Here we have also, of course,

$$\int T d\rho = l,$$

the length of the chain being given.

It is easy to see that this leads, by the usual methods, to the equation $$\frac{d}{ds}\{(Pr + u)\rho'\} - r\nabla P = 0$$

where u is a scalar multiplier.

504.] As a simple case, suppose the chain to be uniform. Then r may be merged in u. Suppose farther that gravity is the only force, then $$P = S\alpha\rho, \qquad \nabla P = -\alpha,$$

and $$\frac{d}{ds}\{(S\alpha\rho + u)\rho'\} + \alpha = 0.$$

Differentiating, and operating by $S\rho'$, we find

$$S.\rho'\left\{\rho'\left(S\alpha\rho' + \frac{du}{ds}\right) + \alpha\right\} = 0;$$

which shews that u is constant, and may therefore be allowed for by change of origin.

The curve lies obviously in a plane parallel to α, and its equation is

$$(S\alpha\rho)^2 + \alpha^2 s^2 = \text{const.},$$

which is a well-known form of the equation of the catenary.

When the quantity Q of § 496 is a vector or a quaternion, we have simply an equation like that there given for each of the constituents.

505.] Suppose P and the constituents of σ to be functions which vanish at the bounding surface of a simply-connected space Σ, or such at least that either P or the constituents vanish there, the others (or other) not becoming infinite.

Then, by § 467,

$$\iiint ds S.\nabla (P\sigma) = \iint ds\, PS\sigma U\nu = 0,$$

if the integrals be taken through and over Σ.

Thus $$\iiint ds S.\sigma\nabla P = -\iiint ds\, PS.\nabla\sigma.$$

By the help of this expression we may easily prove a very remarkable proposition of Thomson (*Cam. and Dub. Math. Journal* Jan. 1848, or *Reprint of Papers on Electrostatics*, § 206.)

To shew that there is one, and but one, solution of the equation

$$S.\nabla (e^2\nabla u) = 4\pi r$$

where r *vanishes at an infinite distance, and* e *is any real scalar what ever, continuous or discontinuous.*

Let v be the potential of a distribution of density r, so that

$$\nabla^2 v = 4\pi r,$$

and consider the integral

$$Q = -\iiint ds \left(e\nabla u - \frac{1}{e}\nabla v\right)^2.$$

That Q may be a minimum as depending on the value of u (which is obviously possible since it cannot be negative, and since it may have any positive value, however large, if only greater than this minimum), we must have

$$0 = \tfrac{1}{2}\delta Q = -\iiint ds S.(e^2\nabla u - \nabla v)\nabla\delta u$$
$$= \iiint ds\, \delta u\, S.\nabla (e^2\nabla u - \nabla v),$$

by the lemma given above,

$$= \iiint ds\, \delta u \{S.\nabla (e^2\nabla u) - 4\pi r\}.$$

Thus any value of u which satisfies the given equation is such as to make Q a minimum.

But there is only one value of u which makes Q a minimum; for, let Q_1 be the value of Q when

$$u_1 = u + \phi$$

is substituted for this value of u, and we have

$$Q_1 = -\iiint ds\left(e\nabla(u+\phi)-\frac{1}{e}\nabla v\right)^2$$
$$= Q - 2\iiint ds S(e^2\nabla u - \nabla v)\nabla\phi - \iiint ds e^2(\nabla\phi)^2.$$

The middle term of this expression may, by the proposition at the beginning of this section, be written

$$2\iiint ds\,\phi\,\{S\nabla(e^2\nabla u) - 4\pi r\},$$

and therefore vanishes. The last term is essentially positive. Thus if u_1 anywhere differ from u (except, of course, by a constant quan tity) it cannot make Q a minimum; and therefore u is a unique solution

---

MISCELLANEOUS EXAMPLES.

1. The expression

$$V\alpha\beta V\gamma\delta + V\alpha\gamma V\delta\beta + V\alpha\delta V\beta\gamma$$

denotes a vector. What vector?

2. If two surfaces intersect along a common line of curvature, they meet at a constant angle.

3. By the help of the quaternion formulae of rotation, translate into a new form the solution (given in § 234) of the problem of inscribing in a sphere a closed polygon the directions of whose sides are given.

4. Express, in terms of the masses, and geocentric vectors of the sun and moon, the sun's vector disturbing force on the moon, and expand it to terms of the second order; pointing out the magnitudes and directions of the separate components.

(Hamilton, *Lectures*, p. 615.)

5. If $q = r^{\frac{1}{2}}$, shew that

$$2\,dq = 2\,dr^{\frac{1}{2}} = \tfrac{1}{2}(dr + Kq\,dr\,q^{-1})Sq^{-1} - \tfrac{1}{2}(dr + q^{-1}dr\,Kq)Sq^{-1}$$
$$= (dr\,q + Kq\,dr)q^{-1}(q+Kq)^{-1} - (dr\,q + Kq\,dr)(r+Tr)^{-1}$$
$$= \frac{dr + Uq^{-1}dr\,Uq^{-1}}{Tq(Uq+Uq^{-1})} = \frac{dr\,Uq + Uq^{-1}dr}{q(Uq+Uq^{-1})} = \frac{q^{-1}(Uq\,dr + dr\,Uq^{-1})}{Uq+Uq^{-1}}$$
$$= \frac{q^{-1}(q\,dr + Tr\,dr\,q^{-1})}{Tq(Uq+Uq^{-1})} = \frac{dr\,Uq + Uq^{-1}dr}{Tq(1+Ur)} = \frac{dr\,Kq^{-1} + q^{-1}dr}{1+Ur}$$

$$2dq = \left\{ dr + V.Vdr\frac{V}{S}q \right\} q^{-1} = \left\{ dr - V.Vdr\frac{V}{S}q^{-1} \right\} q^{-1}$$

$$-\frac{dr}{q} + V.V\frac{dr}{q}\frac{V}{S}q = \frac{dr}{q} - V.V\frac{dr}{q}\frac{V}{S}q^{-1}$$

$$-drq^{-1} + V.Vq^{-1}Vdr\left(1 + \frac{V}{S}q^{-1}\right):$$

and give geometrical interpretations of these varied expressions for the same quantity. (*Ibid.* p. 628.)

6. Shew that the equation of motion of a homogeneous solid of revolution about a point in its axis, which is not its centre of gravity, is

$$BV\rho\ddot{\rho} - A\Omega\dot{\rho} = V\rho\gamma,$$

where Ω is a constant. (*Trans. R. S. E.*, 1869.)

7. Integrate the differential equations

$$(a.)\quad \frac{dq}{dt} + aq = b,$$

$$(b.)\quad \frac{d\rho}{dt} + \phi\rho = a,$$

$$(c.)\quad \frac{d^2\rho}{dt^2} + \phi\left(\frac{d\rho}{dt}\right) + \psi\rho = 0\,;$$

where a and b are given quaternions, and ϕ and ψ given linear and vector functions. (Tait, *Proc. R.S.E.*, 1870–1.)

8. Derive (4) of § 92 directly from (3) of § 91.

9. Find the successive values of the continued fraction

$$u_x = \left(\frac{j}{i+}\right)^x 0,$$

where i and j have their quaternion significations, and x has the values 1, 2, 3, &c. (Hamilton, *Lectures*, p. 645.)

10. If we have

$$u_x = \left(\frac{j}{i+}\right)^x c,$$

where c is a given quaternion, find the successive values.

For what values of c does u become constant? (*Ibid.* p. 652.)

11. Prove that the moment of hydrostatic pressures on the faces of any polyhedron is zero, (*a.*) when the fluid pressure is the same throughout, (*b.*) when it is due to any set of forces which have a potential.

12. What vector is given, in terms of two known vectors, by the relation

$$\rho^{-1} = \tfrac{1}{2}(\alpha^{-1} + \beta^{-1})?$$

Shew that the origin lies on the circle which passes through the extremities of these three vectors.

13. Tait, *Trans. and Proc. R. S. E.*, 1870–3.

With the notation of §§ 467, 477, prove

(*a.*) $\iiint S(\alpha\nabla)\tau ds = \iint \tau S\alpha Uvds.$

(*b.*) If $S(\rho\nabla)\tau = -n\tau,$

$$(n+3)\iiint \tau ds = -\iint \tau S\rho Uvds.$$

(*c.*) With the additional restriction $\nabla^2\tau = 0$

$$\iint S.Uv(2n\rho + (n+3)\rho^2\nabla).\tau ds = 0.$$

(*d.*) Express the value of the last integral over a non-closed surface by a line-integral.

(*e.*) $-\int Td\rho = \iint dsS.Uv\nabla\sigma,$

if $\sigma = Ud\rho$ all round the curve.

(*f.*) For any portion of surface whose bounding edge lies wholly on a sphere with the origin as centre

$$\iint dsS.(U\rho Uv\nabla).\sigma = 0,$$

whatever be the vector σ.

(*g*) $\int Vd\rho\nabla.\sigma = \iint ds(Uv\nabla^2 - S(Uv\nabla)\nabla)\sigma$

whatever be σ.

14. Tait, *Trans. R. S. E.*, 1873.

Interpret the equation

$$d\sigma = uqd\rho q^{-1},$$

and shew that it leads to the following results

$$\nabla^2\sigma = q\nabla u q^{-1},$$

$$\nabla.uq^{-1} = 0,$$

$$\nabla^2.u^{\frac{1}{2}} = 0.$$

Hence shew that the only sets of surfaces which, together, cut space into cubes are planes and their electric images.

15. What problem has its conditions stated in the following six equations, from which ξ, η, ζ are to be determined as scalar functions of x, y, z, or of $\rho = ix + jy + kz$?

$$\nabla^2\xi = 0, \qquad \nabla^2\eta = 0, \qquad \nabla^2\zeta = 0,$$

$$S\nabla\xi\nabla\eta = 0, \qquad S\nabla\eta\nabla\zeta = 0, \qquad S\nabla\zeta\nabla\xi = 0,$$

$$\text{where} \qquad \nabla = i\frac{d}{dx} + j\frac{d}{dy} + k\frac{d}{dz}$$

Shew that they give the farther equations

$$0 = \nabla^2\xi\eta = \nabla^2\eta\zeta = \nabla^2\zeta\xi = \nabla^2.\xi\eta\zeta.$$

Shew that (with a change of origin) the general solution of these equations may be put in the form

$$S\rho(\phi+f)^{-1}\rho = 1,$$

where ϕ is a self-conjugate linear and vector function, and ξ, η, ζ are to be found respectively from the three values of f at any point by relations similar to those in Ex. 24 to Chapter IX. (See Lamé, *Journal de Mathematiques,* 1843.)

16. Shew that, if ρ be a planet's radius vector, the potential P of masses external to the solar system introduces into the equation of motion a term of the form $S(\rho\nabla)\nabla P$.

Shew that this is a self-conjugate linear and vector function of ρ, and that it involves only *five* independent constants.

Supposing the undisturbed motion to be circular, find the chief effects which this disturbance can produce

17. In § 405 above, we have the equations

$$Va(\ddot{\varpi}+n^2\varpi) = 0, \quad Sa\varpi = 0, \quad \dot{a} = \omega Via, \quad Ta = 1,$$

where ω^2 is neglected. Shew that with the assumptions

$$q = i^{\frac{\omega t}{\pi}}, \quad a = q\beta q^{-1}, \quad r = \beta^{\frac{\omega_1 t}{\pi}}, \quad \varpi = qr\tau r^{-1}q^{-1},$$

we have $\dot{\beta} = 0, \quad T\beta = 1, \quad S\beta\tau = 0, \quad V\beta(\ddot{\tau}+n^2\tau) = 0,$

provided $\omega Sia-\omega_1 = 0$. Hence deduce the behaviour of the Foucault pendulum without the x, y, and ξ, η transformations in the text.

Apply analogous methods to the problems proposed at the end of § 401 of the text.

18. Hamilton, *Bishop Law's Premium Examination,* 1862.

(*a.*) If OABP be four points of space, whereof the three first are given, and not collinear; if also OA $= a$, OB $= \beta$, OP $= \rho$; and if, in the equation

$$F\frac{\rho}{a} = F\frac{\beta}{a},$$

the characteristic of operation F be replaced by S, the locus of P is a plane. What plane?

(*b*) In the same general equation, if F be replaced by V, the locus is an indefinite right line. What line?

(*c.*) If F be changed to K, the locus of P is a point. What point?

(*d.*) If F be made $= U$, the locus is an indefinite half-line, or ray. What ray?

(*e.*) If F be replaced by T, the locus is a sphere. What sphere?

(*f.*) If F be changed to TV, the locus is a cylinder of revolution. What cylinder?

(*g.*) If F be made TVU, the locus is a cone of revolution. What cone?

(*h.*) If SU be substituted for F, the locus is one sheet of such a cone. Of what cone? and which sheet?

(*i.*) If F be changed to VU, the locus is a pair of rays. Which pair?

19. Hamilton, *Bishop Law's Premium Examination*, 1863.

(*a.*) The equation $\quad S\rho\rho' + a^2 = 0$

expresses that ρ and ρ' are the vectors of two points P and P′, which are conjugate with respect to the sphere

$$\rho^2 + a^2 = 0;$$

or of which one is on the polar plane of the other.

(*b.*) Prove by quaternions that if the right line PP′, connecting two such points, intersect the sphere, it is cut harmonically thereby.

(*c.*) If P′ be a given external point, the cone of tangents drawn from it is represented by the equation,

$$(V\rho\rho')^2 = a^2(\rho - \rho')^2;$$

and the orthogonal cone, concentric with the sphere, by

$$(S\rho\rho')^2 + a^2\rho^2 = 0.$$

(*d.*) Prove and interpret the equation,

$$T(n\rho - a) = T(\rho - na), \text{ if } T\rho = Ta.$$

(*e.*) Transform and interpret the equation of the ellipsoid,

$$T(\iota\rho + \rho\kappa) = \kappa^2 - \iota^2.$$

(*f.*) The equation

$$(\kappa^2 - \iota^2)^2 = (\iota^2 + \kappa^2)S\rho\rho' + 2S\iota\rho\kappa\rho'$$

expresses that ρ and ρ' are values of conjugate points, with respect to the same ellipsoid.

(*g.*) The equation of the ellipsoid may also be thus written,

$$S\nu\rho = 1, \text{ if } (\kappa^2 - \iota^2)^2\nu = (\iota - \kappa)^2\rho + 2\iota S\kappa\rho + 2\kappa S\iota\rho.$$

(*h.*) The last equation gives also,

$$(\kappa^2 - \iota^2)^2\nu = (\iota^2 + \kappa^2)\rho^2 + 2V\iota\rho\kappa.$$

(*i.*) With the same signification of ν, the differential equations of the ellipsoid and its reciprocal become

$$S\nu d\rho = 0, \qquad S\rho d\nu = 0.$$

(*j.*) Eliminate ρ between the four scalar equations,

$$S\alpha\rho = a, \quad S\beta\rho = b, \quad S\gamma\rho = c, \quad S\epsilon\rho = e.$$

20. Hamilton, *Bishop Law's Premium Examination*, 1864.

(*a.*) Let $A_1B_1, A_2B_2, \ldots A_nB_n$ be any given system of posited right lines, the $2n$ points being all given; and let their vector sum,

$$AB = A_1B_1 + A_2B_2 + \ldots + A_nB_n,$$

be a line which does not vanish. Then a point H, and a scalar h, can be determined, which shall satisfy the quaternion equation,

$$HA_1 . A_1B_1 + \ldots + HA_n . A_nB_n = h . AB;$$

namely by assuming any origin O, and writing,

$$OH = V\frac{OA_1 . A_1B_1 + \ldots + OA_n . A_nB_n}{A_1B_1 + \ldots + A_nB_n},$$

$$h = S\frac{OA_1 . A_1B_1 + \ldots}{A_1B_1 + \ldots}.$$

(*b.*) For any assumed point C, let

$$Q_C = CA_1 . A_1B_1 + \ldots + CA_n . A_nB_n;$$

then this quaternion sum may be transformed as follows,

$$Q_C = Q_H + CH . AB = (h + CH) . AB;$$

and therefore its tensor is

$$TQ_C = (h^2 + \overline{CH}^2)^{\frac{1}{2}} . \overline{AB},$$

in which $\overline{AB}$ and $\overline{CH}$ denote lengths.

(*c.*) The least value of this tensor TQ_C is obtained by placing the point C at H; if then a quaternion be said to be a minimum when its tensor is such, we may write

$$\text{min. } Q_C = Q_H = h . AB;$$

so that this minimum of Q_C is a vector.

(*d.*) The equation

$$TQ_C = c = \text{any scalar constant} > TQ_H$$

expresses that the locus of the variable point C is a spheric surface, with its centre at the fixed point H, and with a radius r, or $\overline{CH}$, such that

$$r . \overline{AB} = (TQ_C{}^2 - TQ_H{}^2)^{\frac{1}{2}} = (c^2 - h^2 . \overline{AB}^2)^{\frac{1}{2}};$$

so that H, as being thus the common centre of a series of concentric spheres, determined by the given system of right lines, may be said to be the *Central Point*, or simply the *Centre*, of that system.

(*e*.) The equation

$$TVQ_C = c_1 = \text{any scalar constant} > TQ_H$$

represents a right cylinder, of which the radius

$$= (c_1^2 - h^2 . AB^2)^{\frac{1}{2}}$$

divided by $\overline{AB}$, and of which the axis of revolution is the line, $VQ_C = Q_H = h . AB$;

wherefore this last right line, as being the common axis of a series of such right cylinders, may be called the *Central Axis* of the system.

(*f*.) The equation

$$SQ_C = C_2 = \text{any scalar constant}$$

represents a plane; and all such planes are parallel to the *Central Plane*, of which the equation is

$$SQ_C = 0.$$

(*g*.) Prove that the central axis intersects the central plane perpendicularly, in the central point of the system.

(*h*.) When the n given vectors $A_1B_1, \ldots A_nB_n$ are parallel, and are therefore proportional to n scalars, $b_1, \ldots b_n$, the scalar h and the vector Q_H vanish; and the centre H is then determined by the equation

$$b_1 . HA_1 + b_2 . HA_2 + \ldots + b_n . HA_n = 0,$$

or by the expression,

$$OH = \frac{b_1 . OA_1 + \ldots + b_n . OA_n}{b_1 + \ldots + b_n},$$

where O is again an arbitrary origin.

21. Hamilton, *Bishop Law's Premium Examination*, 1860.

(*a*.) The normal at the end of the variable vector ρ, to the surface of revolution of the sixth dimension, which is represented by the equation

$$(\rho^2 - \alpha^2)^3 = 27\alpha^2 (\rho - \alpha)^4, \ldots\ldots\ldots\ldots\ldots\ldots \text{(a)}$$

or by the system of the two equations,

$$\rho^2 - \alpha^2 = 3t^2\alpha^2, \quad (\rho - \alpha)^2 = t^3\alpha^2, \ldots\ldots\ldots \text{(a}'\text{)}$$

and the tangent to the meridian at that point, are respectively parallel to the two vectors,

$$\nu = 2(\rho - a) - t\rho,$$

and $$\tau = 2(1-2t)(\rho - a) + t^2\rho;$$

so that they intersect the axis a, in points of which the vectors are, respectively,

$$\frac{2a}{2-t}, \text{ and } \frac{2(1-2t)a}{(2-t)^2-2}.$$

(*b*) If $d\rho$ be in the same meridian plane as ρ, then

$$t(1-t)(4-t)\,d\rho = 3\tau\,dt, \quad \text{and} \quad S\frac{\rho\,dt}{d\rho} = \frac{4-t}{3}.$$

(*c*.) Under the same condition,

$$S\frac{d\nu}{d\rho} = \frac{2}{3}(1-t).$$

(*d*.) The vector of the centre of curvature of the meridian, at the end of the vector ρ, is, therefore,

$$\sigma = \rho - \nu\left(S\frac{d\nu}{d\rho}\right)^{-1} = \rho - \frac{3}{2}\frac{\nu}{1-t} = \frac{6a-(4-t)\rho}{2(1-t)}.$$

(*e*.) The expressions in Example 38 give

$$\nu^2 = a^2t^2(1-t)^2, \quad \tau^2 = a^2t^3(1-t)^2(4-t);$$

hence $$(\sigma - \rho)^2 = \frac{9}{4}a^2t^2, \quad \text{and} \quad d\rho^2 = \frac{9a^2t}{4-t}dt^2;$$

the radius of curvature of the meridian is, therefore,

$$R = T(\sigma - \rho) - \frac{3}{2}tTa;$$

and the length of an element of arc of that curve is

$$ds = Td\rho = 3Ta\left(\frac{t}{4-t}\right)^{\frac{1}{2}}dt.$$

(*f*.) The same expressions give

$$4(Va\rho)^2 = -a^4t^3(1-t)^2(4-t);$$

thus the auxiliary scalar t is confined between the limits 0 and 4, and we may write $t = 2 \operatorname{vers}\theta$, where θ is a real angle, which varies continuously from 0 to 2π; the recent expression for the element of arc becomes, therefore, $$ds = 3Ta.td\theta,$$

and gives by integration

$$s = 6Ta(\theta - \sin\theta),$$

if the arc s be measured from the point, say F, for which $\rho = a$, and which is common to all the meridians; and the total periphery of any one such curve is $= 12\pi Ta$.

(*g.*) The value of σ gives

$$4(\sigma^2-\alpha^2) = 3\alpha^2 t(4-t),\ 16(V\alpha\sigma)^2 = -\alpha^4 t^3(4-t)^3;$$

if, then, we set aside the axis of revolution α, which is *crossed* by all the normals to the surface (a), the surface of centres of curvature which is *touched* by all those normals is represented by the equation,

$$4(\sigma^2-\alpha^2)^3+27\alpha^2(V\alpha\sigma)^2 = 0. \quad\ldots\ldots\ldots\ldots\ldots\ldots\ldots\ldots \text{(b)}$$

(*h.*) The point F is common to the two surfaces (a) and (b), and is a singular point on each of them, being a triple point on (a), and a double point on (b); there is also at it an infinitely sharp cusp on (b), which tends to coincide with the axis α, but a determined tangent plane to (a), which is perpendicular to that axis, and to that cusp; and the point, say F', of which the vector $--\alpha$, is another and an exactly similar cusp on (b), but does not belong to (a).

(*i.*) Besides the *three* universally *coincident* intersections of the surface (a), with *any* transversal, drawn through its triple point F, in *any* given direction β, there are always *three other real intersections*, of which indeed one coincides with F if the transversal be perpendicular to the axis, and for which the following is a general formula:

$$\rho = T\alpha.[U\alpha+\{2SU(\alpha\beta)^{\frac{1}{3}}\}^3 U\beta].$$

(*j.*) The point, say V, of which the vector is $\rho = 2\alpha$, is a double point of (a), near which that surface has a cusp, which coincides nearly with its tangent cone at that point; and the semi-angle of this cone is $-\frac{\pi}{6}$.

Auxiliary Equations·

$$\begin{cases} 2S\rho(\rho-\alpha) = \alpha^2 t^2(3+t), \\ 2S\alpha(\rho-\alpha) = \alpha^2 t^2(3-t). \end{cases}$$

$$\begin{cases} S\nu\rho = -\alpha^2 t(1-t)(1-2t) \\ 2S\nu(\rho-\alpha) = \alpha^2 t^3(1-t). \end{cases}$$

$$\begin{cases} S\rho\tau = \alpha^2 t^2(1-t)(4-t), \\ 2S(\rho-\alpha)\tau = \alpha^2 t^3(1-t)(4-t). \end{cases}$$

CATALOGUE OF

WORKS

PUBLISHED FOR THE SYNDICS

OF THE

Cambridge University Press.

London: C. J. CLAY AND SONS,
CAMBRIDGE UNIVERSITY PRESS WAREHOUSE,
AVE MARIA LANE.

GLASGOW: 263, ARGYLE STREET.

Cambridge: DEIGHTON, BELL AND CO.
Leipzig: F. A. BROCKHAUS.

1000
1/4/87

PUBLICATIONS OF

The Cambridge University Press.

THE HOLY SCRIPTURES, &c.

THE CAMBRIDGE PARAGRAPH BIBLE of the Authorized English Version, with the Text Revised by a Collation of its Early and other Principal Editions, the Use of the Italic Type made uniform, the Marginal References remodelled, and a Critical Introduction prefixed, by F. H. A. SCRIVENER, M.A., LL.D., Editor of the Greek Testament, Codex Augiensis, &c., and one of the Revisers of the Authorized Version. Crown 4to. gilt. 21*s.*

From the *Times.*

"Students of the Bible should be particularly grateful (to the Cambridge University Press) for baving produced, with the able assistance of Dr Scrivener, a complete critical edition of the Authorized Version of the English Bible, an edition such as, to use the words of the Editor, 'would have been executed long ago had this version been nothing more than the greatest and best known of English classics.' Falling at a time when the formal revision of this version has been undertaken by a distinguished company of scholars and divines, the publication of this edition must be considered most opportune."

From the *Athenæum.*

"Apart from its religious importance, the English Bible has the glory, which but few sister versions indeed can claim, of being the chief classic of the language, of having, in conjunction with Shakspeare, and in an immeasurable degree more than he, fixed the language beyond any possibility of important change. Thus the recent contributions to the literature of the subject, by such workers as Mr Francis Fry and Canon Westcott, appeal to a wide range of sympathies; and to these may now be added Dr Scrivener, well known for his labours in the cause of the Greek Testament criticism, who has brought out, for the Syndics of the Cambridge University Press, an edition of the English Bible, according to the text of 1611, revised by a comparison with later issues on principles stated by him in his Introduction. Here he enters at length into the history of the chief editions of the version, and of such features as the marginal notes, the use of italic type, and the changes of orthography, as well as into the most interesting question as to the original texts from which our translation is produced."

From the *Methodist Recorder.*

"This noble quarto of over 1300 pages is in every respect worthy of editor and publishers alike. The name of the Cambridge University Press is guarantee enough for its perfection in outward form, the name of the editor is equal guarantee for the worth and accuracy of its contents. Without question, it is the best Paragraph Bible ever published, and its reduced price of a guinea brings it within reach of a large number of students."

From the *London Quarterly Review.*

"The work is worthy in every respect of the editor's fame, and of the Cambridge University Press. The noble English Version, to which our country and religion owe so much, was probably never presented before in so perfect a form."

THE CAMBRIDGE PARAGRAPH BIBLE. STUDENT'S EDITION, on *good writing paper*, with one column of print and wide margin to each page for MS. notes. This edition will be found of great use to those who are engaged in the task of Biblical criticism. Two Vols. Crown 4to. gilt. 31*s.* 6*d.*

THE AUTHORIZED EDITION OF THE ENGLISH BIBLE (1611), ITS SUBSEQUENT REPRINTS AND MODERN REPRESENTATIVES. Being the Introduction to the Cambridge Paragraph Bible (1873), re-edited with corrections and additions. By F. H. A. SCRIVENER, M.A., D.C.L., LL.D., Prebendary of Exeter and Vicar of Hendon. Crown 8vo. 7*s.* 6*d.*

THE LECTIONARY BIBLE, WITH APOCRYPHA, divided into Sections adapted to the Calendar and Tables of Lessons of 1871. Crown 8vo. 3*s.* 6*d.*

London: C. J. CLAY & SONS, Cambridge University Press Warehouse, Ave Maria Lane.

USHER (*indignantly*)—" Is that cigar lit, sir ? "
MR. SHORTS—" No. Got a match ? "

recently appeared under their auspices."—*Notes and Queries.*

"Cambridge has worthily taken the lead with the Breviary, which is of especial value for that part of the reform of the Prayer-Book which will fit it for the wants of our time . . .

"The editors have done their work excellently, and deserve all praise for their labours in rendering what they justly call 'this most interesting Service-book' more readily accessible to historical and liturgica students."—*Saturday Review.*

FASCICULUS III. In quo continetur PROPRIUM SANCTORUM quod et sanctorale dicitur, una cum accentuario. Demy 8vo. 15*s.*

FASCICULI I. II. III. complete, £2. 2*s.*

GREEK AND ENGLISH TESTAMENT, in parallel Columns on the same page. Edited by J. SCHOLEFIELD, M.A. late Regius Professor of Greek in the University. Small Octavo. New Edition, with the Marginal References as arranged and revised by Dr SCRIVENER. Cloth, red edges. 7*s.* 6*d.*

GREEK AND ENGLISH TESTAMENT. THE STUDENT'S EDITION of the above, on *large writing paper.* 4to. 12*s.*

GREEK TESTAMENT, ex editione Stephani tertia, 1550. Small 8vo. 3*s.* 6*d.*

THE NEW TESTAMENT IN GREEK according to the text followed in the Authorised Version, with the Variations adopted in the Revised Version. Edited by F. H. A. SCRIVENER, M.A., D.C.L., LL.D. Crown 8vo. 6*s.* Morocco boards or limp. 12*s.*

THE PARALLEL NEW TESTAMENT GREEK AND ENGLISH, being the Authorised Version set forth in 1611 Arranged in Parallel Columns with the Revised Version of 1881, and with the original Greek, as edited by F. H. A. SCRIVENER, M.A., D.C.L., LL.D. Prebendary of Exeter and Vicar of Hendon. Crown 8vo. 12*s.* 6*d.* *The Revised Version is the Joint Property of the Universities of Cambridge and Oxford.*

London: C. J. CLAY & SONS, Cambridge University Press Warehouse, Ave Maria Lane.

THE BOOK OF ECCLESIASTES, with Notes and Introduction. By the Very Rev. E. H. PLUMPTRE, D.D., Dean of Wells. Large Paper Edition. Demy 8vo. 7*s.* 6*d.*

"No one can say that the Old Testament is a dull or worn-out subject after reading this singularly attractive and also instructive commentary. Its wealth of literary and historical illustration surpasses anything to which we can point in English exegesis of the Old Testament; indeed, even Delitzsch, whose pride it is to leave no source of illustration unexplored, is far inferior on this head to Dr Plumptre." *Academy*, Sept. 10, 1881.

THE GOSPEL ACCORDING TO ST MATTHEW in Anglo-Saxon and Northumbrian Versions, synoptically arranged: with Collations of the best Manuscripts. By J. M. KEMBLE, M.A. and Archdeacon HARDWICK. Demy 4to. 10*s.*

New Edition. By the Rev. Professor SKEAT. [*Immediately.*

THE GOSPEL ACCORDING TO ST MARK in Anglo-Saxon and Northumbrian Versions, synoptically arranged: with Collations exhibiting all the Readings of all the MSS. Edited by the Rev. W. W. SKEAT, Litt.D., Elrington and Bosworth Professor of Anglo-Saxon. Demy 4to. 10*s.*

THE GOSPEL ACCORDING TO ST LUKE, uniform with the preceding, by the same Editor. Demy 4to. 10*s.*

THE GOSPEL ACCORDING TO ST JOHN, uniform with the preceding, by the same Editor. Demy 4to. 10*s.*

"*The Gospel according to St John, in Anglo-Saxon and Northumbrian Versions:* Edited for the Syndics of the University Press, by the Rev. Walter W. Skeat, M.A., completes an undertaking designed and commenced by that distinguished scholar, J. M. Kemble, some forty years ago. Of the particular volume now before us, we can only say it is worthy of its two predecessors. We repeat that the service rendered to the study of Anglo-Saxon by this Synoptic collection cannot easily be overstated."—*Contemporary Review.*

THE POINTED PRAYER BOOK, being the Book of Common Prayer with the Psalter or Psalms of David, pointed as they are to be sung or said in Churches. Royal 24mo. 1*s.* 6*d.*

The same in square 32mo. cloth. 6*d.*

THE CAMBRIDGE PSALTER, for the use of Choirs and Organists. Specially adapted for Congregations in which the "Cambridge Pointed Prayer Book" is used. Demy 8vo. cloth extra, 3*s.* 6*d.* cloth limp, cut flush. 2*s.* 6*d.*

THE PARAGRAPH PSALTER, arranged for the use of Choirs by BROOKE FOSS WESTCOTT, D.D., Regius Professor of Divinity in the University of Cambridge. Fcap. 4to. 5*s.*

The same in royal 32mo. Cloth 1*s.* Leather 1*s.* 6*d.*

"The Paragraph Psalter exhibits all the care, thought, and learning that those acquainted with the works of the Regius Professor of Divinity at Cambridge would expect to find, and there is not a clergyman or organist in England who should be without this Psalter as a work of reference."—*Morning Post.*

THE MISSING FRAGMENT OF THE LATIN TRANSLATION OF THE FOURTH BOOK OF EZRA, discovered, and edited with an Introduction and Notes, and a facsimile of the MS., by ROBERT L. BENSLY, M.A., Reader in Hebrew, Gonville and Caius College, Cambridge. Demy 4to. 10*s.*

"It has been said of this book that it has added a new chapter to the Bible, and, startling as the statement may at first sight appear, it is no exaggeration of the actual fact, if by the Bible we understand that of the larger size which contains the Apocrypha. and if the Second Book of Esdras can be fairly called a part of the Apocrypha."—*Saturday Review.*

GOSPEL DIFFICULTIES, or the Displaced Section of S. Luke. By the Rev. J. J. HALCOMBE, Rector of Balsham and Rural Dean of North Camps, formerly Reader and Librarian at the Charterhouse. Crown 8vo. 10*s.* 6*d.*

London: C. J. CLAY & SONS, Cambridge University Press Warehouse, Ave Maria Lane.

THEOLOGY—(ANCIENT).

THE GREEK LITURGIES. Chiefly from original Authorities. By C. A. SWAINSON, D.D., Master of Christ's College, Cambridge. Crown 4to. Paper covers. 15*s.*

"Jeder folgende Forscher wird dankbar anerkennen, dass Swainson das Fundament zu einer historisch-kritischen Geschichte der Griechischen Liturgien sicher gelegt hat."—ADOLPH HARNACK, *Theologische Literatur-Zeitung.*

THE PALESTINIAN MISHNA. By W. H. LOWE, M.A., Lecturer in Hebrew at Christ's College, Cambridge. Royal 8vo. 21*s.*

SAYINGS OF THE JEWISH FATHERS, comprising Pirqe Aboth and Pereq R. Meir in Hebrew and English, with Critical and Illustrative Notes. By CHARLES TAYLOR, D.D. Master of St John's College, Cambridge, and Honorary Fellow of King's College, London. Demy 8vo. 10*s.*

"The 'Masseketh Aboth' stands at the head of Hebrew non-canonical writings. It is of ancient date, claiming to contain the dicta of teachers who flourished from B C. 200 to the same year of our era. The precise time of its compilation in its present form is, of course, in doubt. Mr Taylor's explanatory and illustrative commentary is very full and satisfactory." —*Spectator.*

"A careful and thorough edition which does credit to English scholarship, of a short treatise from the Mishna, containing a series of sentences or maxims ascribed mostly to Jewish teachers immediately preceding, or immediately following the Christian era. . ."—*Contemporary Review.*

THEODORE OF MOPSUESTIA'S COMMENTARY ON THE MINOR EPISTLES OF S. PAUL. The Latin Version with the Greek Fragments, edited from the MSS. with Notes and an Introduction, by H. B. SWETE, D.D., Rector of Ashdon, Essex, and late Fellow of Gonville and Caius College, Cambridge. In Two Volumes. Volume I., containing the Introduction, with Facsimiles of the MSS., and the Commentary upon Galatians—Colossians. Demy 8vo. 12*s.*

"In dem oben verzeichneten Buche liegt uns die erste Hälfte einer vollständigen, ebenso sorgfaltig gearbeiteten wie schön ausgestatteten Ausgabe des Commentars mit ausführlichen Prolegomena und reichhaltigen kritischen und erläuternden Anmerkungen vor."—*Literarisches Centralblatt.*

"It is the result of thorough, careful, and patient investigation of all the points bearing on the subject, and the results are presented with admirable good sense and modesty."—*Guardian.*

"Auf Grund dieser Quellen ist der Text bei Swete mit musterhafter Akribie hergestellt. Aber auch sonst hat der Herausgeber mit unermüdlichem Fleisse und eingehendster Sachkenntniss sein Werk mit allen denjenigen Zugaben ausgerüstet, welche bei einer solchen Text-Ausgabe nur irgend erwartet werden können. . . . Von den drei Haupthandschriften . . . sind vortreffliche photographische Facsimile's beigegeben, wie überhaupt das ganze Werk von der *University Press* zu Cambridge mit bekannter Eleganz ausgestattet ist."—*Theologische Literaturzeitung.*

"It is a hopeful sign, amid forebodings which arise about the theological learning of the Universities, that we have before us the first instalment of a thoroughly scientific and painstaking work, commenced at Cambridge and completed at a country rectory."—*Church Quarterly Review* (Jan. 1881).

"Hernu Swete's Leistung ist eine so tüchtige dass wir das Werk in keinen besseren Händen wissen möchten, und mit den sichersten Erwartungen auf das Gelingen der Fortsetzung entgegen sehen."—*Göttingische gelehrte Anzeigen* (Sept. 1881).

VOLUME II., containing the Commentary on 1 Thessalonians—Philemon, Appendices and Indices. 12*s.*

"Eine Ausgabe . . . für welche alle zugänglichen Hülfsmittel in musterhafter Weise benützt wurden . . . eine reife Frucht siebenjährigen Fleisses."—*Theologische Literaturzeitung* (Sept. 23, 1882).

"Mit derselben Sorgfalt bearbeitet die wir bei dem ersten Theile gerühmt haben."—*Literarisches Centralblatt* (July 29, 1882).

"M. Jacobi...commença...une édition du texte. Ce travail a été repris en Angleterre et mené à bien dans les deux volumes que je signale en ce moment...Elle est accompagnée de notes érudites, suivie de divers appendices, parmi lesquels on appréciera surtout un recueil des fragments des oeuvres dogmatiques de Théodore, et précédée d'une introduction où sont traitées à fond toutes les questions d'histoire littéraire qui se rattachent soit au commentaire lui-même, soit à sa version Latine."—*Bulletin Critique*, 1885.

SANCTI IRENÆI EPISCOPI LUGDUNENSIS libros quinque adversus Hæreses, versione Latina cum Codicibus Claromontano ac Arundeliano denuo collata, præmissa de placitis Gnosticorum prolusione, fragmenta necnon Græce, Syriace, Armeniace, commentatione perpetua et indicibus variis edidit W. WIGAN HARVEY, S.T.B. Collegii Regalis olim Socius. 2 Vols. 8vo. 18*s.*

M. MINUCII FELICIS OCTAVIUS. The text newly revised from the original MS., with an English Commentary, Analysis, Introduction, and Copious Indices. Edited by H. A. HOLDEN, LL.D. Examiner in Greek to the University of London. Crown 8vo. 7*s.* 6*d.*

THEOPHILI EPISCOPI ANTIOCHENSIS LIBRI TRES AD AUTOLYCUM edidit, Prolegomenis Versione Notulis Indicibus instruxit GULIELMUS GILSON HUMPHRY, S.T.B. Collegii Sanctiss. Trin. apud Cantabrigienses quondam Socius. Post 8vo. 5*s.*

THEOPHYLACTI IN EVANGELIUM S. MATTHÆI COMMENTARIUS, edited by W. G. HUMPHRY, B.D. Prebendary of St Paul's, late Fellow of Trinity College. Demy 8vo. 7*s.* 6*d.*

TERTULLIANUS DE CORONA MILITIS, DE SPECTACULIS, DE IDOLOLATRIA, with Analysis and English Notes, by GEORGE CURREY, D.D. Preacher at the Charter House, late Fellow and Tutor of St John's College. Crown 8vo. 5*s.*

FRAGMENTS OF PHILO AND JOSEPHUS. Newly edited by J. RENDEL HARRIS, M.A., Fellow of Clare College, Cambridge. With two Facsimiles. Demy 4to. 12*s.* 6*d.*

THE ORIGIN OF THE LEICESTER CODEX OF THE NEW TESTAMENT. By J. RENDEL HARRIS, M.A. [*Nearly ready.*

THEOLOGY—(ENGLISH).

WORKS OF ISAAC BARROW, compared with the Original MSS., enlarged with Materials hitherto unpublished. A new Edition, by A. NAPIER, M.A. of Trinity College, Vicar of Holkham, Norfolk. 9 Vols. Demy 8vo. £3. 3*s.*

TREATISE OF THE POPE'S SUPREMACY, and a Discourse concerning the Unity of the Church, by ISAAC BARROW. Demy 8vo. 7*s.* 6*d.*

PEARSON'S EXPOSITION OF THE CREED, edited by TEMPLE CHEVALLIER, B.D. late Fellow and Tutor of St Catharine's College, Cambridge. New Edition. Revised by R. SINKER, B.D., Librarian of Trinity College. Demy 8vo. 12*s.*

"A new edition of Bishop Pearson's famous work *On the Creed* has just been issued by the Cambridge University Press. It is the well-known edition of Temple Chevallier, thoroughly overhauled by the Rev. R. Sinker, of Trinity College. The whole text and notes have been most carefully examined and corrected, and special pains have been taken to verify the almost innumerable references. These have been more clearly and accurately given in very many places, and the citations themselves have been adapted to the best and newest texts of the several authors—texts which have undergone vast improvements within the last two centuries. The Indices have also been revised and enlarged......Altogether this appears to be the most complete and convenient edition as yet published of a work which has long been recognised in all quarters as a standard one." *Guardian.*

AN ANALYSIS OF THE EXPOSITION OF THE CREED written by the Right Rev. JOHN PEARSON, D.D. late Lord Bishop of Chester, by W. H. MILL, D.D. late Regius Professor of Hebrew in the University of Cambridge. Demy 8vo. 5*s.*

WHEATLY ON THE COMMON PRAYER, edited by G. E. CORRIE, D.D. late Master of Jesus College. Demy 8vo. 7*s*. 6*d*.

TWO FORMS OF PRAYER OF THE TIME OF QUEEN ELIZABETH. Now First Reprinted. Demy 8vo. 6*d*.

"From 'Collections and Notes' 1867—1876, by W. Carew Hazlitt (p. 340), we learn that—'A very remarkable volume, in the original vellum cover, and containing 25 Forms of Prayer of the reign of Elizabeth, each with the autograph of Humphrey Dyson, has lately fallen into the hands of my friend Mr H. Pyne. It is mentioned specially in the Preface to the Parker Society's volume of Occasional Forms of Prayer, but it had been lost sight of for 200 years.' By the kindness of the present possessor of this valuable volume, containing in all 25 distinct publications, I am enabled to reprint in the following pages the two Forms of Prayer supposed to have been lost."—*Extract from the* PREFACE.

CÆSAR MORGAN'S INVESTIGATION OF THE TRINITY OF PLATO, and of Philo Judæus, and of the effects which an attachment to their writings had upon the principles and reasonings of the Fathers of the Christian Church. Revised by H. A. HOLDEN, LL.D. Crown 8vo. 4*s*.

SELECT DISCOURSES, by JOHN SMITH, late Fellow of Queens' College, Cambridge. Edited by H. G. WILLIAMS, B.D. late Professor of Arabic. Royal 8vo. 7*s*. 6*d*.

"The 'Select Discourses' of John Smith, collected and published from his papers after his death, are, in my opinion, much the most considerable work left to us by this Cambridge School [the Cambridge Platonists]. They have a right to a place in English literary history."—Mr MATTHEW ARNOLD, in the *Contemporary Review*.

"Of all the products of the Cambridge School, the 'Select Discourses' are perhaps the highest, as they are the most accessible and the most widely appreciated...and indeed no spiritually thoughtful mind can read them unmoved. They carry us so directly into an atmosphere of divine philosophy, luminous with the richest lights of meditative genius... He was one of those rare thinkers in whom largeness of view, and depth, and wealth of poetic and speculative insight, only served to evoke more fully the religious spirit, and while he drew the mould of his thought from Plotinus he vivified the substance of it from St Paul."—Principal TULLOCH, *Rational Theology in England in the 17th Century*.

"We may instance Mr Henry Griffin Williams's revised edition of Mr John Smith's 'Select Discourses,' which have won Mr Matthew Arnold's admiration, as an example of worthy work for an University Press to undertake."—*Times*.

THE HOMILIES, with Various Readings, and the Quotations from the Fathers given at length in the Original Languages. Edited by G. E. CORRIE, D.D. late Master of Jesus College. Demy 8vo. 7*s*. 6*d*.

DE OBLIGATIONE CONSCIENTIÆ PRÆLECTIONES decem Oxonii in Schola Theologica habitæ a ROBERTO SANDERSON, SS. Theologiæ ibidem Professore Regio. With English Notes, including an abridged Translation, by W. WHEWELL, D.D. late Master of Trinity College. Demy 8vo. 7*s*. 6*d*.

ARCHBISHOP USHER'S ANSWER TO A JESUIT, with other Tracts on Popery. Edited by J. SCHOLEFIELD, M.A. late Regius Professor of Greek in the University. Demy 8vo. 7*s*. 6*d*.

WILSON'S ILLUSTRATION OF THE METHOD OF explaining the New Testament, by the early opinions of Jews and Christians concerning Christ. Edited by T. TURTON, D.D. late Lord Bishop of Ely. Demy 8vo. 5*s*.

LECTURES ON DIVINITY delivered in the University of Cambridge, by JOHN HEY, D.D. Third Edition, revised by T. TURTON, D.D. late Lord Bishop of Ely. 2 vols. Demy 8vo. 15*s*.

S. AUSTIN AND HIS PLACE IN THE HISTORY OF CHRISTIAN THOUGHT. Being the Hulsean Lectures for 1885. By W. CUNNINGHAM, B.D., Chaplain and Birkbeck Lecturer, Trinity College, Cambridge. Demy 8vo. Buckram, 12*s*. 6*d*.

ARABIC, SANSKRIT, SYRIAC, &c.

THE DIVYÂVADÂNA, a Collection of Early Buddhist Legends, now first edited from the Nepalese Sanskrit MSS. in Cambridge and Paris. By E. B. COWELL, M.A., Professor of Sanskrit in the University of Cambridge, and R. A. NEIL, M.A., Fellow and Lecturer of Pembroke College. Demy 8vo. 18*s.*

POEMS OF BEHA ED DIN ZOHEIR OF EGYPT. With a Metrical Translation, Notes and Introduction, by E. H. PALMER, M.A., Barrister-at-Law of the Middle Temple, late Lord Almoner's Professor of Arabic, formerly Fellow of St John's College, Cambridge. 2 vols. Crown 4to.

Vol. I. The ARABIC TEXT. 10*s.* 6*d.*; cloth extra. 15*s.*

Vol. II. ENGLISH TRANSLATION. 10*s.* 6*d.*; cloth extra. 15*s.*

"We have no hesitation in saying that in both Prof Palmer has made an addition to Oriental literature for which scholars should be grateful; and that, while his knowledge of Arabic is a sufficient guarantee for his mastery of the original, his English compositions are distinguished by versatility, command of language, rhythmical cadence, and, as we have remarked, by not unskilful imitations of the styles of several of our own favourite poets, living and dead."—*Saturday Review.*

"This sumptuous edition of the poems of Behá-ed-dín Zobeir is a very welcome addition to the small series of Eastern poets accessible to readers who are not Orientalists."—*Academy.*

THE CHRONICLE OF JOSHUA THE STYLITE, composed in Syriac A.D. 507 with an English translation and notes, by W. WRIGHT, LL.D., Professor of Arabic. Demy 8vo. 10*s.* 6*d.*

"Die lehrreiche kleine Chronik Josuas hat nach Assemani und Martin in Wright einen dritten Bearbeiter gefunden, der sich um die Emendation des Textes wie um die Erklärung der Realien wesentlich verdient gemacht hat . . . Ws. Josua-Ausgabe ist eine sehr dankenswerte Gabe und besonders empfehlenswert als ein Lehrmittel für den syrischen Unterricht; es erscheint auch gerade zur rechten Zeit, da die zweite Ausgabe von Roedigers syrischer Chrestomathie im Buchhandel vollständig vergriffen und diejenige von Kirsch-Bernstein nur noch in wenigen Exemplaren vorhanden ist."—*Deutsche Litteraturzeitung.*

KALILAH AND DIMNAH, OR, THE FABLES OF BIDPAI; being an account of their literary history, together with an English Translation of the same, with Notes, by I. G. N. KEITH-FALCONER, M.A., Lord Almoner's Professor of Arabic in the University of Cambridge. Demy 8vo. 7*s.* 6*d.*

NALOPÀKHYÀNAM, OR, THE TALE OF NALA; containing the Sanskrit Text in Roman Characters, followed by a Vocabulary and a sketch of Sanskrit Grammar. By the late Rev. THOMAS JARRETT, M.A. Trinity College, Regius Professor of Hebrew. Demy 8vo. 10*s.*

NOTES ON THE TALE OF NALA, for the use of Classical Students, by J. PEILE, Litt.D., Fellow and Tutor of Christ's College. Demy 8vo. 12*s.*

CATALOGUE OF THE BUDDHIST SANSKRIT MANUSCRIPTS in the University Library, Cambridge. Edited by C. BENDALL, M.A., Fellow of Gonville and Caius College. Demy 8vo. 12*s.*

"It is unnecessary to state how the compilation of the present catalogue came to be placed in Mr Bendall's hands; from the character of his work it is evident the selection was judicious, and we may fairly congratulate those concerned in it on the result . . . Mr Bendall has entitled himself to the thanks of all Oriental scholars, and we hope he may have before him a long course of successful labour in the field he has chosen."—*Athenæum.*

GREEK AND LATIN CLASSICS, &c.

SOPHOCLES: The Plays and Fragments, with Critical Notes, Commentary, and Translation in English Prose, by R. C. JEBB, Litt.D., LL.D., Professor of Greek in the University of Glasgow.

Part I. Oedipus Tyrannus. Demy 8vo. *New Edition, In the Press.*
Part II. Oedipus Coloneus. Demy 8vo. 12*s.* 6*d.*
Part III. The Antigone. [*In the Press.*

"Of his explanatory and critical notes we can only speak with admiration. Thorough scholarship combines with taste, erudition, and boundless industry to make this first volume a pattern of editing. The work is made complete by a prose translation, upon pages alternating with the text, of which we may say shortly that it displays sound judgment and taste, without sacrificing precision to poetry of expression."—*The Times.*

"This larger edition he has deferred these many years for reasons which he has given in his preface, and which we accept with entire satisfaction, as we have now the first portion of a work composed in the fulness of his powers and with all the resources of fine erudition and laboriously earned experience...We will confidently aver, then, that the edition is neither tedious nor long; for we get in one compact volume such a cyclopædia of instruction, such a variety of helps to the full comprehension of the poet, as not so many years ago would have needed a small library, and all this instruction and assistance given, not in a dull and pedantic way, but in a style of singular clearness and vivacity. In fact, one might take this edition with him on a journey, and, without any other help whatever, acquire with comfort and delight a thorough acquaintance with the noblest production of, perhaps, the most difficult of all Greek poets—the most difficult, yet possessed at the same time of an immortal charm for one who has mastered him, as Mr Jebb has, and can feel so subtly perfection of form and language...We await with lively expectation the continuation, and completion of Mr Jebb's great task, and it is a fortunate thing that his power of work seems to be as great as the style is happy in which the work is done."—*The Athenæum.*

"An edition which marks a definite advance, which is whole in itself, and brings a mass of solid and well-wrought material such as future constructors will desire to adapt, is definitive in the only applicable sense of the term, and such is the edition of Professor Jebb. No man is better fitted to express in relation to Sophocles the mind of the present generation."—*The Saturday Review.*

AESCHYLI FABULAE.—ΙΚΕΤΙΔΕΣ ΧΟΗΦΟΡΟΙ IN LIBRO MEDICEO MENDOSE SCRIPTAE EX VV. DD. CONIECTURIS EMENDATIUS EDITAE cum Scholiis Graecis et brevi adnotatione critica, curante F. A. PALEY, M.A., LL.D. Demy 8vo. 7*s.* 6*d.*

THE AGAMEMNON OF AESCHYLUS. With a Translation in English Rhythm, and Notes Critical and Explanatory. **New Edition Revised.** By BENJAMIN HALL KENNEDY, D.D., Regius Professor of Greek. Crown 8vo. 6*s.*

"One of the best editions of the masterpiece of Greek tragedy."—*Athenæum.*

THE THEÆTETUS OF PLATO with a Translation and Notes by the same Editor. Crown 8vo. 7*s.* 6*d.*

ARISTOTLE.—ΠΕΡΙ ΨΥΧΗΣ. ARISTOTLE'S PSYCHOLOGY, in Greek and English, with Introduction and Notes, by EDWIN WALLACE, M.A., late Fellow and Tutor of Worcester College, Oxford. Demy 8vo. 18*s.*

"The notes are exactly what such notes ought to be,—helps to the student, not mere displays of learning. By far the more valuable parts of the notes are neither critical nor literary, but philosophical and expository of the thought, and of the connection of thought, in the treatise itself. In this relation the notes are invaluable. Of the translation, it may be said that an English reader may fairly master by means of it this great treatise of Aristotle."—*Spectator.*

"Wallace's Bearbeitung der Aristotelischen Psychologie ist das Werk eines denkenden und in allen Schriften des Aristoteles und grösstenteils auch in der neueren Litteratur zu denselben belesenen Mannes... Der schwächste Teil der Arbeit ist der kritische... Aber in allen diesen Dingen liegt auch nach der Absicht des Verfassers nicht der Schwerpunkt seiner Arbeit, sondern."—Prof. Susemihl in *Philologische Wochenschrift.*

ARISTOTLE.—ΠΕΡΙ ΔΙΚΑΙΟΣΥΝΗΣ. THE FIFTH BOOK OF THE NICOMACHEAN ETHICS OF ARISTOTLE. Edited by HENRY JACKSON, Litt.D., Fellow of Trinity College, Cambridge. Demy 8vo. 6*s.*

"It is not too much to say that some of the points he discusses have never had so much light thrown upon them before.... Scholars will hope that this is not the only portion of the Aristotelian writings which he is likely to edit."—*Athenæum.*

London: C. J. CLAY & SONS, Cambridge University Press Warehouse, Ave Maria Lane.

ARISTOTLE. THE RHETORIC. With a Commentary by the late E. M. COPE, Fellow of Trinity College, Cambridge, revised and edited by J. E. SANDYS, Litt.D. With a biographical Memoir by the late H. A. J. MUNRO, Litt.D. 3 Vols., Demy 8vo. **Now reduced to 21s.** (*originally published at 31s. 6d.*)

"This work is in many ways creditable to the University of Cambridge. If an English student wishes to have a full conception of what is contained in the *Rhetoric* of Aristotle, to Mr Cope's edition he must go."—*Academy.*

"Mr Sandys has performed his arduous duties with marked ability and admirable tact. . . . . . In every part of his work—revising, supplementing, and completing—he has done exceedingly well."—*Examiner.*

PINDAR. OLYMPIAN AND PYTHIAN ODES. With Notes Explanatory and Critical, Introductions and Introductory Essays. Edited by C. A. M. FENNELL, Litt. D., late Fellow of Jesus College. Crown 8vo. 9s.

"Mr Fennell deserves the thanks of all classical students for his careful and scholarly edition of the Olympian and Pythian odes. He brings to his task the necessary enthusiasm for his author, great industry, a sound judgment, and, in particular, copious and minute learning in comparative philology."—*Athenæum.*

"Considered simply as a contribution to the study and criticism of Pindar, Mr Fennell's edition is a work of great merit."—*Saturday Review.*

—— THE ISTHMIAN AND NEMEAN ODES. By the same Editor. Crown 8vo. 9s.

". . . As a handy and instructive edition of a difficult classic no work of recent years surpasses Mr Fennell's 'Pindar.'"—*Athenæum.*

"This work is in no way inferior to the previous volume. The commentary affords valuable help to the study of the most difficult of Greek authors, and is enriched with notes on points of scholarship and etymology which could only have been written by a scholar of very high attainments."—*Saturday Review.*

PRIVATE ORATIONS OF DEMOSTHENES, with Introductions and English Notes, by F. A. PALEY, M.A. Editor of Aeschylus, etc. and J. E. SANDYS, Litt.D. Fellow and Tutor of St John's College, and Public Orator in the University of Cambridge.

PART I. Contra Phormionem, Lacritum, Pantaenetum, Boeotum de Nomine, Boeotum de Dote, Dionysodorum. **New Edition.** Crown 8vo. 6s.

"Mr Paley's scholarship is sound and accurate, his experience of editing wide, and if he is content to devote his learning and abilities to the production of such manuals as these, they will be received with gratitude throughout the higher schools of the country. Mr Sandys is deeply read in the German literature which bears upon his author, and the elucidation of matters of daily life, in the delineation of which Demosthenes is so rich, obtains full justice at his hands. . . . We hope this edition may lead the way to a more general study of these speeches in schools than has hitherto been possible."—*Academy.*

PART II. Pro Phormione, Contra Stephanum I. II.; Nicostratum, Cononem, Calliclem. **New Edition.** Crown 8vo. 7s. 6d.

"It is long since we have come upon a work evincing more pains, scholarship, and varied research and illustration than Mr Sandys's contribution to the 'Private Orations of Demosthenes'."—*Saturday Review.*

". . . . . . the edition reflects credit on Cambridge scholarship, and ought to be extensively used."—*Athenæum.*

DEMOSTHENES AGAINST ANDROTION AND AGAINST TIMOCRATES, with Introductions and English Commentary, by WILLIAM WAYTE, M.A., late Professor of Greek, University College, London. Crown 8vo. 7s. 6d.

"These speeches are highly interesting, as illustrating Attic Law, as that law was influenced by the exigences of politics . . . As vigorous examples of the great orator's style, they are worthy of all admiration; and they have the advantage—not inconsiderable when the actual attainments of the average schoolboy are considered—of having an easily comprehended subject matter . . . Besides a most lucid and interesting introduction, Mr Wayte has given the student effective help in his running commentary. We may note, as being so well managed as to form a very valuable part of the exegesis, the summaries given with every two or three sections throughout the speech."—*Spectator.*

PLATO'S PHÆDO, literally translated, by the late E. M. COPE, Fellow of Trinity College, Cambridge, revised by HENRY JACKSON, Litt. D., Fellow of Trinity College. Demy 8vo. 5s.

THE BACCHAE OF EURIPIDES. With Introduction, Critical Notes, and Archæological Illustrations, by J. E. SANDYS, Litt.D. New and Enlarged Edition. Crown 8vo. 12*s*. 6*d*.

"Of the present edition of the *Bacchæ* by Mr Sandys we may safely say that never before has a Greek play, in England at least, had fuller justice done to its criticism, interpretation, and archæological illustration, whether for the young student or the more advanced scholar. The Cambridge Public Orator may be said to have taken the lead in issuing a complete edition of a Greek play, which is destined perhaps to gain redoubled favour now that the study of ancient monuments has been applied to its illustration."—*Saturday Review*.

"The volume is interspersed with well-executed woodcuts, and its general attractiveness of form reflects great credit on the University Press. In the notes Mr Sandys has more than sustained his well-earned reputation as a careful and learned editor, and shows considerable advance in freedom and lightness of style. . . . Under such circumstances it is superfluous to say that for the purposes of teachers and advanced students this handsome edition far surpasses all its predecessors."—*Athenæum*.

"It has not, like so many such books, been hastily produced to meet the momentary need of some particular examination; but it has employed for some years the labour and thought of a highly finished scholar, whose aim seems to have been that his book should go forth ***totus teres atque rotundus***, armed at all points with all that may throw light upon its subject. The result is a work which will not only assist the schoolboy or undergraduate in his tasks, but will adorn the library of the scholar."—*The Guardian*.

THE TYPES OF GREEK COINS. By PERCY GARDNER, Litt. D., F.S.A., Disney Professor of Archæology. With 16 Autotype plates, containing photographs of Coins of all parts of the Greek World. Impl. 4to. Cloth extra, £1. 11*s*. 6*d*.; Roxburgh (Morocco back), £2. 2*s*.

"Professor Gardner's book is written with such lucidity and in a manner so straightforward that it may well win converts, and it may be distinctly recommended to that omnivorous class of readers—'men in the schools'."—*Saturday Review*.

"'The Types of Greek Coins' is a work which is less purely and dryly scientific. Nevertheless, it takes high rank as proceeding upon a truly scientific basis at the same time that it treats the subject of numismatics in an attractive style and is elegant enough to justify its appearance in the drawing-room."—*Athenæum*.

A SELECTION OF GREEK INSCRIPTIONS, with Introductions and Annotations by E. S. ROBERTS, M.A., Fellow and Tutor of Gonville and Caius College. [*Nearly ready*.

ESSAYS ON THE ART OF PHEIDIAS. By C. WALDSTEIN, M.A., Phil. D., Reader in Classical Archæology in the University of Cambridge. Royal 8vo. With numerous Illustrations. 16 Plates. Buckram, 30*s*.

"I acknowledge expressly the warm enthusiasm for ideal art which pervades the whole volume, and the sharp eye Dr Waldstein has proved himself to possess in his special line of study, namely, stylistic analysis, which has led him to several happy and important discoveries. His book will be universally welcomed as a very valuable contribution towards a more thorough knowledge of the style of Pheidias."—*The Academy*.

"'Essays on the Art of Pheidias' form an extremely valuable and important piece of work. . . . Taking it for the illustrations alone, it is an exceedingly fascinating book."—*Times*.

M. TULLI CICERONIS AD. M. BRUTUM ORATOR. A revised text edited with Introductory Essays and with critical and explanatory notes, by J. E. SANDYS, Litt.D. Demy 8vo. 16*s*.

M. TULLI CICERONIS DE FINIBUS BONORUM ET MALORUM LIBRI QUINQUE. The text revised and explained; With a Translation by JAMES S. REID, Litt. D., Fellow and Tutor of Gonville and Caius College. 3 Vols. [*In the Press*.
VOL. III. Containing the Translation. Demy 8vo. 8*s*.

M. T. CICERONIS DE OFFICIIS LIBRI TRES, with Marginal Analysis, an English Commentary, and copious Indices, by H. A. HOLDEN, LL.D., Examiner in Greek to the University of London. **Sixth Edition,** Revised and Enlarged. Crown 8vo. 9*s*.

"Few editions of a classic have found so much favour as Dr Holden's *De Officiis*, and the present revision (sixth edition) makes the position of the work secure."—*American Journal of Philology*.

M. TVLLI CICERONIS PRO C RABIRIO [PERDVELLIONIS REO] ORATIO AD QVIRITES With Notes Introduction and Appendices by W. E. HEITLAND, M.A., Fellow and Tutor of St John's College, Cambridge. Demy 8vo. 7*s*. 6*d*.

London: C. J. CLAY & SONS, Cambridge University Press Warehouse, Ave Maria Lane.

M. TULLII CICERONIS DE NATURA DEORUM Libri Tres, with Introduction and Commentary by JOSEPH B. MAYOR, M.A., together with a new collation of several of the English MSS. by J. H. SWAINSON, M.A.

Vol. I. Demy 8vo. 10*s*. 6*d*. Vol. II. 12*s*. 6*d*. Vol. III. 10*s*.

"Such editions as that of which Prof. Mayor has given us the first instalment will doubtless do much to remedy this undeserved neglect. It is one on which great pains and much learning have evidently been expended, and is in every way admirably suited to meet the needs of the student... The notes of the editor are all that could be expected from his well-known learning and scholarship."—*Academy*.

"Der vorliegende zweite Band enthält N. D. II. und zeigt ebenso wie der erste einen erheblichen Fortschritt gegen die bisher vorhandenen commentirten Ausgaben. Man darf jetzt, nachdem der grösste Theil erschienen ist, sagen, dass niemand, welcher sich sachlich oder kritisch mit der Schrift De Nat. Deor. beschäftigt, die neue Ausgabe wird ignoriren dürfen."—P. SCHWENCKE in *JB. f. cl. Alt.* vol. 35, p. 90 foll.

"Nell' edizione sua è più compiuto, che in qualunque altra edizione anteriore, e in parte nuove, non meno l' apparato critico dal testo che l' esame ed il commento del contenuto del libro."—R. BONGHI in *Nuova Antologia*, Oct. 1881, pp. 717—731.

P. VERGILI MARONIS OPERA cum Prolegomenis et Commentario Critico edidit B. H. KENNEDY, S.T.P., Graecae Linguae Prof. Regius. Extra Fcap. 8vo. 5*s*.

See also Pitt Press Series, pp. 24—27.

MATHEMATICS, PHYSICAL SCIENCE, &c.

MATHEMATICAL AND PHYSICAL PAPERS. By Sir W. THOMSON, LL.D., D.C.L., F.R.S., Professor of Natural Philosophy in the University of Glasgow. Collected from different Scientific Periodicals from May 1841, to the present time. Vol. I. Demy 8vo. 18*s*. Vol. II. 15*s*. [Volume III. *In the Press*.

"Wherever exact science has found a follower Sir William Thomson's name is known as a leader and a master. For a space of 40 years each of his successive contributions to knowledge in the domain of experimental and mathematical physics has been recognized as marking a stage in the progress of the subject. But, unhappily for the mere learner, he is no writer of text-books. His eager fertility overflows into the nearest available journal... The papers in this volume deal largely with the subject of the dynamics of heat. They begin with two or three articles which were in part written at the age of 17, before the author had commenced residence as an undergraduate in Cambridge."—*The Times*.

"We are convinced that nothing has had a greater effect on the progress of the theories of electricity and magnetism during the last ten years than the publication of Sir W. Thomson's reprint of papers on electrostatics and magnetism, and we believe that the present volume is destined in no less degree to further the advancement of physical science."—*Glasgow Herald*.

MATHEMATICAL AND PHYSICAL PAPERS, by GEORGE GABRIEL STOKES, M.A., D.C.L., LL.D., F.R.S., Fellow of Pembroke College, and Lucasian Professor of Mathematics in the University of Cambridge. Reprinted from the Original Journals and Transactions, with Additional Notes by the Author. Vol. I. Demy 8vo. 15*s*. Vol. II. 15*s*. [Volume III. *In the Press*.

"...The same spirit pervades the papers on pure mathematics which are included in the volume. They have a severe accuracy of style which well befits the subtle nature of the subjects, and inspires the completest confidence in their author."—*The Times*.

A HISTORY OF THE THEORY OF ELASTICITY AND OF THE STRENGTH OF MATERIALS, from Galilei to the present time. VOL. I. Galilei to Saint-Venant, 1639-1850. By the late I. TODHUNTER, D. Sc., F.R.S., edited and completed by Professor KARL PEARSON, M.A. Demy 8vo. 25*s*.

A TREATISE ON GEOMETRICAL OPTICS. By R. S. HEATH, M.A., Professor of Mathematics in Mason Science College, Birmingham. Demy 8vo. [*Nearly ready*.

THE SCIENTIFIC PAPERS OF THE LATE PROF. J. CLERK MAXWELL. Edited by W. D. NIVEN, M.A. In 2 vols. Royal 4to. [*Nearly ready*.

A TREATISE ON NATURAL PHILOSOPHY. By Sir W. THOMSON, LL.D., D.C.L., F.R.S., and P. G. TAIT, M.A., Professor of Natural Philosophy in the University of Edinburgh. **Part I.** Demy 8vo. 16*s.* **Part II.** Demy 8vo. 18*s.*

ELEMENTS OF NATURAL PHILOSOPHY. By Professors Sir W. THOMSON and P. G. TAIT. Demy 8vo. *Second Edition.* 9*s.*

AN ATTEMPT TO TEST THE THEORIES OF CAPILLARY ACTION by FRANCIS BASHFORTH, B.D., and J. C. ADAMS, M.A., F.R.S. Demy 4to. £1. 1*s.*

A TREATISE ON THE THEORY OF DETERMINANTS and their applications in Analysis and Geometry, by R. F. SCOTT, M.A., Fellow of St John's College. Demy 8vo. 12*s.*

HYDRODYNAMICS, a Treatise on the Mathematical Theory of the Motion of Fluids, by HORACE LAMB, M.A., formerly Fellow of Trinity College, Cambridge. Demy 8vo. 12*s.*

THE ANALYTICAL THEORY OF HEAT, by JOSEPH FOURIER. Translated, with Notes, by A. FREEMAN, M.A., Fellow of St John's College, Cambridge. Demy 8vo. 16*s.*

PRACTICAL WORK AT THE CAVENDISH LABORATORY. HEAT. Edited by W. N. SHAW, M.A., Fellow and Lecturer of Emmanuel College. Demy 8vo. 3*s.*

THE ELECTRICAL RESEARCHES OF THE Hon. H. CAVENDISH, F.R.S. Written between 1771 and 1781. Edited from the original MSS. in the possession of the Duke of Devonshire, K. G., by the late J. CLERK MAXWELL, F.R.S. Demy 8vo. 18*s.*

"Every department of editorial duty appears to have been most conscientiously performed; and it must have been no small satisfaction to Prof. Maxwell to see this goodly volume completed before his life's work was done."—*Athenæum.*

AN ELEMENTARY TREATISE ON QUATERNIONS. By P. G. TAIT, M.A. *Second Edition.* Demy 8vo. 14*s.*

THE MATHEMATICAL WORKS OF ISAAC BARROW, D.D. Edited by W. WHEWELL, D.D. Demy 8vo. 7*s.* 6*d.*

COUNTERPOINT. A Practical Course of Study, by Professor Sir G. A. MACFARREN, M.A., Mus. Doc. New Edition, revised. Crown 4to. 7*s.* 6*d.*

A TREATISE ON THE GENERAL PRINCIPLES OF CHEMISTRY, by M. M. PATTISON MUIR, M.A., Fellow and Prælector in Chemistry of Gonville and Caius College. Demy 8vo. 15*s.*

"The value of the book as a digest of the historical developments of chemical thought is immense."—*Academy.*

"Theoretical Chemistry has moved so rapidly of late years that most of our ordinary text books have been left far behind. German students, to be sure, possess an excellent guide to the present state of the science in 'Die Modernen Theorien der Chemie' of Prof. Lothar Meyer; but in this country the student has had to content himself with such works as Dr Tilden's 'Introduction to Chemical Philosophy', an admirable book in its way, but rather slender. Mr Pattison Muir having aimed at a more comprehensive scheme, has produced a systematic treatise on the principles of chemical philosophy which stands far in advance of any kindred work in our language. It is a treatise that requires for its due comprehension a fair acquaintance with physical science, and it can hardly be placed with advantage in the hands of any one who does not possess an extended knowledge of descriptive chemistry. But the advanced student whose mind is well equipped with an array of chemical and physical facts can turn to Mr Muir's masterly volume for unfailing help in acquiring a knowledge of the principles of modern chemistry."—*Athenæum.*

ELEMENTARY CHEMISTRY. I. PRINCIPLES. By M. M. PATTISON MUIR, M.A., and CHARLES SLATER, M.A., M.B. II. COURSE OF LABORATORY WORK. By M. M. PATTISON MUIR, M.A., and D. J. CARNEGIE, B.A. [*In the Press.*

NOTES ON QUALITATIVE ANALYSIS. Concise and Explanatory. By H. J. H. FENTON, M.A., F.I.C., Demonstrator of Chemistry in the University of Cambridge. Cr. 4to. *New Edition.* 6*s.*

LECTURES ON THE PHYSIOLOGY OF PLANTS, by S. H. VINES, D.Sc., Fellow of Christ's College. Demy 8vo. With Illustrations. 21*s.*

"To say that Dr Vines' book is a most valuable addition to our own botanical literature is but a narrow meed of praise: it is a work which will take its place as cosmopolitan: no more clear or concise discussion of the difficult chemistry of metabolism has appeared.... In erudition it stands alone among English books, and will compare favourably with any foreign competitors."—*Nature.*

"It has long been a reproach to English science that the works in most general use in this country for higher botanical teaching have been of foreign origin....This is not as it should be; and we welcome Dr Vines' Lectures on the Physiology of Plants as an important step towards the removal of this reproach....The work forms an important contribution to the literature of the subject....It will be eagerly welcomed by all students, and must be in the hands of all teachers."—*Academy.*

A SHORT HISTORY OF GREEK MATHEMATICS. By J. GOW, Litt.D., Fellow of Trinity College. Demy 8vo. 10*s.* 6*d.*

DIOPHANTOS OF ALEXANDRIA; a Study in the History of Greek Algebra. By T. L. HEATH, M.A., Fellow of Trinity College, Cambridge. Demy 8vo. 7*s.* 6*d.*

"This study in the history of Greek Algebra is an exceedingly valuable contribution to the history of mathematics."—*Academy.*

"The most thorough account extant of Diophantus's place, work, and critics. . . . [The classification of Diophantus's methods of solution taken in conjunction with the invaluable abstract, presents the English reader with a capital picture of what Greek algebraists had really accomplished.]"—*Athenæum.*

THE FOSSILS AND PALÆONTOLOGICAL AFFINITIES OF THE NEOCOMIAN DEPOSITS OF UPWARE AND BRICKHILL with Plates, being the Sedgwick Prize Essay for the Year 1879. By W. KEEPING, M.A., F.G.S. Demy 8vo. 10*s.* 6*d.*

A CATALOGUE OF BOOKS AND PAPERS ON PROTOZOA, CŒLENTERATES, WORMS, and certain smaller groups of animals, published during the years 1861—1883, by D'ARCY W. THOMPSON, B.A. Demy 8vo. 12*s.* 6*d.*

ASTRONOMICAL OBSERVATIONS made at the Observatory of Cambridge by the late Rev. JAMES CHALLIS, M.A., F.R.S., F.R.A.S. For various Years, from 1846 to 1860.

ASTRONOMICAL OBSERVATIONS from 1861 to 1865. Vol. XXI. Royal 4to. 15*s.* From 1866 to 1869. Vol. XXII. Royal 4to. [*Nearly ready.*

A CATALOGUE OF THE COLLECTION OF BIRDS formed by the late H. E. STRICKLAND, now in the possession of the University of Cambridge. By O. SALVIN, M.A. Demy 8vo. £1. 1*s.*

A CATALOGUE OF AUSTRALIAN FOSSILS, Stratigraphically and Zoologically arranged, by R. ETHERIDGE, Jun., F.G.S. Demy 8vo. 10*s.* 6*d.*

ILLUSTRATIONS OF COMPARATIVE ANATOMY, VERTEBRATE AND INVERTEBRATE, for the Use of Students in the Museum of Zoology and Comparative Anatomy. Second Edition. Demy 8vo. 2*s.* 6*d.*

A SYNOPSIS OF THE CLASSIFICATION OF THE BRITISH PALÆOZOIC ROCKS, by the Rev. ADAM SEDGWICK, M.A., F.R.S., and FREDERICK MCCOY, F.G.S. One vol., Royal 4to. Plates, £1. 1*s.*

A CATALOGUE OF THE COLLECTION OF CAMBRIAN AND SILURIAN FOSSILS contained in the Geological Museum of the University of Cambridge, by J. W. SALTER, F.G.S. With a Portrait of PROFESSOR SEDGWICK. Royal 4to. 7*s.* 6*d.*

CATALOGUE OF OSTEOLOGICAL SPECIMENS contained in the Anatomical Museum of the University of Cambridge. Demy 8vo. 2*s.* 6*d.*

LAW.

A SELECTION OF CASES ON THE ENGLISH LAW OF CONTRACT. By GERARD BROWN FINCH, M.A., of Lincoln's Inn, Barrister at Law; Law Lecturer and late Fellow of Queens College, Cambridge. Royal 8vo. 28*s.*

"An invaluable guide towards the best method of legal study."—*Law Quarterly Review.*

THE INFLUENCE OF THE ROMAN LAW ON THE LAW OF ENGLAND. Being the Yorke Prize Essay for 1884. By T. E. SCRUTTON, M.A. Demy 8vo. 10*s.* 6*d.*

"Legal work of just the kind that a learned University should promote by its prizes."—*Law Quarterly Review.*

LAND IN FETTERS. Being the Yorke Prize Essay for 1885. By T. E. SCRUTTON, M.A. Demy 8vo. 7*s.* 6*d.*

AN ANALYSIS OF CRIMINAL LIABILITY. By E. C. CLARK, LL.D., Regius Professor of Civil Law in the University of Cambridge, also of Lincoln's Inn, Barrister-at-Law. Crown 8vo. 7*s.* 6*d.*

"Prof. Clark's little book is the substance of lectures delivered by him upon those portions of Austin's work on jurisprudence which deal with the "operation of sanctions" . . . Students of jurisprudence will find much to interest and instruct them in the work of Prof. Clark."—*Athenæum.*

PRACTICAL JURISPRUDENCE, a Comment on AUSTIN. By E. C. CLARK, LL.D. Regius Professor of Civil Law. Crown 8vo. 9*s.*

"Damit schliesst dieses inhaltreiche und nach allen Seiten anregende Buch über Practical Jurisprudence."—König. *Centralblatt für Rechtswissenschaft.*

A SELECTION OF THE STATE TRIALS. By J. W. WILLIS-BUND, M.A., LL.B., Barrister-at-Law, Professor of Constitutional Law and History, University College, London. Crown 8vo. Vols. I. and II. In 3 parts. **Now reduced to 30*s.*** (*originally published at* 46*s.*)

"This work is a very useful contribution to that important branch of the constitutional history of England which is concerned with the growth and development of the law of treason, as it may be gathered from trials before the ordinary courts. The author has very wisely distinguished these cases from those of impeachment for treason before Parliament, which he proposes to treat in a future volume under the general head 'Proceedings in Parliament.'" —*The Academy.*

"This is a work of such obvious utility that the only wonder is that no one should have undertaken it before . . . In many respects therefore, although the trials are more or less abridged, this is for the ordinary student's purpose not only a more handy, but a more useful work than Howell's."—*Saturday Review.*

"But, although the book is most interesting to the historian of constitutional law, it is also not without considerable value to those who seek information with regard to procedure and the growth of the law of evidence. We should add that Mr Willis-Bund has given short prefaces and appendices to the trials, so as to form a connected narrative of the events in history to which they relate. We can thoroughly recommend the book."—*Law Times.*

"To a large class of readers Mr Willis-Bund's compilation will thus be of great assistance, for he presents in a convenient form a judicious selection of the principal statutes and the leading cases bearing on the crime of treason . . . For all classes of readers these volumes possess an indirect interest, arising from the nature of the cases themselves, from the men who were actors in them, and from the numerous points of social life which are incidentally illustrated in the course of the trials."—*Athenæum.*

THE FRAGMENTS OF THE PERPETUAL EDICT OF SALVIUS JULIANUS, collected, arranged, and annotated by BRYAN WALKER, M.A., LL.D., Law Lecturer of St John's College, and late Fellow of Corpus Christi College, Cambridge. Crown 8vo. 6*s.*

"In the present book we have the fruits of the same kind of thorough and well-ordered study which was brought to bear upon the notes to the Commentaries and the Institutes . . . Hitherto the Edict has been almost inaccessible to the ordinary English student, and such a student will be interested as well as perhaps surprised to find how abundantly the extant fragments illustrate and clear up points which have attracted his attention in the Commentaries, or the Institutes, or the Digest."—*Law Times.*

AN INTRODUCTION TO THE STUDY OF JUSTINIAN'S DIGEST. Containing an account of its composition and of the Jurists used or referred to therein. By HENRY JOHN ROBY, M.A., formerly Prof. of Jurisprudence, University College, London. Demy 8vo. 9*s*.

JUSTINIAN'S DIGEST. Lib. VII., Tit. I. De Usufructu with a Legal and Philological Commentary. By H. J. ROBY, M.A. Demy 8vo. 9*s*.

Or the Two Parts complete in One Volume. Demy 8vo. 18*s*.

"Not an obscurity, philological, historical, or legal, has been left unsifted. More informing aid still has been supplied to the student of the Digest at large by a preliminary account, covering nearly 300 pages, of the mode of composition of the Digest, and of the jurists whose decisions and arguments constitute its substance. Nowhere else can a clearer view be obtained of the personal succession by which the tradition of Roman legal science was sustained and developed. Roman law, almost more than Roman legions, was the backbone of the Roman commonwealth. Mr Roby, by his careful sketch of the sages of Roman law, from Sextus Papirius, under Tarquin the Proud, to the Byzantine Bar, has contributed to render the tenacity and durability of the most enduring polity the world has ever experienced somewhat more intelligible."—*The Times*.

THE COMMENTARIES OF GAIUS AND RULES OF ULPIAN. With a Translation and Notes, by J. T. ABDY, LL.D., Judge of County Courts, late Regius Professor of Laws in the University of Cambridge, and BRYAN WALKER, M.A., LL.D., Law Lecturer of St John's College, Cambridge, formerly Law Student of Trinity Hall and Chancellor's Medallist for Legal Studies. New Edition by BRYAN WALKER. Crown 8vo. 16*s*.

"As scholars and as editors Messrs Abdy and Walker have done their work well . . . For one thing the editors deserve special commendation. They have presented Gaius to the reader with few notes and those merely by way of reference or necessary explanation. Thus the Roman jurist is allowed to speak for himself, and the reader feels that he is really studying Roman law in the original, and not a fanciful representation of it."—*Athenæum*.

THE INSTITUTES OF JUSTINIAN, translated with Notes by J. T. ABDY, LL.D., and BRYAN WALKER, M.A., LL.D. Crown 8vo. 16*s*.

"We welcome here a valuable contribution to the study of jurisprudence. The text of the *Institutes* is occasionally perplexing, even to practised scholars, whose knowledge of classical models does not always avail them in dealing with the technicalities of legal phraseology. Nor can the ordinary dictionaries be expected to furnish all the help that is wanted. This translation will then be of great use. To the ordinary student, whose attention is distracted from the subject-matter by the difficulty of struggling through the language in which it is contained, it will be almost indispensable."—*Spectator*.

"The notes are learned and carefully compiled, and this edition will be found useful to students."—*Law Times*.

SELECTED TITLES FROM THE DIGEST, annotated by B. WALKER, M.A., LL.D. Part I. Mandati vel Contra. Digest XVII. I. Crown 8vo. 5*s*.

"This small volume is published as an experiment. The author proposes to publish an annotated edition and translation of several books of the Digest if this one is received with favour. We are pleased to be able to say that Mr Walker deserves credit for the way in which he has performed the task undertaken. The translation, as might be expected, is scholarly." —*Law Times*.

—— Part II. De Adquirendo rerum dominio and De Adquirenda vel amittenda possessione. Digest XLI. I and II. Crown 8vo. 6*s*.

—— Part III. De Condictionibus. Digest XII. I and 4—7 and Digest XIII. I—3. Crown 8vo. 6*s*.

GROTIUS DE JURE BELLI ET PACIS, with the Notes of Barbeyrac and others; accompanied by an abridged Translation of the Text, by W. WHEWELL, D.D. late Master of Trinity College. 3 Vols. Demy 8vo. 12*s*. The translation separate, 6*s*.

HISTORY.

LIFE AND TIMES OF STEIN, OR GERMANY AND PRUSSIA IN THE NAPOLEONIC AGE, by J. R. SEELEY, M.A., Regius Professor of Modern History in the University of Cambridge, with Portraits and Maps. 3 Vols. Demy 8vo. 30*s*.

"DR BUSCH'S volume has made people think and talk even more than usual of Prince Bismarck, and Professor Seeley's very learned work on Stein will turn attention to an earlier and an almost equally eminent German statesman. It has been the good fortune of Prince Bismarck to help to raise Prussia to a position which she had never before attained, and to complete the work of German unification. The frustrated labours of Stein in the same field were also very great, and well worthy to be taken into account. He was one, perhaps the chief, of the illustrious group of strangers who came to the rescue of Prussia in her darkest hour, about the time of the inglorious Peace of Tilsit, and who laboured to put life and order into her dispirited army, her impoverished finances, and her inefficient Civil Service. Stein strove, too,—no man more,—for the cause of unification when it seemed almost folly to hope for success. Englishmen will feel very pardonable pride at seeing one of their countrymen undertake to write the history of a period from the investigation of which even laborious Germans are apt to shrink."—*Times*.

"In a notice of this kind scant justice can be done to a work like the one before us; no short *résumé* can give even the most meagre notion of the contents of these volumes, which contain no page that is superfluous, and none that is uninteresting.... To understand the Germany of to-day one must study the Germany of many yesterdays, and now that study has been made easy by this work, to which no one can hesitate to assign a very high place among those recent histories which have aimed at original research."—*Athenæum*.

"We congratulate Cambridge and her Professor of History on the appearance of such a noteworthy production. And we may add that it is something upon which we may congratulate England that on the especial field of the Germans, history, on the history of their own country, by the use of their own literary weapons, an Englishman has produced a history of Germany in the Napoleonic age far superior to any that exists in German."—*Examiner*.

THE DESPATCHES OF EARL GOWER, English Ambassador at the court of Versailles from June 1790 to August 1792, to which are added the Despatches of Mr Lindsay and Mr Munro, and the Diary of Lord Palmerston in France during July and August 1791. Edited by OSCAR BROWNING, M.A., Fellow of King's College, Cambridge. Demy 8vo. 15*s*.

THE GROWTH OF ENGLISH INDUSTRY AND COMMERCE. By W. CUNNINGHAM, B.D., late Deputy to the Knightbridge Professor in the University of Cambridge. With Maps and Charts. Crown 8vo. 12*s*.

"Mr Cunningham is not likely to disappoint any readers except such as begin by mistaking the character of his book. He does not promise, and does not give, an account of the dimensions to which English industry and commerce have grown. It is with the process of growth that he is concerned; and this process he traces with the philosophical insight which distinguishes between what is important and what is trivial."—*Guardian*.

CHRONOLOGICAL TABLES OF GREEK HISTORY. Accompanied by a short narrative of events, with references to the sources of information and extracts from the ancient authorities, by CARL PETER. Translated from the German by G. CHAWNER, M.A., Fellow of King's College, Cambridge. Demy 4to. 10*s*.

CHRONOLOGICAL TABLES OF ROMAN HISTORY. By the same. [*Preparing*.

KINSHIP AND MARRIAGE IN EARLY ARABIA, by W. ROBERTSON SMITH, M.A., LL.D., Fellow of Christ's College and University Librarian. Crown 8vo. 7*s*. 6*d*.

"It would be superfluous to praise a book so learned and masterly as Professor Robertson Smith's; it is enough to say that no student of early history can afford to be without *Kinship in Early Arabia*."—*Nature*.

"It is clearly and vividly written, full of curious and picturesque material, and incidentally throws light, not merely on the social history of Arabia, but on the earlier passages of Old Testament history.... We must be grateful to him for so valuable a contribution to the early history of social organisation."—*Scotsman*.

TRAVELS IN NORTHERN ARABIA IN 1876 AND 1877. By CHARLES M. DOUGHTY, of Gonville and Caius College. With Illustrations. Demy 8vo. [*In the Press.*

HISTORY OF NEPAL, translated by MUNSHĪ SHEW SHUNKER SINGH and PANDIT SHRĪ GUNĀNAND; edited with an Introductory Sketch of the Country and People by Dr D. WRIGHT, late Residency Surgeon at Kāthmāndū, and with facsimiles of native drawings, and portraits of Sir JUNG BAHĀDUR, the KING OF NEPĀL, &c. Super-royal 8vo. 10*s.* 6*d.*

"The Cambridge University Press have done well in publishing this work. Such translations are valuable not only to the historian but also to the ethnologist; . . . Dr Wright's Introduction is based on personal inquiry and observation, is written intelligently and candidly, and adds much to the value of the volume"—*Nature.*

A JOURNEY OF LITERARY AND ARCHÆOLOGICAL RESEARCH IN NEPAL AND NORTHERN INDIA, during the Winter of 1884-5. By CECIL BENDALL, M.A., Fellow of Gonville and Caius College, Cambridge; Professor of Sanskrit in University College, London. Demy 8vo. 10*s.*

THE UNIVERSITY OF CAMBRIDGE FROM THE EARLIEST TIMES TO THE ROYAL INJUNCTIONS OF 1535, by J. B. MULLINGER, M.A., Lecturer on History and Librarian to St John's College. Part I. Demy 8vo. (734 pp.), 12*s.*

Part II. From the Royal Injunctions of 1535 to the Accession of Charles the First. Demy 8vo. 18*s.*

"That Mr Mullinger's work should admit of being regarded as a continuous narrative, in which character it has no predecessors worth mentioning, is one of the many advantages it possesses over annalistic compilations, even so valuable as Cooper's, as well as over *Athenae.*"—Prof. A. W. Ward in the *Academy.*

"Mr Mullinger's narrative omits nothing which is required by the fullest interpretation of his subject. He shews in the statutes of the Colleges, the internal organization of the University, its connection with national problems, its studies, its social life, and the activity of its leading members. All this he combines in a form which is eminently readable."—PROF. CREIGHTON in *Cont. Review.*

"Mr Mullinger has succeeded perfectly in presenting the earnest and thoughtful student with a thorough and trustworthy history."—*Guardian.*

"The entire work is a model of accurate and industrious scholarship. The same qualities that distinguished the earlier volume are again visible, and the whole is still conspicuous for minuteness and fidelity of workmanship and breadth and toleration of view."—*Notes and Queries.*

"Mr Mullinger displays an admirable thoroughness in his work. Nothing could be more exhaustive and conscientious than his method: and his style...is picturesque and elevated."—*Times.*

HISTORY OF THE COLLEGE OF ST JOHN THE EVANGELIST, by THOMAS BAKER, B.D., Ejected Fellow. Edited by JOHN E. B. MAYOR, M.A. Two Vols. Demy 8vo. 24*s.*

"To antiquaries the book will be a source of almost inexhaustible amusement, by historians it will be found a work of considerable service on questions respecting our social progress in past times; and the care and thoroughness with which Mr Mayor has discharged his editorial functions are creditable to his learning and industry."—*Athenæum.*

"The work displays very wide reading, and it will be of great use to members of the college and of the university, and, perhaps, of still greater use to students of English history, ecclesiastical, political, social, literary and academical, who have hitherto had to be content with 'Dyer.'"—*Academy.*

SCHOLAE ACADEMICAE: some Account of the Studies at the English Universities in the Eighteenth Century. By CHRISTOPHER WORDSWORTH, M.A., Fellow of Peterhouse. Demy 8vo. 10*s.* 6*d.*

"Mr Wordsworth has collected a great quantity of minute and curious information about the working of Cambridge institutions in the last century, with an occasional comparison of the corresponding state of things at Oxford. . . . To a great extent it is purely a book of reference, and as such it will be of permanent value for the historical knowledge of English education and learning."—*Saturday Review.*

"Of the whole volume it may be said that it is a genuine service rendered to the study of University history, and that the habits of thought of any writer educated at either seat of learning in the last century will, in many cases, be far better understood after a consideration of the materials here collected."—*Academy.*

THE ARCHITECTURAL HISTORY OF THE UNIVERSITY OF CAMBRIDGE AND OF THE COLLEGES OF CAMBRIDGE AND ETON, by the late ROBERT WILLIS, M.A. F.R.S., Jacksonian Professor in the University of Cambridge. Edited with large Additions and brought up to the present time by JOHN WILLIS CLARK, M.A., formerly Fellow of Trinity College, Cambridge. Four Vols. Super Royal 8vo. £6. 6*s.*

Also a limited Edition of the same, consisting of 120 numbered Copies only, large paper Quarto; the woodcuts and steel engravings mounted on India paper; price Twenty-five Guineas **net** each set.

MISCELLANEOUS.

A CATALOGUE OF ANCIENT MARBLES IN GREAT BRITAIN, by Prof. ADOLF MICHAELIS. Translated by C. A. M. FENNELL, Litt. D., late Fellow of Jesus College. Royal 8vo. Roxburgh (Morocco back), £2. 2*s.*

"The object of the present work of Michaelis is to describe and make known the vast treasures of ancient sculpture now accumulated in the galleries of Great Britain, the extent and value of which are scarcely appreciated, and chiefly so because there has hitherto been little accessible information about them. To the loving labours of a learned German the owners of art treasures in England are for the second time indebted for a full description of their rich possessions. Waagen gave to the private collections of pictures the advantage of his inspection and cultivated acquaintance with art, and now Michaelis performs the same office for the still less known private hoards of antique sculptures for which our country is so remarkable. The book is beautifully executed, and with its few handsome plates, and excellent indexes, does much credit to the Cambridge Press. It has not been printed in German, but appears for the first time in the English translation. All lovers of true art and of good work should be grateful to the Syndics of the University Press for the liberal facilities afforded by them towards the production of this important volume by Professor Michaelis." —*Saturday Review.*

"Professor Michaelis has achieved so high a fame as an authority in classical archæology that it seems unnecessary to say how good a book this is."—*The Antiquary.*

RHODES IN ANCIENT TIMES. By CECIL TORR, M.A. With six plates. Demy 8vo. 10*s.* 6*d.*

RHODES IN MODERN TIMES. By the same Author. [*Nearly ready.*

CHAPTERS ON ENGLISH METRE. By Rev. JOSEPH B. MAYOR, M.A. Demy 8vo. 7*s.* 6*d.*

THE WOODCUTTERS OF THE NETHERLANDS during the last quarter of the Fifteenth Century. In three parts. I. History of the Woodcutters. II. Catalogue of their Woodcuts. III. List of the Books containing Woodcuts. By WILLIAM MARTIN CONWAY. Demy 8vo. 10*s.* 6*d.*

A GRAMMAR OF THE IRISH LANGUAGE. By Prof. WINDISCH. Translated by Dr NORMAN MOORE. Crown 8vo. 7*s.* 6*d.*

LECTURES ON TEACHING, delivered in the University of Cambridge in the Lent Term, 1880. By J. G. FITCH, M.A., LL.D. Her Majesty's Inspector of Training Colleges. Cr. 8vo. New Edit. 5*s.*

"As principal of a training college and as a Government inspector of schools, Mr Fitch has got at his fingers' ends the working of primary education, while as assistant commissioner to the late Endowed Schools Commission he has seen something of the machinery of our higher schools . . . Mr Fitch's book covers so wide a field and touches on so many burning questions that we must be content to recommend it as the best existing *vade mecum* for the teacher." —*Pall Mall Gazette.*

"Therefore, without reviewing the book for the second time, we are glad to avail ourselves of the opportunity of calling attention to the re-issue of the volume in the five-shilling form, bringing it within the reach of the rank and file of the profession. We cannot let the occasion pass without making special reference to the excellent section on 'punishments' in the lecture on 'Discipline.'"—*School Board Chronicle.*

For other books on Education, see Pitt Press Series, pp. 30, 31.

FROM SHAKESPEARE TO POPE: an Inquiry into the causes and phenomena of the rise of Classical Poetry in England. By EDMUND GOSSE, M.A., Clark Lecturer in English Literature at Trinity College, Cambridge. Crown 8vo. 6*s*.

THE LITERATURE OF THE FRENCH RENAISSANCE. An Introductory Essay. By A. A. TILLEY, M.A., Fellow and Tutor of King's College, Cambridge. Crown 8vo. 6*s*.

STUDIES IN THE LITERARY RELATIONS OF ENGLAND WITH GERMANY IN THE SIXTEENTH CENTURY. By C. H. HERFORD, M.A. Crown 8vo. 9*s*.

CATALOGUE OF THE HEBREW MANUSCRIPTS preserved in the University Library, Cambridge. By Dr S. M. SCHILLER-SZINESSY. Volume I. containing Section I. *The Holy Scriptures;* Section II. *Commentaries on the Bible*. Demy 8vo. 9*s*.
Volume II. *In the Press*.

A CATALOGUE OF THE MANUSCRIPTS preserved in the Library of the University of Cambridge. Demy 8vo. 5 Vols. 10*s*. each. INDEX TO THE CATALOGUE. Demy 8vo. 10*s*.

A CATALOGUE OF ADVERSARIA and printed books containing MS. notes, preserved in the Library of the University of Cambridge. 3*s*. 6*d*.

THE ILLUMINATED MANUSCRIPTS IN THE LIBRARY OF THE FITZWILLIAM MUSEUM, Catalogued with Descriptions, and an Introduction, by W. G. SEARLE, M.A., late Fellow of Queens' College, Cambridge Demy 8vo. 7*s*. 6*d*

A CHRONOLOGICAL LIST OF THE GRACES, Documents, and other Papers in the University Registry which concern the University Library. Demy 8vo. 2*s*. 6*d*.

CATALOGUS BIBLIOTHECÆ BURCKHARDTIANÆ. Demy 4to. 5*s*.

GRADUATI CANTABRIGIENSES: SIVE CATALOGUS exhibens nomina eorum quos ab Anno Academico Admissionum MDCCC usque ad octavum diem Octobris MDCCCLXXXIV gradu quocunque ornavit Academia Cantabrigiensis, e libris subscriptionum desumptus. Cura HENRICI RICHARDS LUARD S. T. P. Coll. SS. Trin. Socii atque Academiæ Registrarii. Demy 8vo. 12*s*. 6*d*.

STATUTES OF THE UNIVERSITY OF CAMBRIDGE and for the Colleges therein, made published and approved (1878—1882) under the Universities of Oxford and Cambridge Act, 1877. With an Appendix. Demy 8vo. 16*s*.

STATUTES OF THE UNIVERSITY OF CAMBRIDGE. With some Acts of Parliament relating to the University. Demy 8vo. 3*s*. 6*d*.

ORDINANCES OF THE UNIVERSITY OF CAMBRIDGE. Demy 8vo., cloth. 7*s*. 6*d*.

TRUSTS, STATUTES AND DIRECTIONS affecting (1) The Professorships of the University. (2) The Scholarships and Prizes. (3) Other Gifts and Endowments. Demy 8vo. 5*s*.

COMPENDIUM OF UNIVERSITY REGULATIONS, for the use of persons in Statu Pupillari. Demy 8vo. 6*d*.

The Cambridge Bible for Schools and Colleges.

GENERAL EDITOR: THE VERY REVEREND J. J. S. PEROWNE, D.D., DEAN OF PETERBOROUGH.

"It is difficult to commend too highly this excellent series, the volumes of which are now becoming numerous."—*Guardian*.

"The modesty of the general title of this series has, we believe, led many to misunderstand its character and underrate its value. The books are well suited for study in the upper forms of our best schools, but not the less are they adapted to the wants of all Bible students who are not specialists. We doubt, indeed, whether any of the numerous popular commentaries recently issued in this country will be found more serviceable for general use."—*Academy*.

"One of the most popular and useful literary enterprises of the nineteenth century."—*Baptist Magazine*.

"Of great value. The whole series of comments for schools is highly esteemed by students capable of forming a judgment. The books are scholarly without being pretentious: information is so given as to be easily understood."—*Sword and Trowel*.

The Very Reverend J. J. S. PEROWNE, D.D., Dean of Peterborough, has undertaken the general editorial supervision of the work, assisted by a staff of eminent coadjutors. Some of the books have been already edited or undertaken by the following gentlemen:

Rev. A. CARR, M.A., *late Assistant Master at Wellington College.*
Rev. T. K. CHEYNE, M.A., D.D., *late Fellow of Balliol College, Oxford.*
Rev. S. COX, *Nottingham.*
Rev. A. B. DAVIDSON, D.D., *Professor of Hebrew, Edinburgh.*
The Ven. F. W. FARRAR, D.D., *Archdeacon of Westminster.*
Rev. C. D. GINSBURG, LL.D.
Rev. A. E. HUMPHREYS, M.A., *late Fellow of Trinity College, Cambridge.*
Rev. A. F. KIRKPATRICK, M.A., *Fellow of Trinity College, Regius Professor of Hebrew.*
Rev. J. J. LIAS, M.A., *late Professor at St David's College, Lampeter.*
Rev. J. R. LUMBY, D.D., *Norrisian Professor of Divinity.*
Rev. G. F. MACLEAR, D.D., *Warden of St Augustine's College, Canterbury.*
Rev. H. C. G. MOULE, M.A., *late Fellow of Trinity College, Principal of Ridley Hall, Cambridge.*
Rev. W. F. MOULTON, D.D., *Head Master of the Leys School, Cambridge.*
Rev. E. H. PEROWNE, D.D., *Master of Corpus Christi College, Cambridge.*
The Ven. T. T. PEROWNE, B.D., *Archdeacon of Norwich.*
Rev. A. PLUMMER, M.A., D.D., *Master of University College, Durham.*
The Very Rev. E. H. PLUMPTRE, D.D., *Dean of Wells.*
Rev. W. SIMCOX, M.A., *Rector of Weyhill, Hants.*
W. ROBERTSON SMITH, M.A., *Fellow of Christ's College, and University Librarian.*
Rev. H. D. M. SPENCE, M.A., *Dean of Gloucester.*
Rev. A. W. STREANE, M.A., *Fellow of Corpus Christi College, Cambridge.*

London: C. J. CLAY & SONS, Cambridge University Press Warehouse, Ave Maria Lane.

THE CAMBRIDGE BIBLE FOR SCHOOLS & COLLEGES.

Continued.

Now Ready. Cloth, Extra Fcap. 8vo.

THE BOOK OF JOSHUA. By the Rev. G. F. MACLEAR, D.D. With 2 Maps. 2*s*. 6*d*.

THE BOOK OF JUDGES. By the Rev. J. J. LIAS, M.A. With Map. 3*s*. 6*d*.

THE FIRST BOOK OF SAMUEL. By the Rev. Professor KIRKPATRICK, M.A. With Map. 3*s*. 6*d*.

THE SECOND BOOK OF SAMUEL. By the Rev. Professor KIRKPATRICK, M.A. With 2 Maps. 3*s*. 6*d*.

THE FIRST BOOK OF KINGS. By the Rev. Prof. LUMBY, D.D. 3*s*. 6*d*.

THE BOOK OF JOB. By the Rev. A. B. DAVIDSON, D.D. 5*s*.

THE BOOK OF ECCLESIASTES. By the Very Rev. E. H. PLUMPTRE, D.D., Dean of Wells. 5*s*.

THE BOOK OF JEREMIAH. By the Rev. A. W. STREANE, M.A. With Map. 4*s*. 6*d*.

THE BOOK OF HOSEA. By Rev. T. K. CHEYNE, M.A., D.D. 3*s*.

THE BOOKS OF OBADIAH AND JONAH. By Archdeacon PEROWNE. 2*s*. 6*d*.

THE BOOK OF MICAH. By Rev. T. K. CHEYNE, D.D. 1*s*. 6*d*.

THE BOOKS OF HAGGAI AND ZECHARIAH. By Archdeacon PEROWNE. 3*s*.

THE GOSPEL ACCORDING TO ST MATTHEW. By the Rev. A. CARR, M.A. With 2 Maps. 2*s*. 6*d*.

THE GOSPEL ACCORDING TO ST MARK. By the Rev. G. F. MACLEAR, D.D. With 4 Maps. 2*s*. 6*d*.

THE GOSPEL ACCORDING TO ST LUKE. By Archdeacon F. W. FARRAR. With 4 Maps. 4*s*. 6*d*.

THE GOSPEL ACCORDING TO ST JOHN. By the Rev. A. PLUMMER, M.A., D.D. With 4 Maps. 4*s*. 6*d*.

THE ACTS OF THE APOSTLES. By the Rev. Professor LUMBY, D.D. With 4 Maps. 4*s*. 6*d*.

THE EPISTLE TO THE ROMANS. By the Rev. H. C. G. MOULE, M.A. 3*s*. 6*d*.

THE FIRST EPISTLE TO THE CORINTHIANS. By the Rev. J. J. LIAS, M.A. With a Map and Plan. 2*s*.

THE SECOND EPISTLE TO THE CORINTHIANS. By the Rev. J. J. LIAS, M.A. 2*s*.

THE EPISTLE TO THE EPHESIANS. By the Rev. H. C. G. MOULE, M.A. 2*s*. 6*d*.

THE EPISTLE TO THE HEBREWS. By Arch. FARRAR. 3*s*. 6*d*.

THE GENERAL EPISTLE OF ST JAMES. By the Very Rev. E. H. PLUMPTRE, D.D., Dean of Wells. 1*s*. 6*d*.

THE EPISTLES OF ST PETER AND ST JUDE. By the same Editor. 2*s*. 6*d*.

THE EPISTLES OF ST JOHN. By the Rev. A. PLUMMER, M.A., D.D. 3*s*. 6*d*.

THE CAMBRIDGE BIBLE FOR SCHOOLS & COLLEGES.

Continued.

Preparing.

THE BOOK OF GENESIS. By the Very Rev. the DEAN OF PETERBOROUGH.

THE BOOKS OF EXODUS, NUMBERS AND DEUTERONOMY. By the Rev. C. D. GINSBURG, LL.D.

THE SECOND BOOK OF KINGS. By the Rev. Prof. LUMBY, D.D.

THE BOOK OF PSALMS. By the Rev. Prof. KIRKPATRICK, M.A.

THE BOOK OF ISAIAH. By W. ROBERTSON SMITH, M.A.

THE BOOK OF EZEKIEL. By the Rev. A. B. DAVIDSON, D.D.

THE EPISTLE TO THE GALATIANS. By the Rev. E. H. PEROWNE, D.D.

THE EPISTLES TO THE PHILIPPIANS, COLOSSIANS AND PHILEMON. By the Rev. H. C. G. MOULE, M.A.

THE EPISTLES TO THE THESSALONIANS. By the Rev. W. F. MOULTON, D.D.

THE BOOK OF REVELATION. By the Rev. W. SIMCOX, M.A.

THE CAMBRIDGE GREEK TESTAMENT

FOR SCHOOLS AND COLLEGES,

with a Revised Text, based on the most recent critical authorities, and English Notes, prepared under the direction of the General Editor,

THE VERY REVEREND J. J. S. PEROWNE, D.D.

Now Ready.

THE GOSPEL ACCORDING TO ST MATTHEW. By the Rev. A. CARR, M.A. With 4 Maps. 4*s.* 6*d.*

"Copious illustrations, gathered from a great variety of sources, make his notes a very valuable aid to the student. They are indeed remarkably interesting, while all explanations on meanings, applications, and the like are distinguished by their lucidity and good sense."—*Pall Mall Gazette.*

THE GOSPEL ACCORDING TO ST MARK. By the Rev. G. F. MACLEAR, D.D. With 3 Maps. 4*s.* 6*d.*

"The Cambridge Greek Testament, of which Dr Maclear's edition of the Gospel according to St Mark is a volume, certainly supplies a want. Without pretending to compete with the leading commentaries, or to embody very much original research, it forms a most satisfactory introduction to the study of the New Testament in the original . . . Dr Maclear's introduction contains all that is known of St Mark's life, with references to passages in the New Testament in which he is mentioned; an account of the circumstances in which the Gospel was composed, with an estimate of the influence of St Peter's teaching upon St Mark; an excellent sketch of the special characteristics of this Gospel; an analysis, and a chapter on the text of the New Testament generally The work is completed by three good maps."—*Saturday Review.*

THE GOSPEL ACCORDING TO ST LUKE. By Archdeacon FARRAR. With 4 Maps. 6*s.*

THE GOSPEL ACCORDING TO ST JOHN. By the Rev. A. PLUMMER, M.A., D.D. With 4 Maps. 6*s.*

"A valuable addition has also been made to 'The Cambridge Greek Testament for Schools,' Dr Plummer's notes on 'the Gospel according to St John' are scholarly, concise, and instructive, and embody the results of much thought and wide reading."—*Expositor.*

THE ACTS OF THE APOSTLES. By the Rev. Prof. LUMBY, D.D., with 4 Maps. 6*s.*

THE FIRST EPISTLE TO THE CORINTHIANS. By the Rev. J. J. LIAS, M.A. 3*s.*

THE EPISTLE TO THE HEBREWS. By Archdeacon FARRAR. [*In the Press.*

THE EPISTLES OF ST JOHN. By the Rev. A. PLUMMER, M.A., D.D. 4*s.*

THE PITT PRESS SERIES.

[*Copies of the Pitt Press Series may generally be obtained bound in two parts for Class use, the text and notes in separate volumes.*]

I. GREEK.

SOPHOCLES.—OEDIPUS TYRANNUS. School Edition, with Introduction and Commentary, by R. C. JEBB, Litt. D., LL.D., Professor of Greek in the University of Glasgow. 4*s.* 6*d.*

XENOPHON.—ANABASIS, BOOKS I. III. IV. and V. With a Map and English Notes by ALFRED PRETOR, M.A., Fellow of St Catharine's College, Cambridge. 2*s.* each.

"In Mr Pretor's edition of the Anabasis the text of Kühner has been followed in the main, while the exhaustive and admirable notes of the great German editor have been largely utilised. These notes deal with the minutest as well as the most important difficulties in construction, and all questions of history, antiquity, and geography are briefly but very effectually elucidated."—*The Examiner.*

"We welcome this addition to the other books of the *Anabasis* so ably edited by Mr Pretor. Although originally intended for the use of candidates at the university local examinations, yet this edition will be found adapted not only to meet the wants of the junior student, but even advanced scholars will find much in this work that will repay its perusal."—*The Schoolmaster.*

"Mr Pretor's 'Anabasis of Xenophon, Book IV.' displays a union of accurate Cambridge scholarship, with experience of what is required by learners gained in examining middle-class schools. The text is large and clearly printed, and the notes explain all difficulties. . . . Mr Pretor's notes seem to be all that could be wished as regards grammar, geography, and other matters."—*The Academy.*

BOOKS II. VI. and VII. By the same Editor. 2*s.* 6*d.* each.

"Another Greek text, designed it would seem for students preparing for the local examinations, is 'Xenophon's Anabasis,' Book II., with English Notes, by Alfred Pretor, M.A. The editor has exercised his usual discrimination in utilising the text and notes of Kuhner, with the occasional assistance of the best hints of Schneider, Vollbrecht and Macmichael on critical matters, and of Mr R. W. Taylor on points of history and geography. . . When Mr Pretor commits himself to Commentator's work, he is eminently helpful. . . Had we to introduce a young Greek scholar to Xenophon, we should esteem ourselves fortunate in having Pretor's text-book as our chart and guide."—*Contemporary Review.*

XENOPHON.—ANABASIS. By A. PRETOR, M.A., Text and Notes, complete in two Volumes. 7*s.* 6*d.*

XENOPHON.—AGESILAUS. The Text revised with Critical and Explanatory Notes, Introduction, Analysis, and Indices. By H. HAILSTONE, M.A., late Scholar of Peterhouse. 2*s.* 6*d.*

XENOPHON.—CYROPAEDEIA. BOOKS I. II. With Introduction, Notes and Map. By Rev. H. A. HOLDEN, M.A., LL.D. 2 vols. Vol. I. Text. Vol. II. Notes. 6*s.*

—— —— BOOKS III., IV. By the same Editor. [*In the Press.*

ARISTOPHANES—RANAE. With English Notes and Introduction by W. C. GREEN, M.A., late Assistant Master at Rugby School. 3*s.* 6*d.*

ARISTOPHANES—AVES. By the same Editor. *New Edition.* 3*s.* 6*d.*

"The notes to both plays are excellent. Much has been done in these two volumes to render the study of Aristophanes a real treat to a boy instead of a drudgery, by helping him to understand the fun and to express it in his mother tongue. —*The Examiner.*

ARISTOPHANES—PLUTUS. By the same Editor. 3*s.* 6*d.*

PLATONIS APOLOGIA SOCRATIS. With Introduction, Notes and Appendices by J. ADAM, B.A., Fellow and Classical Lecturer of Emmanuel College. 3*s.* 6*d.*

HERODOTUS, BOOK VIII., CHAPS. 1—90. Edited with Notes and Introduction by E. S. SHUCKBURGH, M.A., late Fellow of Emmanuel College. [*Immediately.*

EURIPIDES. HERCULES FURENS. With Introductions, Notes and Analysis. By A. GRAY, M.A., Fellow of Jesus College, and J. T. HUTCHINSON, M.A., Christ's College. New Edition, with additions. 2s.

"Messrs Hutchinson and Gray have produced a careful and useful edition."—*Saturday Review.*

EURIPIDES. HERACLEIDÆ. With Introduction and Critical Notes by E. A. BECK, M.A., Fellow of Trinity Hall. 3s. 6d.

LUCIANI SOMNIUM CHARON PISCATOR ET DE LUCTU, with English Notes by W. E. HEITLAND, M.A., Fellow of St John's College, Cambridge. New Edition, with Appendix. 3s. 6d.

PLUTARCH'S LIVES OF THE GRACCHI. With Introduction, Notes and Lexicon by Rev. HUBERT A. HOLDEN, M.A., LL.D., Examiner in Greek to the University of London. 6s.

PLUTARCH'S LIFE OF SULLA. With Introduction, Notes, and Lexicon. By the Rev. HUBERT A. HOLDEN, M.A., LL.D. 6s.

OUTLINES OF THE PHILOSOPHY OF ARISTOTLE. Edited by E. WALLACE, M.A. (See p. 31.)

II. LATIN.

M. T. CICERONIS DE AMICITIA. Edited by J. S. REID, Litt. D., Fellow and Tutor of Gonville and Caius College. New Edition, with Additions. 3s. 6d.

"Mr Reid has decidedly attained his aim, namely, 'a thorough examination of the Latinity of the dialogue.' . . . . The revision of the text is most valuable, and comprehends sundry acute corrections. . . . This volume, like Mr Reid's other editions, is a solid gain to the scholarship of the country."—*Athenæum.*

"A more distinct gain to scholarship is Mr Reid's able and thorough edition of the *De Amicitiâ* of Cicero, a work of which, whether we regard the exhaustive introduction or the instructive and most suggestive commentary, it would be difficult to speak too highly. . . . When we come to the commentary, we are only amazed by its fulness in proportion to its bulk. Nothing is overlooked which can tend to enlarge the learner's general knowledge of Ciceronian Latin or to elucidate the text."—*Saturday Review.*

M. T. CICERONIS CATO MAJOR DE SENECTUTE. Edited by J. S. REID, Litt. D. Revised Edition. 3s. 6d.

"The notes are excellent and scholarlike, adapted for the upper forms of public schools, and likely to be useful even to more advanced students."—*Guardian.*

M. T. CICERONIS ORATIO PRO ARCHIA POETA. Edited by J. S. REID, Litt. D. Revised Edition. 2s.

"It is an admirable specimen of careful editing. An Introduction tells us everything we could wish to know about Archias, about Cicero's connexion with him, about the merits of the trial, and the genuineness of the speech. The text is well and carefully printed. The notes are clear and scholar-like. . . . No boy can master this little volume without feeling that he has advanced a long step in scholarship."—*The Academy.*

M. T. CICERONIS PRO L. CORNELIO BALBO ORATIO. Edited by J. S. REID, Litt. D. 1s. 6d.

"We are bound to recognize the pains devoted in the annotation of these two orations to the minute and thorough study of their Latinity, both in the ordinary notes and in the textual appendices."—*Saturday Review.*

M. T. CICERONIS PRO P. CORNELIO SULLA ORATIO. Edited by J. S. REID, Litt. D. 3s. 6d.

"Mr Reid is so well known to scholars as a commentator on Cicero that a new work from him scarcely needs any commendation of ours. His edition of the speech *Pro Sulla* is fully equal in merit to the volumes which he has already published . . . It would be difficult to speak too highly of the notes. There could be no better way of gaining an insight into the characteristics of Cicero's style and the Latinity of his period than by making a careful study of this speech with the aid of Mr Reid's commentary . . . Mr Reid's intimate knowledge of the minutest details of scholarship enables him to detect and explain the slightest points of distinction between the usages of different authors and different periods . . . The notes are followed by a valuable appendix on the text, and another on points of orthography; an excellent index brings the work to a close."—*Saturday Review.*

M. T. CICERONIS PRO CN. PLANCIO ORATIO. Edited by H. A. HOLDEN, LL.D., Examiner in Greek to the University of London. Second Edition. 4*s.* 6*d.*

"As a book for students this edition can have few rivals. It is enriched by an excellent introduction and a chronological table of the principal events of the life of Cicero; while in its appendix, and in the notes on the text which are added, there is much of the greatest value. The volume is neatly got up, and is in every way commendable."—*The Scotsman.*

M. T. CICERONIS IN Q. CAECILIUM DIVINATIO ET IN C. VERREM ACTIO PRIMA. With Introduction and Notes by W. E. HEITLAND, M.A., and HERBERT COWIE, M.A., Fellows of St John's College, Cambridge. 3*s.*

M. T. CICERONIS ORATIO PRO L. MURENA, with English Introduction and Notes. By W. E. HEITLAND, M.A., Fellow and Classical Lecturer of St John's College, Cambridge. **Second Edition, carefully revised.** 3*s.*

"Those students are to be deemed fortunate who have to read Cicero's lively and brilliant oration for L. Murena with Mr Heitland's handy edition, which may be pronounced 'four-square' in point of equipment, and which has, not without good reason, attained the honours of a second edition."—*Saturday Review.*

M. T. CICERONIS IN GAIUM VERREM ACTIO PRIMA. With Introduction and Notes. By H. COWIE, M.A., Fellow of St John's College, Cambridge. 1*s.* 6*d.*

M. T. CICERONIS ORATIO PRO T. A. MILONE, with a Translation of Asconius' Introduction, Marginal Analysis and English Notes. Edited by the Rev. JOHN SMYTH PURTON, B.D., late President and Tutor of St Catharine's College. 2*s.* 6*d.*

"The editorial work is excellently done."—*The Academy.*

M. T. CICERONIS SOMNIUM SCIPIONIS. With Introduction and Notes. By W. D. PEARMAN, M.A., Head Master of Potsdam School, Jamaica. 2*s.*

M. TULLI CICERONIS ORATIO PHILIPPICA SECUNDA. With Introduction and Notes by A. G. PESKETT, M.A., Fellow of Magdalene College. 3*s.* 6*d.*

P. OVIDII NASONIS FASTORUM LIBER VI. With a Plan of Rome and Notes by A. SIDGWICK, M.A., Tutor of Corpus Christi College, Oxford. 1*s.* 6*d.*

"Mr Sidgwick's editing of the Sixth Book of Ovid's *Fasti* furnishes a careful and serviceable volume for average students. It eschews 'construes' which supersede the use of the dictionary, but gives full explanation of grammatical usages and historical and mythical allusions, besides illustrating peculiarities of style, true and false derivations, and the more remarkable variations of the text."—*Saturday Review.*

"It is eminently good and useful. . . . The Introduction is singularly clear on the astronomy of Ovid, which is properly shown to be ignorant and confused; there is an excellent little map of Rome, giving just the places mentioned in the text and no more; the notes are evidently written by a practical schoolmaster."—*The Academy.*

M. ANNAEI LUCANI PHARSALIAE LIBER PRIMUS, edited with English Introduction and Notes by W. E. HEITLAND, M.A. and C. E. HASKINS, M.A., Fellows and Lecturers of St John's College, Cambridge. 1*s.* 6*d.*

"A careful and scholarlike production."—*Times.*

"In nice parallels of Lucan from Latin poets and from Shakspeare, Mr Haskins and Mr Heitland deserve praise."—*Saturday Review.*

GAI IULI CAESARIS DE BELLO GALLICO COMMENT. I. II. III. With Maps and English Notes by A. G. PESKETT, M.A., Fellow of Magdalene College, Cambridge. 3*s*.

"In an unusually succinct introduction he gives all the preliminary and collateral information that is likely to be useful to a young student; and, wherever we have examined his notes, we have found them eminently practical and satisfying. . . The book may well be recommended for careful study in school or college."—*Saturday Review.*

"The notes are scholarly, short, and a real help to the most elementary beginners in Latin prose."—*The Examiner.*

—— COMMENT. IV. AND V. AND COMMENT. VII. by the same Editor. 2*s*. each.

—— COMMENT. VI. AND COMMENT. VIII. by the same Editor. 1*s*. 6*d*. each.

P. VERGILI MARONIS AENEIDOS LIBRI I., II., III., IV., V., VI., VII., VIII., IX., X., XI., XII. Edited with Notes by A. SIDGWICK, M.A., Tutor of Corpus Christi College, Oxford. 1*s*. 6*d*. each.

"Much more attention is given to the literary aspect of the poem than is usually paid to it in editions intended for the use of beginners. The introduction points out the distinction between primitive and literary epics, explains the purpose of the poem, and gives an outline of the story." —*Saturday Review.*

"Mr Arthur Sidgwick's 'Vergil, Aeneid, Book XII.' is worthy of his reputation, and is distinguished by the same acuteness and accuracy of knowledge, appreciation of a boy's difficulties and ingenuity and resource in meeting them, which we have on other occasions had reason to praise in these pages."—*The Academy.*

"As masterly in its clearly divided preface and appendices as in the sound and independent character of its annotations. . . . There is a great deal more in the notes than mere compilation and suggestion. . . . No difficulty is left unnoticed or unhandled."—*Saturday Review.*

BOOKS IX. X. in one volume. 3*s*.

BOOKS X., XI., XII. in one volume. 3*s*. 6*d*.

P. VERGILI MARONIS GEORGICON LIBRI I. II. By the same Editor. 2*s*.

—— —— Libri III. IV. By the same Editor. 2*s*.

QUINTUS CURTIUS. A Portion of the History. (ALEXANDER IN INDIA.) By W. E. HEITLAND, M.A., Fellow and Lecturer of St John's College, Cambridge, and T. E. RAVEN, B.A., Assistant Master in Sherborne School. 3*s*. 6*d*.

"Equally commendable as a genuine addition to the existing stock of school-books is *Alexander in India*, a compilation from the eighth and ninth books of Q. Curtius, edited for the Pitt Press by Messrs Heitland and Raven. . . . The work of Curtius has merits of its own, which, in former generations, made it a favourite with English scholars, and which still make it a popular text-book in Continental schools. . . . . . The reputation of Mr Heitland is a sufficient guarantee for the scholarship of the notes, which are ample without being excessive, and the book is well furnished with all that is needful in the nature of maps, indices, and appendices." —*Academy.*

BEDA'S ECCLESIASTICAL HISTORY, BOOKS III., IV., the Text from the very ancient MS. in the Cambridge University Library, collated with six other MSS. Edited, with a life from the German of EBERT, and with Notes, &c. by J. E. B. MAYOR, M.A., Professor of Latin, and J. R. LUMBY, D.D., Norrisian Professor of Divinity. Revised edition. 7*s*. 6*d*.

"To young students of English History the illustrative notes will be of great service, while the study of the texts will be a good introduction to Mediæval Latin."—*The Nonconformist.*

"In Bede's works Englishmen can go back to *origines* of their history, unequalled for form and matter by any modern European nation. Prof. Mayor has done good service in rendering a part of Bede's greatest work accessible to those who can read Latin with ease. He has adorned this edition of the third and fourth books of the 'Ecclesiastical History' with that amazing erudition for which he is unrivalled among Englishmen and rarely equalled by Germans. And however interesting and valuable the text may be, we can certainly apply to his notes the expression, *La sauce vaut mieux que le poisson.* They are literally crammed with interesting information about early English life. For though ecclesiastical in name, Bede's history treats of all parts of the national life, since the Church had points of contact with all."—*Examiner.*

BOOKS I. and II. *In the Press.*

III. FRENCH.

LA CANNE DE JONC. By A. DE VIGNY. Edited with Notes by Rev. H. A. BULL, M.A., late Master at Wellington College. 2*s.*

BATAILLE DE DAMES. By SCRIBE and LEGOUVÉ. Edited by Rev. H. A. BULL, M.A. 2*s.*

JEANNE D'ARC by A. DE LAMARTINE. With a Map and Notes Historical and Philological and a Vocabulary by Rev. A. C. CLAPIN, M.A., St John's College, Cambridge, and Bachelier-ès-Lettres of the University of France. 2*s.*

LE BOURGEOIS GENTILHOMME, Comédie-Ballet en Cinq Actes. Par J.-B. POQUELIN DE MOLIÈRE (1670). With a life of Molière and Grammatical and Philological Notes. By the same Editor. 1*s.* 6*d.*

LA PICCIOLA. By X. B. SAINTINE. The Text, with Introduction, Notes and Map, by the same Editor. 2*s.*

LA GUERRE. By MM. ERCKMANN-CHATRIAN. With Map, Introduction and Commentary by the same Editor. 3*s.*

LAZARE HOCHE—PAR ÉMILE DE BONNECHOSE. With Three Maps, Introduction and Commentary, by C. COLBECK, M.A., late Fellow of Trinity College, Cambridge. 2*s.*

LE VERRE D'EAU. A Comedy, by SCRIBE. With a Biographical Memoir, and Grammatical, Literary and Historical Notes. By the same Editor. 2*s.*

"It may be national prejudice, but we consider this edition far superior to any of the series which hitherto have been edited exclusively by foreigners. Mr Colbeck seems better to understand the wants and difficulties of an English boy. The etymological notes especially are admirable. . . . The historical notes and introduction are a piece of thorough honest work."—*Journal of Education.*

HISTOIRE DU SIÈCLE DE LOUIS XIV PAR VOLTAIRE. Part I. Chaps. I.—XIII. Edited with Notes Philological and Historical, Biographical and Geographical Indices, etc. by GUSTAVE MASSON, B. A. Univ. Gallic., Assistant Master of Harrow School, and G. W. PROTHERO, M.A., Fellow and Tutor of King's College, Cambridge. 2*s.* 6*d.*

—— Part II. Chaps. XIV.—XXIV. With Three Maps of the Period. By the same Editors. 2*s.* 6*d.*

—— Part III. Chap. XXV. to the end. By the same Editors. 2*s.* 6*d.*

M. DARU, par M. C. A. SAINTE-BEUVE, (Causeries du Lundi, Vol. IX.). With Biographical Sketch of the Author, and Notes Philological and Historical. By GUSTAVE MASSON. 2*s.*

LA SUITE DU MENTEUR. A Comedy in Five Acts, by P. CORNEILLE. Edited with Fontenelle's Memoir of the Author, Voltaire's Critical Remarks, and Notes Philological and Historical. By GUSTAVE MASSON. 2*s.*

LA JEUNE SIBÉRIENNE. LE LÉPREUX DE LA CITÉ D'AOSTE. Tales by COUNT XAVIER DE MAISTRE. With Biographical Notice, Critical Appreciations, and Notes. By G. MASSON. 2*s.*

LE DIRECTOIRE. (Considérations sur la Révolution Française. Troisième et quatrième parties.) Par MADAME LA BARONNE DE STAËL-HOLSTEIN. With a Critical Notice of the Author, a Chronological Table, and Notes Historical and Philological, by G. MASSON, B.A., and G. W. PROTHERO, M.A. Revised and enlarged Edition. *2s.*

"Prussia under Frederick the Great, and France under the Directory, bring us face to face respectively with periods of history which it is right should be known thoroughly, and which are well treated in the Pitt Press volumes. The latter in particular, an extract from the world-known work of Madame de Staël on the French Revolution, is beyond all praise for the excellence both of its style and of its matter."—*Times.*

DIX ANNÉES D'EXIL. LIVRE II. CHAPITRES 1—8. Par MADAME LA BARONNE DE STAËL-HOLSTEIN. With a Biographical Sketch of the Author, a Selection of Poetical Fragments by Madame de Staël's Contemporaries, and Notes Historical and Philological. By GUSTAVE MASSON and G. W. PROTHERO, M.A. Revised and enlarged edition. *2s.*

FRÉDÉGONDE ET BRUNEHAUT. A Tragedy in Five Acts, by N. LEMERCIER. Edited with Notes, Genealogical and Chronological Tables, a Critical Introduction and a Biographical Notice. By GUSTAVE MASSON. *2s.*

LE VIEUX CÉLIBATAIRE. A Comedy, by COLLIN D'HARLEVILLE. With a Biographical Memoir, and Grammatical, Literary and Historical Notes. By the same Editor. *2s.*

LA METROMANIE, A Comedy, by PIRON, with a Biographical Memoir, and Grammatical, Literary and Historical Notes. By the same Editor. *2s.*

LASCARIS, OU LES GRECS DU XVE. SIÈCLE, Nouvelle Historique, par A. F. VILLEMAIN, with a Biographical Sketch of the Author, a Selection of Poems on Greece, and Notes Historical and Philological. By the same Editor. *2s.*

LETTRES SUR L'HISTOIRE DE FRANCE (XIII—XXIV.). Par AUGUSTIN THIERRY. By GUSTAVE MASSON, B.A. and G. W. PROTHERO, M.A. With Map. *2s. 6d.*

IV. GERMAN.

SELECTED FABLES. LESSING and GELLERT. Edited with Notes by KARL HERMANN BREUL, M.A., Lecturer in German at the University of Cambridge. *3s.*

DIE KARAVANE von WILHELM HAUFF. Edited with Notes by A. SCHLOTTMANN, Ph. D. *3s. 6d.*

CULTURGESCHICHTLICHE NOVELLEN, von W. H. RIEHL, with Grammatical, Philological, and Historical Notes, and a Complete Index, by H. J. WOLSTENHOLME, B.A. (Lond.). *4s. 6d.*

ERNST, HERZOG VON SCHWABEN. UHLAND. With Introduction and Notes. By H. J. WOLSTENHOLME, B.A. (Lond.), Lecturer in German at Newnham College, Cambridge. *3s. 6d.*

ZOPF UND SCHWERT. Lustspiel in fünf Aufzügen von KARL GUTZKOW. With a Biographical and Historical Introduction, English Notes, and an Index. By the same Editor. *3s. 6d.*

"We are glad to be able to notice a careful edition of K. Gutzkow's amusing comedy 'Zopf and Schwert' by Mr H. J. Wolstenholme. . . . These notes are abundant and contain references to standard grammatical works."—*Academy.*

Goethe's Knabenjahre. (1749—1759.) GOETHE'S BOYHOOD: being the First Three Books of his Autobiography. Arranged and Annotated by WILHELM WAGNER, Ph. D., late Professor at the Johanneum, Hamburg. *2s.*

HAUFF. DAS WIRTHSHAUS IM SPESSART. Edited by A. SCHLOTTMANN, Ph. D., late Assistant Master at Uppingham School. 3*s*. 6*d*.

DER OBERHOF. A Tale of Westphalian Life, by KARL IMMERMANN. With a Life of Immermann and English Notes, by WILHELM WAGNER, Ph.D., late Professor at the Johanneum, Hamburg. 3*s*.

A BOOK OF GERMAN DACTYLIC POETRY. Arranged and Annotated by the same Editor. 3*s*.

Der erste Kreuzzug (THE FIRST CRUSADE), by FRIEDRICH VON RAUMER. Condensed from the Author's 'History of the Hohenstaufen', with a life of RAUMER, two Plans and English Notes. By the same Editor. 2*s*.

"Certainly no more interesting book could be made the subject of examinations. The story of the First Crusade has an undying interest. The notes are, on the whole, good."—*Educational Times.*

A BOOK OF BALLADS ON GERMAN HISTORY. Arranged and Annotated by the same Editor. 2*s*.

"It carries the reader rapidly through some of the most important incidents connected with the German race and name, from the invasion of Italy by the Visigoths under their King Alaric, down to the Franco-German War and the installation of the present Emperor. The notes supply very well the connecting links between the successive periods, and exhibit in its various phases of growth and progress, or the reverse, the vast unwieldy mass which constitutes modern Germany." —*Times.*

DER STAAT FRIEDRICHS DES GROSSEN. By G. FREYTAG. With Notes. By the same Editor. 2*s*.

GOETHE'S HERMANN AND DOROTHEA. With an Introduction and Notes. By the same Editor. Revised edition by J. W. CARTMELL, M.A. 3*s*. 6*d*.

"The notes are among the best that we know, with the reservation that they are often too abundant."—*Academy.*

Das Jahr 1813 (THE YEAR 1813), by F. KOHLRAUSCH. With English Notes. By W. WAGNER. 2*s*.

V. ENGLISH.

COWLEY'S ESSAYS. With Introduction and Notes. By the Rev. J. RAWSON LUMBY, D.D., Norrisian Professor of Divinity; Fellow of St Catharine's College. 4*s*.

SIR THOMAS MORE'S UTOPIA. With Notes by the Rev. J. RAWSON LUMBY, D.D. 3*s*. 6*d*.

"To Dr Lumby we must give praise unqualified and unstinted. He has done his work admirably. . . . . Every student of history, every politician, every social reformer, every one interested in literary curiosities, every lover of English should buy and carefully read Dr Lumby's edition of the 'Utopia.' We are afraid to say more lest we should be thought extravagant, and our recommendation accordingly lose part of its force."—*The Teacher.*

"It was originally written in Latin and does not find a place on ordinary bookshelves. A very great boon has therefore been conferred on the general English reader by the managers of the *Pitt Press Series*, in the issue of a convenient little volume of *More's Utopia* not in the original Latin, but in the quaint *English Translation thereof made by Raphe Robynson*, which adds a linguistic interest to the intrinsic merit of the work. . . . All this has been edited in a most complete and scholarly fashion by Dr J. R. Lumby, the Norrisian Professor of Divinity, whose name alone is a sufficient warrant for its accuracy. It is a real addition to the modern stock of classical English literature."—*Guardian.*

BACON'S HISTORY OF THE REIGN OF KING HENRY VII. With Notes by the Rev. J. RAWSON LUMBY, D.D. 3*s*.

MORE'S HISTORY OF KING RICHARD III. Edited with Notes, Glossary and Index of Names. By J. RAWSON LUMBY, D.D. to which is added the conclusion of the History of King Richard III. as given in the continuation of Hardyng's Chronicle, London, 1543. 3*s*. 6*d*.

THE TWO NOBLE KINSMEN, edited with Introduction and Notes by the Rev. Professor SKEAT, Litt.D., formerly Fellow of Christ's College, Cambridge. 3*s*. 6*d*.

"This edition of a play that is well worth study, for more reasons than one, by so careful a scholar as Mr Skeat, deserves a hearty welcome."—*Athenæum*.

"Mr Skeat is a conscientious editor, and has left no difficulty unexplained."—*Times*.

LOCKE ON EDUCATION. With Introduction and Notes by the Rev. R. H. QUICK, M.A. 3*s*. 6*d*.

"The work before us leaves nothing to be desired. It is of convenient form and reasonable price, accurately printed, and accompanied by notes which are admirable. There is no teacher too young to find this book interesting; there is no teacher too old to find it profitable."—*The School Bulletin, New York*.

MILTON'S TRACTATE ON EDUCATION. A facsimile reprint from the Edition of 1673. Edited, with Introduction and Notes, by OSCAR BROWNING, M.A., Senior Fellow of King's College, Cambridge, and University Lecturer. 2*s*.

"A separate reprint of Milton's famous letter to Master Samuel Hartlib was a desideratum, and we are grateful to Mr Browning for his elegant and scholarly edition, to which is prefixed the careful *résumé* of the work given in his 'History of Educational Theories.'"—*Journal of Education*.

THEORY AND PRACTICE OF TEACHING. By the Rev. EDWARD THRING, M.A., Head Master of Uppingham School, late Fellow of King's College, Cambridge. New Edition. 4*s*. 6*d*.

"Any attempt to summarize the contents of the volume would fail to give our readers a taste of the pleasure that its perusal has given us."—*Journal of Education*.

LECTURES ON THE TEACHING OF MODERN LANGUAGES. By C. COLBECK, M.A., Assistant Master of Harrow School. [*In the Press*.

GENERAL AIMS OF THE TEACHER, AND FORM MANAGEMENT. Two Lectures delivered in the University of Cambridge in the Lent Term, 1883, by F. W. FARRAR, D.D. Archdeacon of Westminster, and R. B. POOLE, B.D. Head Master of Bedford Modern School. 1*s*. 6*d*.

THREE LECTURES ON THE PRACTICE OF EDUCATION. Delivered in the University of Cambridge in the Easter Term, 1882, under the direction of the Teachers' Training Syndicate. 2*s*.

JOHN AMOS COMENIUS, Bishop of the Moravians. His Life and Educational Works, by S. S. LAURIE, A.M., F.R.S.E., Professor of the Institutes and History of Education in the University of Edinburgh. Second Edition, revised. 3*s*. 6*d*.

OUTLINES OF THE PHILOSOPHY OF ARISTOTLE. Compiled by EDWIN WALLACE, M.A., LL.D. (St Andrews), late Fellow of Worcester College, Oxford. Third Edition Enlarged. 4*s*. 6*d*.

"A judicious selection of characteristic passages, arranged in paragraphs, each of which is preceded by a masterly and perspicuous English analysis."—*Scotsman*.

"Gives in a comparatively small compass a very good sketch of Aristotle's teaching."—*Sat. Review*.

A SKETCH OF ANCIENT PHILOSOPHY FROM THALES TO CICERO, by JOSEPH B. MAYOR, M.A. 3*s*. 6*d*.

"Professor Mayor contributes to the Pitt Press Series *A Sketch of Ancient Philosophy* in which he has endeavoured to give a general view of the philosophical systems illustrated by the genius of the masters of metaphysical and ethical science from Thales to Cicero. In the course of his sketch he takes occasion to give concise analyses of Plato's Republic, and of the Ethics and Politics of Aristotle; and these abstracts will be to some readers not the least useful portions of the book."—*The Guardian*.

[*Other Volumes are in preparation.*]

University of Cambridge.

LOCAL EXAMINATIONS.

Examination Papers, for various years, with the *Regulations for the Examination.* Demy 8vo. 2*s.* each, or by Post, 2*s.* 2*d.*

Class Lists, for various years, Boys 1*s.*, Girls 6*d.*

Annual Reports of the Syndicate, with Supplementary Tables showing the success and failure of the Candidates. 2*s.* each, by Post 2*s.* 3*d.*

HIGHER LOCAL EXAMINATIONS.

Examination Papers for various years, *to which are added the Regulations for the Examination.* Demy 8vo. 2*s.* each, by Post 2*s.* 2*d.*

Class Lists, for various years. 1*s.* By post, 1*s.* 2*d.*

Reports of the Syndicate. Demy 8vo. 1*s.*, by Post 1*s.* 2*d.*

LOCAL LECTURES SYNDICATE.

Calendar for the years 1875—80. Fcap. 8vo. *cloth.* 2*s.*; **for 1880—81.** 1*s.*

TEACHERS' TRAINING SYNDICATE.

Examination Papers for various years, *to which are added the Regulations for the Examination.* Demy 8vo. 6*d.*, by Post 7*d.*

CAMBRIDGE UNIVERSITY REPORTER.

Published by Authority.

Containing all the Official Notices of the University, Reports of Discussions in the Schools, and Proceedings of the Cambridge Philosophical, Antiquarian, and Philological Societies. 3*d.* weekly.

CAMBRIDGE UNIVERSITY EXAMINATION PAPERS.

These Papers are published in occasional numbers every Term, and in volumes for the Academical year.

VOL. XIII.	Parts 177 to 195.	PAPERS for the Year	1883—84, 15*s.* *cloth.*
VOL. XIV.	„ 1 to 20.	„ „	1884—85, 15*s.* *cloth*
VOL. XV.	„ 21 to 43.	„ „	1885—86, 15*s.* *cloth*

Oxford and Cambridge Schools Examinations.

Papers set in the Examination for Certificates, July, 1885. 2*s.* 6*d.*

List of Candidates who obtained Certificates at the Examinations held in 1885 and 1886; and Supplementary Tables. 6*d.*

Regulations of the Board for 1887. 9*d.*

Report of the Board for the year ending Oct. 31, 1886. 1*s.*

Studies from the Morphological Laboratory in the University of Cambridge. Edited by ADAM SEDGWICK, M.A., Fellow and Lecturer of Trinity College, Cambridge. Vol. II. Part I. Royal 8vo. 10*s.* Vol. II. Part II. 7*s.* 6*d.* Vol. III. Part I. 7*s.* 6*d.*

London: C. J. CLAY AND SONS,
CAMBRIDGE UNIVERSITY PRESS WAREHOUSE,
AVE MARIA LANE.
GLASGOW: 263, ARGYLE STREET.

CAMBRIDGE: PRINTED BY C. J. CLAY, M.A. AND SONS, AT THE UNIVERSITY PRESS.

52

PLEASE DO NOT REMOVE
CARDS OR SLIPS FROM THIS POCKET

UNIVERSITY OF TORONTO LIBRARY

P&A Sci.